ISBN 978-1-330-39764-0
PIBN 10051363

FB &c Ltd, Dalton House, 60 Windsor Avenue, London, SW19 2RR.
Company number 08720141. Registered in England and Wales.

QUARTERLY JOURNAL

OF

MICROSCOPICAL SCIENCE:

EDITED BY

EDWIN LANKESTER, M.D., F.R.S., &c.,

AND

E. RAY LANKESTER, B.A. Oxon., F.R.M.S.

VOLUME X.—New Series.

With Illustrations on Wood and Stone.

LONDON
JOHN CHURCHILL AND SONS, NEW BURLINGTON STREET
1870.

LONDON:
J. E. ADLARD, PRINTER. BARTHOLOMEW CLOSE.

MEMOIRS.

Notes on SPONGES.—1. *On* HYALONEMA MIRABILIS, *Gray.* —2. *On* APHROCALLISTES BOCAGEI, sp. nov.—3. *On a new* GENUS *and species of* DEEP SEA SPONGE. By ED. PERCEVAL WRIGHT, A.M., M.D., F.L.S., Professor of Botany, Dublin University. (Plates I, II, and III.)

1. *Hyalonema mirabilis*, Gray. (Plate III.)

So much is now known of a deep sea fauna that one is very apt to forget how little was known about it just eighteen months ago. It is quite true that here and there stray facts were to be met with that most distinctly showed that animal life was to be found at very great depths of the ocean, not to mention the important researches of Sir John and Sir James Ross, or of Dr. Wallich, there were also records of the occurrence of zoantharian corals, at depths of 300 to 400 fathoms, and of fishes taken at depths of 312 fathoms. Indeed, to collect all the scattered observations bearing on this subject would be a work requiring considerable research. But some way or other the true bearing and the extreme importance of all these facts were in a great measure, and by a great many, overlooked; and the fact that great depths act as no bar to the existence of animal life, however known to a few, and however much it should have been known to all, has only during the last year and a half been fully recognised.

Accustomed to dredge in what will be now considered the trifling depths of from 80 to 100 fathoms, I learnt with some surprise of the deep sea fishing for sharks at Setubal in depths of from 300 to 400 fathoms; and when my friend Professor Bocage, of Lisbon, told me of the discovery of quite recent specimens of *Hyalonema mirabilis* in these same depths, I had a great wish to go and investigate the fact for myself. The prevalent opinion, however, at the time about the discovery of the Hyalonema off the coast of Portugal was, that the specimens taken by the Setubal fishermen were stray specimens thrown overboard from some vessel trading between Japan and Lisbon, and that it would be a perfectly hopeless task to look for living specimens off the west coast of Portugal; and one great friend of mine, whose opinion on the subject had very great weight with me, believed this so firmly that I yielded to his arguments on the point, and abandoned the idea of going to Setubal in the spring of 1868. This same friend has since, by suggesting the deep sea expeditions of 1868 and 1869, and by the amount of work that he has accomplished in connection with these expeditions, opened up to the student of nature quite a new world. It so happened, too, that at the very time that I was dredging up specimens of Hyalonema in the tranquil waters of Setubal he was dredging it, and a host of other glorious species in the more stormy seas of the north, for hearing again from Professor Bocage in July, 1868, that the Hyalonema had been actually taken *in situ*, I lost no time in going out to Lisbon in 1868, the time when the shark fishery season commenced. I have in another place[1] given a short account of my excursion in Portugal, and will only here refer to it for the purpose of stating my firm conviction that, though so many new and rare species have been taken by the several deep sea expeditions of the Swedish, British, and American Governments; yet I believe much more remains to be done, and I would suggest that the deep ground off Setubal is well worthy of investigation, as it lies within a distance of from ten to thirty miles of the shore; and as the sea there is, as a rule, peculiarly tranquil during the months of August and September, it would be possible to make a very thorough investigation of it without even the assistance, most valuable though such assistance be, of a man-of-war or a Government survey vessel. The present King of Portugal is in every way entitled to take his place in the ranks of science; and the national museum at Lisbon is already indebted to him for most important aid, and for many valuable collections; perhaps Professor Bocage might induce him to follow the example set by Britain, and persuade him to allow a Government survey vessel to spend a fortnight or three weeks on the ground I refer to; the collections that would surely be made would form a most desirable addition to the museum at Lisbon, as well as be most valuable for exchanges. It was my object on my return home to make a report on the structure of *Hyalonema mirabilis*, of which I had succeeded in taking living specimens, I had also every opportunity given me by Professor Bocage of studying the magnificent series of specimens preserved by

[1] 'Annals and Magazine of Natural History,' December, 1868.

him in alcohol in his museum. Finding, however, that Professor Wyville Thomson had taken a large number of specimens during his cruise in the "Lightning," and that he intended publishing a memoir on the genus, I contented myself with a simple record of its occurrence in a living state; of its mode of growth, viz. as Loven suggested, with its siliceous stem anchored in the mud, and with expressing my opinion that the stem was truly a part of the sponge-mass, and that the Polythoa was simply parasitic upon the stem. Nor do I here intend to do more than call attention to one or two peculiarities which it strikes me are to be met with in the specimens that I have examined from Portugal, and which do not seem to exist; at least, not exactly after the same fashion in the specimens taken in the "Lightning" and "Porcupine" expeditions. My knowledge of these latter is based upon a very casual examination of the specimens taken by Professor Wyville Thomson; and upon a more careful examination of a beautiful little specimen, about an inch and a half in length, most kindly given to me by Professor Thomson. And in these remarks I do not mean to anticipate at all the memoir on this genus which is so impatiently expected, but rather to state what I know about the differences between the specimens taken off Portugal, and those off the west coast of Great Britain and Ireland.

Some of the Setubal specimens are of very great size; the stems of several measuring nearly two feet in length. In one very perfect specimen the head consists of a large somewhat oval mass, about eight inches broad in its long diameter, and four inches across in its short diameter; it is cup-shaped, resembling somewhat the ordinary shape of a common toilet sponge, and, like it, it is hollow on the inner surface or on that portion where the "glass rope" ends. The outer surface has been somewhat worn off by either lying on the mud or from rough handling, and presents that appearance of wet brown paper that must be familiar to all who have examined specimens of Hyalonema with the sponge mass attached from Japan. On opening out the sponge, the interior concave surface appears to have remained uninjured, and here will be seen a delicate network of spicules and sarcode, lining the concavity and passing into the texture of the sponge. A number of irregular large openings (oscula) are also seen, and these are covered over with a delicate open sarcode network, the edges of the meshes of which are thickly lined by the spicules called 'spiculate cruciform spicules' by Dr. Bowerbank. These spicules are met with all through the sponge, but almost always lining the cavities or hollow passages of the mass. They are likewise to be found as a lining all over the surface of the sponge, but in no place are they to be met with arranged in so regular a fashion as on the meshes of the network covering the oscula. From the peculiar way in which they are placed on the edges of the meshes, and from the fact that the barbs on the stem of the spicules all point in the one direction, it is possible that while it would be easy to glide over the slimy sarcode down into an osculum, return would be no easy task, as any solid body would be at once caught and retained by the barbs. From the manner in which the cruciform basal portion of these spicules is inserted in or attached to the sarcode, I make no doubt but that they are subject to being moved up and down and to and fro, and that on the contraction of an osculum, and on the consequent discharge of water from the oscular cavity, the spicules are pushed outwards and upwards, falling down again on the expansion of the osculum. In all the numerous writings on the structure of Hyalonema, I cannot find that the exact position of these spicules in the living sponge has been determined. I have, therefore, thought it advisable to give the accompanying illustration (Plate III), for which I am indebted to Mr. Lens Aldous. It represents one of the oscula removed from a specimen of *H. mirabilis* in the Lisbon Museum. The spiculate cruciform spicules which line the edges of the sarcode network are very easily displaced, and but comparatively few of them were on the specimen drawn by Mr. Aldous, but in a living state they line, packed in a close row, the edges of the sarcode mesh; they differ slightly from any of those figured by Dr. Bowerbank or Max Schultze. One other subject I should like here to allude to. The oscula of *H. mirabilis* being now discovered and described, and they being found to be just those that one would have expected and just in the position in which one would have looked for them, it scarcely requires my statement that I saw the little parasitic Polythoa in a living state on the siliceous axis of the Hyalonema, and that I watched them expand their tentacles, after the fashion of any other zoantharian, to prove that though they have mouths these mouths are their own, and not at the service directly or indirectly of the Hyalonema. Is it too much to expect to settle the last lingering doubt that may still exist in some minds as to the nature of these independent though parasitic organisms?

2. *Aphrocallistes Bocagei*, sp. nov. (Plate I.)

Sponge fistulous, erect, branching somewhat irregularly;

skeleton siliceo-fibrous, more or less symmetrically radial; radii short and stout on the outer surface, and somewhat longer and thinner on the inner surface of the skeleton, forming a series of hexagonal spaces, which are nearly all of the same dimensions, central umbo of the ray giving origin on its inner surface, often on both surfaces, to a long spine. These spines, generally long, sharp-pointed, sometimes knob-headed. Spicules, acerate; retentive verticillately spined; attenuated rectangulated hexradiate, and subfusiformi cylindrical entirely spinous. Main tube closed by an irregular siliceous network, which is deeply concave. Pores and dermal system unknown.

Habitat.—Cape de Verde Islands, in museum of Lisbon, in British Museum, London, and in my own collection (Sept., 1868) also off south-west coast of Ireland in deep water; Professor Thomson, " Porcupine " Expedition, 1869.

Dr. J. E. Gray established the genus Aphrocallistes in 1858 for a very beautiful sponge from Malacca (' Proc. Zool. Soc.,' London, 1858, p. 115, Pl. XI, Radiata), *A. beatrix*. Dr. Bowerbank having identified the *Iphiteon panicea* of Valencienne in the museum of the Jardin des Plantes, Paris, as belonging to the same genus as *A. beatrix* adopts Valenciennes's name. As, however there were never any descriptive characters of the genus Iphiteon published until 1869 (it was affixed to the specimen in the Paris Museum in 1800), and was described as Aphrocallistes in 1858, I have no hesitation in assigning the priority to Dr. Gray's name and in adopting it for those sponges, with a siliceo-fibrous skeleton in which the reticulations are symmetrical. It is true that by an accident Dr. Gray described the genus as having calcareous instead of siliceous spicules—an error which he afterwards corrected. But this mistake could not for a moment mislead when the rest of the diagnosis and the beautiful figure by Mr. Ford were taken into account; indeed, such a figure with a name attached would amount to a publication.

It is confessedly unsatisfactory to describe a sponge from a dead and bleached specimen; for if in any group of the animal kingdom, surely here we require all the assistance it is possible to have from an examination of all the structures of the organism. It is, therefore, not without an apology that I publish the above beautiful form as a new species. When examining the very interesting collection at the Museum of Lisbon in September, 1868, I discovered three or four specimens of this sponge, which I immediately regarded as a new species of this genus. Professor Bocage, with his accustomed liberality, at once gave me the specimen figured on Plate I, and accorded me permission to describe it. In dedicating it to my friend I take this opportunity of thanking him for the many kindnesses which he showed me while in Portugal. The memory of a delightful Sunday spent with him at charming Cintra will ever remain with me. The museum under his care is one of the most interesting in Europe ; the more especially interesting on account of the fine collection of native species brought together by the persevering energy of Professor Bocage and his admirable assistant Sig. Capello. It is also rich in species from the Portuguese settlements abroad, and this sponge formed but one of a fine series of invertebrata from the Cape de Verd Islands.

The nearest ally of this species is undoubtedly *A. beatrix*, Gray, and it is quite possible when we know more about both forms, and when they have both been examined in a living state, that they may prove but varieties of the same species. This is is possible, but for the moment I think not probable. There is a certain regularity of form in the sponges which have a non-elastic siliceo-fibrous skeleton, which I venture to think will be found to be in a measure characteristic of the species. But apart even from this consideration, the areas forming the skeleton in *A. Bocagei* are much more regularly hexagonal than those in *A. beatrix*. The spines on the bosses are very much longer in the former than in the latter species; in it too the central cavity is larger. The reticulated network-like lid is much more radial in its composition than in *A. beatrix*. The bosses of the rays of the body-skeleton are often knobbed, and there is an apparent absence of porrecto multiradiate spicules so characteristic of *A. beatrix*. This fact I do not lay much stress upon, as it may arise from an error of observation. I have, however, met with these spicules in every specimen that I have examined of *A. beatrix*, and never in the many specimens examined of *A. Bocagei*. This latter, too, is a much more erect form than the former, and I should expect that when the sarcode layer of both species is known that the spicules of this layer may be somewhat different in both. I am indebted to Mr. Ford for the accompanying drawing which he made for me in January, 1869. Circumstances have prevented me from publishing a description of it sooner. Professor Wyville Thomson has kindly forwarded to me portions of this sponge taken in the recent cruise of H.M.S. "Porcupine." He informs me that it was dredged living off the south-west coast of Ireland at a great depth. The portion sent to me is a fragment of a dead specimen. But Professor Thomson thinks that living specimens were met with, and that they are somewhere among the vast stores of good things collected during the expedition. Professor Alexander Agassiz also recognises Mr. Ford's drawing as that of a species taken by Count Pourtales in his last expedition, and informs me that all the sponges taken in the course of the coast survey expeditions of America have been forwarded to Professor Oscar Schmidt for description; perhaps, therefore, we may expect still another supplement to that most important and useful work "Die Spongien des Adriätischen Meeres." If so, I hope he will agree with me in considering this species a good one, and that from better specimens and with his great powers of drawing he will still further describe and illustrate it.

3. *On a new Genus and Species of Sponge from the Deep Sea.* (Plate II).

In March, 1869, my friend Dr. Wallich, so well known by his botanical and zoological writings, as well as by his researches into the deep sea fauna, gave me a small portion of a minute sponge, of which three specimens had been brought up from the great depth of 1913 fathoms, with the request that I should describe it. I have to apologise to Dr. Wallich for letting the summer pass over without fulfilling the promise that I made to him. But there were two difficulties in my way. One was to have the most perfect of the three specimens discovered drawn. This specimen had been presented by Dr. Wallich along with a vast collection of Foraminifera, Polycystina, Diatomaceæ, and Desmidiaceæ, to the Royal Microscopical Society of London. The other difficulty was to find out where to place the species when described. My first difficulty has been surmounted—thanks to the Council of the Royal Microscopical Society and their assistant-secretary Mr. Reeves—by Mr. C. Stewart, F.L.S., of St. Thomas' Hospital, to whose friendship I am indebted for the accompanying very characteristic, faithful, and beautiful drawing. My second difficulty I am not so sure of having as yet clearly seen my way through. But to this I will allude more particularly a little further on. By the help of the enlarged figure on Plate II, and the following description, I hope this earliest discovered (October, 1860) of all the deep sea sponges will be easily recognised.

Wyvillethomsonia, gen. nov.

Sponge body subspherical, attached by a stem. In the summit of the sponge, *i.e.* the end farthest from the stem, there is one large osculum, which is fringed by long, delicate, biacerate spicules. The interior of the sponge body consists of several cavities which open into the osculum. The stem is prolonged through the body as an axis, and consists of numerous biacerate spicules somewhat more robust than those fringing the osculum, and mixed with these are a number of anchoring spicules (fusiformi-recurvo-ternate of Bowerbank), the recurved end being always directed to the point of attachment (which in this case is a small stone). The body is composed of a large number of spicules (furcated attenuato-patento-ternate of Dr. Bowerbank), the radii of the ternate spicules meeting each other as they lie on the surface of the sponge, form a remarkable loose network-like pattern; the long pointed process from the central boss projecting inwards towards the axis of the sponge.

The whole of the body of the sponge and of the stem is covered by a thin sarcode layer which abounds in stellate spicules varying much in size. One remarkable spicule (bifurcated expando-ternate) seems to terminate the axis in the centre of the large osculum.

W. Wallichii, sp. nov. (Plate II, figs. 1 to 6.)

Habitat.—Dredged from a depth of 1913 fathoms, October, 1860, in lat. 58° 23′ N., long. 48° 50′ W., by Dr. Wallich, who was then Acting Naturalist to H.M.S. "Bulldog," Sir F. L. McClintock, R.N., commander.

I name this genus and species after my friends Professor Wyville Thomson and Dr. Wallich. The name of Wallich has been long since employed in botany as a generic term, otherwise I should have employed it as such here. Those who may object to the length of my generic name I may remind of the precedent I have in Vaughanthomsonia. I could not commemorate two more original workers than Drs. Wallich and Thomson. There can be little doubt but that the three specimens dredged of this species are in a very young condition, but from what we know of sponges generally I think it is fair to assume that a mere question of size of specimen is of very little consequence in determining a species; indeed, once the sponge arrives at that stage of its existence that it forms all its characteristic spicules, neither the form of these nor their general arrangement in the sponge structure is very much altered by growth; hence the diminutive size of the specimens examined by me, seeing that they appear complete in all their parts, is not a sufficient reason for this species remaining undescribed. At the next meeting (15th April, 1869) of the Dublin Microscopical Club, after Dr. Wallich had given me the small portions of the third specimen above referred to, I exhibited a series of the spicules, and stated it as my impression that the species belonged to the section of sponges with siliceo-fibrous skeleton and hexradiate spicules called Vitrea by Wyville Thomson. In this I was led astray by some siliceous network, like that met with in Aphrocallistes which was entangled by the body spicules of the little sponge, and I have now little hesitation in referring it to the Corticatæ of Oscar Schmidt, suggesting that its affinities are to the genus *Stellata*, Sdt. I do this for the following reasons: The number of stellate spicules in the outer sarcode layer, which on some portions of the sponge body are so tightly packed together as to form quite a hard layer of silex; the prevalence of the large furcate ternate spicules, which are certainly most important in the structure of the sponge mass—such spicules (No. 850—51 of Bowerbank) are to be met with in *Pachymatisma Listeri*, Bowk. MS. in *Stellata discophora*, Sdt., *S. Helleri*, Sdt., *S. mamillaris*, Sdt., and *S. mucronata*, Sdt. So far as I know these two forms of spicules are only met with in the same sponge when that sponge belongs to the division Corticatæ of O. Schmidt. The genus, however, cannot be easily confounded with any of those placed among the corticates. In some specimens of mud, taken from the same locality by Dr. Wallich, spicules (furcate ternate) occur seven eighths of an inch in length, proving the existence of some enormous specimens of some sponge of this group. Professor Wyville Thomson, who was present at the meeting of our Dublin Microscopical Club at which I exhibited this species, stated that he had taken this species, or at least one very closely allied to it, in the same ground that he had taken *Holtenia Carpenteri*, W. Thomson.

On Certain Imperfections *and* Tests *of* Object-glasses.

By G. W. Royston-Pigott, M.D., M.R.C.P., M.A., F.R.A.S., late Fellow of St. Peter's College, Cambridge. (Received Sept., 1869.)

The actual diameter of the least circle of aberration caused by lenticular vision is the real gist of the much-debated question of "aberration" and imperfect definition. The performance of the eye-piece is altogether secondary and inferior to that of the objective, whose errors are multiplied by it, as well as the ratio of the distances of the final conjugate foci from the posterior lenses of the objective.

This circle or ring, being the smallest space through which the focal pencil passes, is seldom so reduced as to leave no traces in the highly developed image presented to the eye.

Omitting here to dilate upon the terms now so vaguely used, as "resolution," "penetration," and "definition," I may be permitted to enumerate a few points worthy of consideration, as they have occurred to me during the last twenty-five years.

Under the use of very high power every free edge of an object, and every isolated point, exhibits an umbra and penumbra exactly representing the diameter of the least circle of aberration generated by the final objective refractions.

Every object being an assemblage of such points exhibits, under high power, similar aberrating shadow, principally visible at the sharp borders and edges; this shadow or penumbra depending upon the aberration, and being independent of the size of the object considered as an assemblage of points. After a multitude of experiments, I conclude that—

This shadow can be considerably diminished—

(*a*) By limiting the aperture of the illuminating pencil.

(*b*) By reducing the aperture of the objective.

(*c*) By further correction and better approximation to aplanatism of the objective itself.

(*d*) By viewing objects directly, without a covering glass, properly adjusting the position of the front lenses.

(*e*) By the use of immersion lenses, destroying the aberrating effects of a plate of air.

(*f*) By carefully searching the axis of the instrument for a position of minimum aberration of the conjugate foci.

(*g*) By employing direct rays, from a radiant point, free

from circles of confusion and chromatic dispersion, such as a fine pencil of rays admitted through an exceedingly fine aperture from the direct solar beam.

(*h*) By analysing the aberration of the annular surfaces of the objective, and selecting such areas and annuli as are the more perfect in their operation.

(*i*) Lastly, by abstention from pressing the powers of the objective beyond distinct vision, a canon universally adopted by astronomers, but too much neglected by microscopists. In telescopes every inch of diameter is generally considered to barely admit a power of 100, a 10-inch objective scarcely allowing a power of 1000. In microscopes the power may be similarly estimated for useful effects by taking 100 times the reciprocal focal length, one eighth giving 800 diameters. There are two other points to which I beg to direct the attention of microscopists, as worthy of their best efforts and scientific research.

(*k*) The spherical aberration, both lateral and longitudinal; which will be improved—

(*l*) By extending the visual focal distance of deep objectives, and so withdrawing the face-glass from its extreme and dangerous proximity to the "covering glass."

(*m*) By greatly increasing the depth of focal vision, and calculating its amount. With a 3-inch objective it is possible to view, at one and the same instance, both surfaces of a thin covering glass; and the marvellous and delightful *perspective* view into deeper parts of insects given by Ross's 4-inch objective illustrates the same principle of the advantages of increasing focal depth. These points having occupied my attention for many years, I hope shortly to communicate the methods adopted to accomplish these desiderata.

There is a great deal of interest in the subject of definition, because it is one common to the sister sciences of the astronomer and the microscopist. To the former "dividing power," "definition," giving fixed stars a round disc, resembling a bright spangle placed upon black velvet, and "diffraction rings," and lastly "nebulosity" and haze, have all their special, though humbler, representatives in the microscopic field. And in order to obtain precise ideas, it is wise to proceed from the known to the unknown; I therefore beg to suggest the study of these points in the images formed of given and known objects by minute lenses and high microscopic power. For the information of those interested in this point I have calculated the diameters of the circles of least aberration for parallel rays for minute lenses of the following dimensions, the index of refraction being taken at $1 \cdot 50 = \mu$ *for plate-glass :*

Focal length . .	$\frac{1}{20}$th inch	= ·05″
Aperture . . .	$\frac{2}{100}$ths ,,	= ·02″

The diameters of the aberrating penumbra of a point—

	Diameter of least aberration.		
Plano-convex	. ·0009″	=	$\frac{1}{1111}$th of an inch nearly.
Equiconvex	. ·00033″	=	$\frac{1}{3000}$th ,,
Convexo-plane	. ·00023″	=	$\frac{1}{4300}$th ,,
Crossed lens	. ·000214″	=	$\frac{1}{4650}$th ,,

If an equiconvex lens be used where the diameter of circle of least aberration $= \cdot 835 \times \frac{\text{cube of } \frac{1}{2} \text{ aperture}}{\text{square of focal length}}$, the aberration will be a minimum when the object and image are equidistant from the lens, and each at twice its focal length from it, or $v = 2f$ and $u = 2f$.

Now, by a combination of many glasses, objectives may be *corrected* to show scarcely any penumbral aberration, but, unfortunately, nearly all opticians using the Podura markings as an unequalled standard, all the best glasses are corrected to show what is absolutely false and delusive, and the result is a misplaced belief. Till the explosion of this creed, every one, being taught to look upon the Podura spectral markings as the *ne plus ultra* of objective accuracy, is satisfied with that ignorance which is bliss. The glasses are constructed on purpose to show this supposed standard appearance; but opticians will be obliged, at some future day, to elect a new standard. The same glasses which show this admirable note (!!!) fail in the higher tests, such as will now be described.

I have found in the best objectives a residuary aberration, and some of it, when the glasses are pressed with too high an amplification by deep DD eye-pieces, will in some degree always remain to put a stop to further research, except with deeper objectives still; and these objectives, rising already to the 50th of an inch, will probably soon reach their limit owing to manipulative defects in their manufacture. The precious stones, such as the topaz and sapphire, and lenses formed *by accident* or fortunate working, may perhaps assist a further development of power or detection of error.

For forming beautiful images of bright discs, to imitate close double stars, I recommend the use of two minute crossed lenses, set with their deepest convexities in contact; but a fine *objective analyser* may be formed of two lenses in contact, made of plate glass of 1·5 index of refraction; the first lens being equiconvex, and the second concavo-convex, whose radii are as 1 to 101—which combination for forming an image of a distant object—as a silvered-glass ball or illuminated globe or lamp—is absolutely aplanatic for plate glass of refractive power 1·5. I propose to name this instrument the aplanatometer.

Armed with such an analyser amateurs will experience little difficulty in examining the performance of their objectives. The appearance of black points on a white disc, or a white disc upon a black ground, as imaged by minute lenses, affords one of the most instructive lessons in microscopic aberration to students of this difficult subject. Familiarity with these optical phenomena of *known objects* enables the observer to detect the errors of observation with fidelity.

When a thorough acquaintance has been made with these signs of defective correction, objectives required to be examined will be found either to increase or diminish the glaring aberration of these minute images, and increase, or even destroy, the penumbra of circular aberration.

Still more difficult of definition is the minute lenticular image of a brilliant point, or of a minute brilliant aperture, or reflecting mercurial globule, placed sufficiently far from the lens to produce a minute image. Thus, if the lens is the 20th of an inch in focal length, the image of the globule will become smaller and smaller as it recedes.

At a distance of forty inches the image of a globule of mercury 100th of an inch in diameter will be 80,000th of an inch + 64,000,000th, very nearly. By this plan, therefore, we possess a means of forming images of known diameter, and with more or less aberration, according to the kind of lenses employed.

The Podura is, according to the high-class objectives made by Messrs. Powell and Lealand, a scale covered with ribbing on each side of the basic membrane; each ribbing is separated by a clear interval, through which the second set of beads placed on the other side are visible. When the scale is folded over, so as to expose the underneath side, a similar ribbing is seen. Mr. Wenham states[1] that he cannot see any signs of structure at the folded edge, but the ribs and intervals are clearly displayed when the objective corrections are exquisitely performed. These ribs are composed of beads, varying from the 30,000th to the 150,000th of an inch in diameter, according to the size and character of the scale.

[1] 'Microscopical Trausac.,' July, 1869.

The upper and lower sets of beaded ribbing cross each other at an acute angle, and their general direction is somewhat wavy. The markings are caused by the cross intersections of the ribbing, and exactly imitate the effect of watered silk caused by pressing two pieces powerfully together.

Minute aplanatic lenses being employed on the stage—

A. Image of a watch, 8 feet distance. Some objectives show the image enveloped in yellow fog; the time cannot be distinguished. No focusing or "collar adjustment" gets rid of the nebulosity.

A small bright aperture appears nebulous and radiating; either the brightest part of the radiant cone is *before* or *behind* the most distinct image. The images of two contiguous bright apertures coalesce and cannot be divided as double stars.

B. The divisions of a micrometer imaged by the minute lens can scarcely be discerned in a haze of nebulous light: the micrometer was placed on the plane mirror illuminated by a bright cloud.

C. Rows of beads appear cylindrical bodies unless the aberration is finely corrected.

D. The image of a church clock, 200 feet distance, can be formed by the plane mirror, and the stage lenses will give the details of the face, the small dots for the minute hand, and, of course, the figures are beautifully displayed, sharply cut and defined or lost in mist according to the size of the stage lenses and the power employed, and the corrections of the objectives.

Other Tests.

Podura scales. As I had the honour of communicating to the Royal Microscopical Society in May last an account of the podura beads existing on both sides of the basic membrane, I may be permitted to draw attention to the nearest approach made to the definition of this difficult object as given by Mr. Richard Beck, who has figured the cylindrical bands of the Podura, but described them as out of focus and a false appearance, whereas I always find these bands to present their provisional appearance as heralding the development of their beading.

Lepisma saccharina. The exceedingly beautiful striæ of this scale are also shown, in the same work, as perfectly transparent and clear, like threads of glass. These cylindrical ribs are also composed of spherical beads, and they exist on both sides of the scale, the lower set radiating from the quill being much smaller than the upper set.

The finest and most resplendent definition of the diatoms may certainly be seen with the immersion lens and a 1-16th, which it converts into a 1-20th objective. The structure of the *P. formosum* has been an object of careful study for many years, as one of the easiest forms for definition. Separated spherules generally characterise this diatom. The beading of this object is brilliant in the extreme, *never grey*. The exquisite beauty of these minute gems of nature rival the most glorious tints of the diamond, ruby, or sapphire; but a power of 7000 diameters begins to develop shadow and haze. *Between the spaces of the upper beading* another strueture is discernible, but whether the interspaces are *crossed by* a deeper set of beading or the upper set are superimposed upon the lower I cannot at present decide, but I strongly incline to the latter supposition.

P. strigosum. Here the upper set appear to hide a parallel lower set of beads, like row upon row of cannon shot. But always do I perceive the two sets of different colours, one row pink-red, the intervening row violet or blue; probably the colours are produced by the *dispersion of position,* and may be good evidence of the sets existing in different planes.

P. hippocampus. Similar phenomena are observed. Rows of beading appear to cross in different planes at right angles to each other.

A severe test is the appearance of minute hairs 1-50,000th of an inch diameter. *A fine definition shows a hair to bear two black borders and a central line of light,* with scarcely any penumbra under the 1-16th and immersion lens. Hairs of antennæ of male gnat were employed.

Glittering particles of gold leaf. Some of these may be found 1-50,000th of an inch in diameter; brilliant illumination, if the corrections be not good, shows four to five diffraction rings. I have seen them diminish to one.

Crystalline surface of metal recently broken. The glare is universal unless the aberration is very finely corrected.

On Testing the Magnifying Power.

Sir John Herschel's definition of the power of a lens, as *unity divided by the focal length,* is, perhaps, the best that can be employed. But the real magnifying power varies with different persons, as with short-sighted and long-sighted observers. As a standard, the power of a lens forcing an image to the eye at ten inches may be 10 ÷ focal length. On this principle an inch objective should form an image ten times larger than the object at a screen ten inches distant, and a 1-12th objective should magnify 120 times. If the eye-piece is equivalent to 1-10th single lens, it again enlarges the object 100 times, and the total power will be 12,000 diameters. The magnifying power of a microscope whose objective has a focal length f and eye-piece e, will be conveniently expressed by—

$$M = \frac{10}{f} \times \frac{10}{e} = \frac{100}{ef}$$

I have several times tested this formula against the camera lucida, and have found it extremely accurate for a standard of ten inches' distance of the eye-lens from the objective.

Examples.—With a C eye-piece, Powell and Lealand's 1-8th objective magnifies 800 times, and its focal length is one inch. The 1-12th, with the same eye-piece, gives 1200 diameters, and the 1-50th 5000 diameters. The formula produces identical results, precisely expressing their printed tables :

(1) $\frac{1}{8}$th. $e = 1, f = \frac{1}{8}$.

$$M = \frac{100}{1 \times \frac{1}{8}} = 800.$$

(2) $\frac{1}{12}$th. $e = 1, f = \frac{1}{12}$.

$$M = \frac{100}{1 \times \frac{1}{12}} = 1200.$$

(3). $\frac{1}{50}$th objective. $e = 1, f = \frac{1}{50}$.

$$M = \frac{100}{1 \times \frac{1}{50}} = 5000.$$

If the draw-tube be used to increase the distance by the 1-5th or 1-6th the magnifying power will, for high objectives, increase proportionately. Thus, if the draw-tube give twelve inches instead of ten, and an eye-piece similar to Browning's G achromatic of ·1000 focal length, and a 1-12th objective—

$$M = \frac{100}{\cdot 1000 \times \frac{1}{12}} \times \frac{12}{10} = \frac{100 \times 12 \times 12 \times 10}{10} = 14{,}400.$$

By the formulæ the focal length of double D eye-piece would, if the 1-12th gives 12,000 diameters with twelve inches of tubing instead of ten, be given by the equation—

$$M = 12{,}000 = \frac{100}{e \times \frac{1}{12}} \times \frac{12}{10};$$

$$\text{Therefore, } e \text{ or DD eye-piece} = \frac{100 \times 12}{12000 \times \frac{1}{12} \times 10} = \frac{3}{25},$$

or rather less than the 8th of an inch, or eight times deeper than Powell and Lealand's third eye-piece C, which is generally preferred to all higher powers for accurate definition.

When objectives are pressed to this extreme amplification the jewelled brilliance of the translucent and radiant beading vanishes; they take a grey and sombre hue, and all intervening structure is lost in an indefinable haze of exaggerated aberration. The spherical beads of diatoms are brilliantly transparent (being formed of pure silica?), and they behave themselves in almost every respect as spherical lenses, showing crescentic shadows, blending and mingling with each other in endless variety, according as the interspherical refractions develop single or multiple shadows. I cannot refrain from expressing my convictions that all diatoms of this character possess a double set of beading, set in different planes.

On some Freshwater Rhizopoda, New *or* Little-Known. By William Archer.

(Continued from vol. ix, N.S., p. 397.)

Pleurophrys spherica (Clap. et Lachm.)? *Pl. ? amphitremoides* (sp. nov.), and *Pl. ? fulva* (sp. nov.)

Having endeavoured in the foregoing to give some account of the remarkable form, *Diaphoropodon mobile*, I pass on to the three others, which in the present series seem next related thereto, figured on Pl. XX, figs. 1, 2, 3 (vol. ix), and which I would (provisionally at least) identify as above. I regret, however, that I have it in my power, at present at least, to do but very little more than refer to the figures and the accompanying explanation of the plate.

It will be seen by the appended notes of interrogation that it is not without some amount of uncertainty as yet that I refer these forms to Claparède and Lachmann's genus, which is thus characterised:—"Body covered by a test furnished with a single opening, and formed of foreign substances agglutinated by means of an organic cement," to which diagnosis is prefixed the remark that this genus is to the Actinophryans as the Difflugiæ are to the Amœbæ.[1]

From this we are to understand that in Pleurophrys the sarcode body emits slender linear, unbranched pseudopodia, through the single aperture of a test formed of agglutinated foreign particles. Nor must the figure given[2] be misunder-

[1] Claparède et Lachmann, 'Les Infusoires et les Rhizopodes,' p. 451.
[2] Op. cit., pl. xxii, fig. 3.

stood, because the pseudopodia are depicted as radiating all around, for, as the explanation of the plate states, the figure is drawn as seen from above, and hence the opening, whence emanates the pencil of pseudopodia, must be below, and these, seen from that point of view, naturally appear to project in all directions.

Three circumstances seem to render the identity of the form of which I have tried to convey an idea by fig. 1 with *Pl. spherica* somewhat doubtful. The first is that this latter form is depicted as possessing linear pseudopodia quite *unbranched*, whereas our form shows its pseudopodia very distinctly, but not indeed very copiously, branched. The second point is that the outer covering is represented by Claparède and Lachmann as formed of irregular arenaceous-looking particles, whereas in our form the test appears to be formed of certain problematic linear or bacillar bodies, along with minute indescribable granules agglutinated in a single stratum by an intervening, indeed seemingly organic, cement into a more or less flexible test. The third point is that our animal appears to be notably larger than Claparède and Lachmann's. That those observers are silent as to the presence of a nucleus may not bear upon the immediate question, because it may have been present in their form, though concealed by the opacity of the outer covering.

But it might almost become a further question if any of the three forms I figure belong rightly to the genus Pleurophrys by reason of the decidedly branched character of the pseudopodia; for though no absolutely strict distinction can be drawn from pseudopodia in this regard, still, as is well known, these, in many forms, maintain a great amount of constancy in their individual character. For instance, compare the pseudopodia of the two forms I have designated in this paper as *Cystophrys Haeckeliana* and *C. oculea* (Pl. XVII, figs. 1 and 3), as well as of others.

The test not being membranous, but formed of foreign and miscellaneous particles, excludes my three forms from Gromia (Duj.), not to speak of the nature of the pseudopodia, which seems to me to be very distinct from those characteristic of that genus. Admitting that their linear but branched character would be compatible with the genus Pleurophrys, they appear to be quite different from the pseudopodia in a Gromia by their comparatively rigid nature, and clear, non-granular, and tufted shrub-like appearance, without any evident current or reticulated arrangement. They are long, and comparatively straight, clear, and silvery, so to say, in appearance, the branches given off more or less dichotomously at an acute angle, and do not again appear to inosculate with their neighbours. They alter in appearance, or position, or ramification but slowly—the change which, indeed, is speediest of accomplishment is that of retraction, leaving the oval body a brown, inert, not then readily recognisable mass. All this is quite unlike the behaviour of a Gromia which, when quiet some time upon a slide, pours forth an overflow of a clouded, fluid sarcode, which gives off at all points irregularly branched prolongations, copiously anastomosing and carrying a vigorous flow of granules in a current almost like a system of vessels. It need not be remarked that the pseudopodia in our form are quite a distinct sort of thing from the finger-like, ever fitful pseudopodia of a Difflugian. That these forms, apart from the tests, are quite distinct in themselves from Difflugia on the one hand and from Gromia on the other, I could have little doubt; whether they rightly fall under Pleurophrys, as I have indicated, may be a question. For some time I imagined, as I now believe erroneously, that these forms might fall under the genus propounded by Schlumberger[1]—Pseudodifflugia,—but that type is described as having a membranous test, which, so far as I can see, would exclude the present forms. I would be disposed to suggest that Pseudodifflugia (Schlumb.) might really come nearer such a form as the so-called *Gromia Dujardinii* (Schultze).

Further, it appears to me that these differ in character quite from the pseudopodia of Euglypha or Cyphoderia, not to speak of the quite distinct kind of test. In these genera and their allies the pseudopodia are few, unbranched, extremely slender and very lively, thrown out comparatively rapidly, sometimes waved with no little vigour, drawn in hurriedly, often showing a knob-like expansion or clavate end during the act—all quite unlike the inert, comparatively persistent, shrub-like tuft of thicker pseudopodia shown by our forms. I think those who may be inclined to consider the importance here attributed to the kind of pseudopodia, and their appearance and behaviour, as too overdrawn, would be at least obliged to admit, on looking over these forms in a living state upon a slide for a length of time, that they possess at least remarkable idiosyncrasies.

The form shown in Pl. XX, fig. 1, is, I think, not uncommon; but it is easy to pass it over, though, as will be seen, by no means minute, as it may be taken almost for a pellet of some kind of excrementitious matter, and, moreover, it is shy of protruding its pseudopodia. The test, as mentioned, formed of a number of elongate and granular

[1] 'Annales des Sciences Naturelles,' 3rd ser., tom. iii, p. 256.

particles agglutinated by an intervening brownish substance, sometimes presents itself of an irregular, or lobed, or distorted shape, though a broad elliptic figure seems to be more usual; in so far as this goes it is further against its presumed identity with *Pleurophrys spherica.* The body of the creature is suspended within this test, from which it mostly stands off a notable distance, and it is granular. I have not noticed a vacuole, but a large orbicular nucleus is often readily to be perceived, which by some management can be extruded intact.

The form represented by fig. 2 (*Pl. ? amphitremoides*) is much smaller, and prone to cover itself with various diatomaceous frustules, and in the specimens I have met with contained numerous chlorophyll granules, but I could not see a nucleus. The ramifications of the pseudopodia seem more copious as compared with the preceding, hence the tuft appears more shrub-like. A smaller and nearly orbicular form, seemingly the same, is met with without diatomaceous frustules, but covered by large arenaceous particles, this perhaps, after all, equally likely to be the same as *Pl. spherica* (Clap. et Lachm.)

The third form, which I have named from its colour *Pleurophrys fulva,* is far smaller than either the preceding, and is characterised by its tawny hue and the pellucid character of its rough test, owing to its use of clear quartzose granules. These must be impacted on a basis of that colour which gives it the tawny hue. I have not seen a nucleus. This is the only one of these forms I have seen "conjugated," in which position pairs are not unfrequently met.

The character of the pseudopodia in these forms is very like that of the pseudopodia of *Diaphoropodon mobile.* I do not mean the long, extravagantly drawn-out ones, but the tufted ones at the sides or after the retraction of the very long ones. The nucleus seen in Fig. 1, too, is very like that of fig. 6, and hence the consideration of the three forms just drawn attention to naturally follows after the latter.

Amphitrema Wrightianum (gen. et. sp. nov.)

The two drawings presented (fig. 4 and 5), taken along with the explanation of the plate already given, convey, I might say, all I am able to offer upon this curious little form. This rhizopod possesses an elliptic compressed test, bearing impacted thereon a number of granular foreign particles, each face being comparatively free from these, which are most crowded along the margins and edges. At each opposite end there exists a rounded aperture, through which emanates a dense shrub-like tuft of more or less branched, linear, pseudopodia of quite the same character as that of those belonging to the forms drawn attention to above in the preceding section. One of these tufts is always notably larger and more drawn out than the other. It sometimes happens that one of the apertures becomes stuffed up by foreign granular particles clustered in a heap, so much so as seemingly to prevent the emission of any pseudopodia at all, or only a few straggling ones make their way out from amongst the particles. Indeed, it is quite rare to get examples in which the short neck-like margin of the aperture can be seen in either, not to speak of both ends, for the same kind of foreign particles which abound at the margin have a tendency to be retained around the aperture, obscuring its margin, and rendering it hard to be proven that a neck-like border exists. The body assumes generally a narrower form than the test, thus usually leaving a space at each side, though it sometimes appears to completely fill the test. It is always densely loaded with chlorophyll granules, along with which occur other brownish coloured particles. I have not seen a nucleus, nor have I seen crude food in its interior.

This seems a sufficiently remarkable form, inasmuch as I do not know of the existence of another rhizopod of its affinity with a single chamber, and with two large apertures for the emission of pseudopodia, if we except *Diplophrys Archeri* (Barker).[1] To the seeming distinctions of these two forms in a generic point of view I shall allude below; in an individual or specific point of view no two forms could be more distinct, though the descriptions of each to a certain extent, no doubt, coincide.

Having thus, in the foregoing sections of this communication, passed in review the various new Rhizopoda figured on the present occasion (with the exception of *Difflugia carinata*), the account of the forms described must naturally conclude with their short diagnostic, generic, and specific characters.

However, as regards locating them first in their due position in higher, more comprehensive groups, it does not appear to me that the known forms of Rhizopoda inhabiting the fresh-waters are yet sufficiently numerous or sufficiently understood to enable us to classify them otherwise than approxi-

[1] 'Quart. Jour. Mic. Sci.,' vol. xvi, p. 123, in "Proceedings of Dublin Microscopical Club," 19th December, 1867.

mately or provisionally under received Orders or Families. Thus the three Pleurophrys (?) shown in Plate XX, figs. 1, 2, 3, and Amphitrema (figs. 4, 5), and Diaphoropodon (fig. 6), would come under the Order Proteina, as adopted by Claparède and Lachmann, but would seem possibly to make a Family intermediate in character between Amœbina and Actinophryna. *Gromia socialis* (figs. 7—11) surely belongs to Gromida, whilst Acanthocystis and Raphidiophrys would appertain to Echinocystida by reason of having siliceous spicules; but Heterophrys, Pompholyxophrys, and Cystophrys have no spicules, yet are, no doubt, closely related, but yet according to the characters given by Claparède and Lachmann they would require a new Order.

Endeavouring, again, to arrange them after the system laid down by Haeckel in his "Radiolaria,"[1] and acquiescing that the forms here drawn attention to, referable to Pleurophrys (Clap. et Lachm.) do not possess a contractile vacuole, and if it be conceded that the marginal, pulsating vacuole shown by Diaphoropodon is entitled to come under that designation, then the latter genus must be placed near Difflugia and far apart from Pleurophrys, whereas I believe it cannot be doubted but that they are really closely allied. Again, the forms I have comprised under the new genera Raphidiophrys, Heterophrys, Pompholyxophrys, Cystophrys—not one of them, in my own opinion, possesses a central capsule, nor so far as I see even an analogue of that part of the organization of Haeckel's marine forms. *A priori* then they would fall under Haeckel's Order Acyttaria and Family Athalamia, where he would, when he then wrote, place *Actinophrys sol.* But I think there can be but small doubt that they have a far stronger claim to admission into the Order Cytophora, Family Radiolaria, notwithstanding the want of a "central capsule."

Again, on having recourse to the system laid down by Carpenter,[2] the forms referable to the genera Raphidiophrys, Acanthocystis, Heterophrys, Pompholyxophrys, would, of course, fall under his Radiolaria. The new Gromia (*G. socialis*) must be placed to Reticulosa; but then what of the form I have called *Cystophrys Haeckeliana,* with its subarborescent pseudopodia, which often coalesce, more or less, in a reticulose manner, whilst the other form referred to the same genus as yet (*C. oculea*) has linear and non-coalescent pseudopodia, clearly placing *it* under Carpenter's Radiolaria. What too of the forms on Pl. XX referred by me to Pleu-

[1] 'Die Radiolarien,' p. 212.
[2] 'Introduction to the Study of the Foraminifera,' 1862, p. 17.

rophrys (Cl. et Lachm.), Diaphoropodon (mihi) and Amphitrema (mihi)? These almost seem to hover between Carpenter's Radiolaria and Reticulosa.

Again, looking to Wallich's proposed classification,[1] *Pompholyxophrys punicea* and *Heterophrys myriopoda*, with no definite nucleus, and no contractile vesicle, and *no* spicula, (confined by the diagnosis) must come under one division of the Order "Herpnemata," whilst Raphidiophrys, likewise with no definite nucleus and no contractile vesicle, but *with* spicula, would come under another division of the same Order; yet I fancy it could hardly be doubted but that, although none of those named show a "nucleus," these forms come most naturally in alongside of the solid-skeletoned "Protodermata" (Wallich). I never saw a contractile vacuole in the little *Gromia socialis*, but it has a nucleus, thus seemingly with only half a right to a place under Wallich's Order Proteina, where, however, seemingly clearly should come Pleurophrys and Diaphoropodon.

In neither Raphidiophrys, Heterophrys, nor Pompholyxophrys, can I see anything capable of being called a nucleus, such as occur, for instance, in Actinospherium. If we admit the validity of that class, are they then "Monera" (Haeckel[2])? Even some of *those* can secrete siliceous structures during a developmental condition. The nature of the sarcode of the body-mass in the new form here described, which, as we have seen, is far indeed from being "structureless and homogeneous"—a characteristic in the diagnosis of the Class given by Haeckel, placed even before the absence of a "nucleus,"—would in itself, no doubt, place them above the "Monera" in the system. But Haeckel would now claim *Actinophrys sol* as a Moneron, but that form could hardly be placed far from such as *Heterophrys Fockii*, which indeed, like *A. sol*, does not show a nucleus, and they agree in possessing a marginal pulsating vacuole.

It, therefore, seems to me that an attempt to place the forms I have brought forward in this paper in their Classes or Orders is not yet to be accomplished. We must just for the present consider the forms on Plates XVI and XVII as annexing themselves most closely to Radiolaria (Haeck.), forming a Heliozoan subgroup, wanting a central capsule, of which *Actinophrys sol* is the very simplest representative; whilst the forms upon Plate XX, even including Diaphoro-

[1] Wallich "On the Structure and Affinities of the Polycystina," in 'Quart. Journ. of Micr. Science,' vol. v, page 57.

[2] 'Monographie der Moneren.' loc. cit.

podon (of course leaving *Difflugia carinata*, fig. 6, out of view), appertain more closely to Gromida.

The foregoing view of the case as regards forms such as those on Plates XVI and XVII, seems to receive a very strong confirmation from certain freshwater Rhizopoda brought forward only a few weeks ago by Greef, in a paper full of close observation and close reasoning founded thereon,[1] some of which forms are identical with mine and certain others very closely related, and which latter do actually seem to possess a true central capsule. I allude to the forms falling under his new genus Astrodisculus (*A. ruber*, *A. flavescens*, *A. flavocapsulatus*, Greef). This valuable and interesting paper of Greef's opens up some new points, and, as is seen, brings to notice some instructive new forms.

Upon a species appertaining to one of the genera—Acanthocystis (Carter), a valuable paper has likewise previously appeared by Grenacher,[2] with whom, so far as his observations are concerned, Greef seems to coincide. Perhaps it would be of advantage to English readers briefly to allude to the points made out in both.

As, however, some of the species Greef himself either describes or alludes to are actually identical with certain of those described by myself in the preceding sections of this communication, before the appearance of his paper (as well as previously in a cursory manner in the 'Proceedings of the Dublin Microscopical Club'), I shall, perhaps, most fittingly allude to his account of them and of the forms most immediately related in a paragraph following the diagnosis of each of the species themselves under the head of "Affinities and Differences."

Not essaying, then, to place our forms under headings of any of the proposed Classes or Orders, I shall simply attempt to give their generic and specific character, with a view to facilitate their more speedy recognition when encountered by other observers.

But before proceeding to do so, I would draw attention to the circumstance that whilst partaking with other forms of a common "heliozoan" or "actinophryan" nature, there is, moreover, one additional character or point of structure in which the forms located in the three genera now sought

[1] "Ueber Radiolarien und Radiolarien-artige Rhizopoden des süssen Wassers," in Schultze's 'Archiv für mikroskopische Anatomie," Bd. v, Heft 4, Oct. 1869.

[2] 'Zeitschrift für wissensch. Zoologie,' Bd. xix, p 288, T. 2, fig 25.

to be constructed by me, Raphidiophrys, Heterophrys, Pompholyxophrys, along with the genus Acanthocystis (as well as, possibly (?), Greef's new genus Astrodisculus), all mutually coincide; I mean, at least, as it appears to my own observation. I think, then, that in the forms included under the genera just named, the body-mass is in all composed of *two distinct regions or strata of sarcode, sharply marked off, one within the other, the outer surrounding the inner as a complete investment.* These two sarcode regions are distinguished by a sharp line of demarcation, and present considerable difference of appearance, colour, and consistency. The outer region appears to me to be the softer, and more plastic and polymorphous, the inner to be more consistent, and in itself (unacted on by outward forces) the less changeable in form. When spicula are characteristic of the species, it is the outer region which bears them—when colouring granules (green, red, or yellow) are characteristic of the species, it is the inner region which contains them. It is the inner region which projects the true pseudopodia, these passing directly through and through the outer investing region; and further, it is the inner region which receives and digests the food in such forms as have shown crude objects incepted. But sharply marked off as may be the inner globular portion from the outer investing stratum, there appears no evidence of a "capsule" including the former; nor does it appear, therefore, that the outer can be regarded as homologous with the "extra-capsular," and the inner as corresponding to the "intra-capsular" region of a true or typical Radiolarian.

It is true that ere now in bringing forward before our Club *Heterophrys Fockii* for the first time, I have suggested a different view,[1] and it is also true that Greef proposed, in his recent paper, the same view which I then suggestively put forward, of the presence in this rhizopod (which I think I see portrayed in his fig. 35), of a "central capsule" represented by the definite outline of the inner globular body. But I do not any longer see the justice of the assumption, when compared with the seemingly *true* "central capsule" of the forms contained in his new genus Astrodisculus and *Acanthocystis spinifera,* for the very same sharp line of demarcation exists in those forms; because they contain *within* the globular central portion a capsule-like structure, it is unnecessary to assume the outer boundary of the inner region as the representative of the "central capsule," and yet the globular inner

[1] "Proc. Dub. Micr. Club," 19th Sept. 1867, in 'Quart. Journ. Micr. Science.'

region seems to me to be in both quite alike in this regard. But it further appears to me to be indisputable that (in *H. Fockii*, or *Pompholyxophrys punicea*, for instance), there is no capsule enclosing the globular central portion—first, because it cannot be seen when crushed or otherwise—secondly, because the margin of the inner globular region sometimes shows pulsating vacuoles (in *H. Fockii*), like those of Actinophrys or Actinosphærium—and thirdly, because crude food is incepted into it—all which could not happen were there a rigid 'capsule' (even though minutely 'perforated' for the passage of pseudopodia) enclosing it.

If this outward boundary of the inner globe in *Heterophrys Fockii* be truly the homologue of the "central capsule," with however delicate a wall, then the same part in Astrodisculus must have a like signification, and in that case must not, surely, the central globe represent rather the "inner vesicle" ("Binnenblase")?

Nor can this outer region, as it seems to me, be supposed to be truly the same thing, that is homologous with, that sarcode-layer, attributed by Wallich to the Polycystina—the "chitonosarc" Wallich—which film of sarcode he supposes to be formed by the universal coalescence of the bases of the pseudopodia themselves.[1] In the forms now described, this layer is not formed by the coalescence of the pseudopodia, for these, taking origin from the inner region, pass out directly through the outer, without the least appearance of any confluence.

The *other* heliozoan genera forming the subject of Greef's and my own communications (along with *Actinophrys sol*) do not present the character of two differentiated portions of the sarcode body. This appears to be similar throughout, but, as the case may be, in the new genera enclosing or surrounded by characteristic structures.

The following, then, may serve as diagnostic characters of the several forms brought forward:—

Genus, *Acanthocystis* (Carter.)

Generic Characters.—Rhizopod composed of two distinct sarcode regions, the inner dense, hyaline and with or without colouring granules, of a globular and somewhat rigid figure, the outer colourless, soft, and delicate (sometimes difficult to be discerned), bearing a number of more or less elongate siliceous spicula discoid at the base and arranged in close approxima-

[1] Wallich "On the Polycystina," in 'Quart. Journ. of Micr. Science,' vol. v, N. S. page 71.

tion vertically upon the periphery of the inner sarcode body, which gives off, through the outer region, and reaching beyond the radiating spicula, a variable number of very slender, delicate, non-coalescing granuliferous pseudopodia.

It will be seen that, in the generic characters, I have claimed for Acanthocystis, as in some other new genera, two differentiated strata of the sarcode mass; the inner one is, of course, that within the cavity, the outer is not so perceptible, yet I think can be readily seen with close examination as a pale, colourless, granular, rather plastic, investment to the radial spines. It is not, I think, conceivable how these could originate, merely touched to the surface of the inner body-mass; they would not grow like a plant just in contact and no more, or even slightly immersed in the upper surface of the inner body: they must be deposited by sarcode. This outer sarcode region, I believe I readily see in living examples; and though I do not find it insisted on by Grenacher or Greef, I think I see it well depicted in the figures of the latter. I think that in Acanthocystis this differentiated outer sarcode region exists just as truly and as marked as in Raphidiophrys, Heterophrys, or Pompholyxophrys.

It will be further seen that I have left out any allusion in the foregoing characters to the assumed more or less curved tangental spicula, described by Carter as characteristic, inasmuch as I now conceive they are not of any essentially different nature, nor, indeed, even necessarily, present. Even when present, I should now, upon re-examination of the two Irish forms, quite coincide with Grenacher's suggestion that they represent but the discoid bases of certain of the spines, whose shafts have not become developed. If they were truly linear or arcuate, or crescentiform spicula, as Carter represented, and as I had myself long thought, though always puzzled about it, they should naturally be apparent on the upper portion of an example when focussed by an observer, and yet I could not perceive them; they, in fact, only appear linear when seen *edgeways* at the periphery of a specimen. These apparently distinct spicula, that is, the bases of the spines, are, of course, really circular, and appear so when viewed at the near or upper portion of a specimen, but they are very pellucid, and hence hard to be made out, even with accurate focussing, and this seems to me to account for the puzzle that the assumed linear spicula could be seen *only* at and towards the periphery.

So far, then, as I can see, the foregoing diagnosis lays down all that can be as yet absolutely stated as appertaining to the genus of which Carter's *A. turfacea* is the type.

Further accounts of its structure internally are given by Grenacher and by Greef in the papers already alluded to, but they are open to confirmation, and it may, therefore, be here useful as briefly as I can, to presently try to convey their observations.

The form described by Grenacher and called by him *Acanthocystis viridis*, is regarded by him as identical with *Actinophrys viridis* (Ehr.), but as distinct from *A. turfacea* (Carter). Greef, on the other hand, considers he latter and Grenacher's form to be identical, and truly none else but *Actinophrys viridis* (Ehr.). I would myself still venture to hold a different opinion, and regard it as not proven that *Acanthocystis turfacea* (Carter), is actually the same as *Actinophrys viridis* (Ehr.). That form is figured by Ehrenberg as densely fringed by the pseudopodia, which are very short, say not more than one third or one half of the diameter of the body. Even assuming that the radial processes are really siliceous spines (the true pseudopodia overlooked), they are thus so far quite unlike either Grenacher's or Greef's or our Irish form in these respects, for in these the spicula are comparatively fine and long, the longer ones quite equal in length to the diameter of the body (if not, indeed, longer), not to speak of their occurring of two distinct lengths. In fact, Ehrenberg's figure shows the radial processes only about equal in length to the shorter series of spines of *A. turfacea.* In the latter form, too, they are far less numerous and less crowded—could be without much difficulty counted—they do not form the *dense* fringe-like border shown by Ehrenberg. Hence it appears to me to be still a matter of doubt that Ehrenberg's *Actinophrys viridis* is trulythe same as *Acanthocystis turfacea* (Carter), or *A. viridis* (Grenacher), provided the two latter forms are distinct; until it should be proved that they indeed are truly identical, Carter's name should for the present, at least, and unless *A. viridis* (Ehr.) could be proved to be the same thing, maintain its currency.

The observations of Greef on the form *A. turfacea* (Carter), (*A. viridis*, Greef), are, as I shall now briefly try to narrate: The body is of a globular figure, frequently densely filled with green granules, showing apparent vacuoles, not pulsating. From the circumference of this globular body there stand off in a radial manner the closely posed siliceous spicula, these of two lengths. The longer are hollow spines expanded at the base in a discoid manner, and minutely forked at the apex. These are said to be by Greef in length about two thirds of the diameter of the body ; I myself think they mostly attain in length as much at least as the full diameter of the body ; and, indeed, Greef's figures so depict them. Amongst these long radial spines there occur a fewer number of others, not attaining half their length (in Irish examples say about one third), still more slender, also discoid at the base, but notably more widely furcate at the apex. The pseudopodia are long, delicate, colourless, and granuliferous, and by both Greef and Grenacher described as possessing an axis like those of Actinophrys. For so far all this is apparent on even a superficial examination, and it accords, too, with Grenacher's description of his form, the main difference being that he represents the bifurcation of the apex of the longer spines as less pronounced and less divergent than in Carter's, and they are said to reach in length only half of the diameter of the body ; thus, indeed, more approaching in this regard *Actinophrys viridis* (Ehr.), but still they are far fewer and far less crowded than in Ehrenberg's figure. A new point brought forward by Grenacher is the existence in the centre of the globular sarcode mass, of a little pale body or cavity from which proceed in an everywhere radiant manner, from the very centre numerous pale, delicate threads or lines showing an agreement in appearance with the *axes* of the pseudopodia ; and the author assumes, though he could not satisfy himself, that these lines radiating from the common centre were, in truth, carried on directly through and through the body mass, reappearing as the axes of the pseudopodia. Greef confirms the account given by Grenacher, as he was able to extrude by pressure a vesicle containing a solid "nuclear mass" (Kernmasse), which he regards as the common central starting-point from which radiate the fine threads, and he thinks he can recognise in the extruded vesicle with its contained corpuscle, the doubly bounded space which occupies the centre of an uninjured example, but which he could never see of that sharply bounded stellate figure depicted by Grenacher. He (Greef) states he has been able to follow the central radiating lines even to under the surface of the body, where they get lost, and incapable of being directly followed outwards into the pseudopodia.

Greef alludes to two kinds of probable reproductive processes—a direct self-division of the total animal into two, and a resting or "encysted" state. The latter consists in the withdrawal of the sarcode body-mass from the inner boundary formed by the union of the bases of the radial spines, leaving a rather wide empty border, and its becoming invested by a double coat, a firm inner one, when empty dotted, as if perforated, and an outer hyaline one. Of any further development of this state Greef has not seen any indication.

It is difficult to condense his long and interesting account so as to do fair justice to the matter, but the foregoing may, I trust, convey an epitome of his principal observations.

I myself would argue as Greef does in one place that in *Acanthocystis turfacea*, just as little as in *A. Pertyana*, is there anything around the central sarcode body which might be denominated a special membrane. The so-called "lorica"—"biegsamer Panzer"—"verdichtete Rindenschicht"—can be but the expression of the mutually approximated discoid bases of the radial spines being held together with a considerable amount of coherence by means of some intervening bond not readily perceptible. One meets not unfrequently portions of the periphery of a defunct Acanthocystis, perhaps as much as a fourth or fifth of a circle (nay, whole globes sometimes), all the sarcode clean gone and nothing but a greater or less number of radial spines left, and those still cohering by their discoid bases, the shafts radiating in the same manner, just as they would if they stood at the periphery of a still living example. See also Greef's fig. 15 of an "encysted" state of this form, where the peripheral system of spines, still mutually coherent, stand off a distance from the contracted inner sarcode body, the latter now surrounded by its special investments. But Greef's views expressed in his account of this outer boundary (page 488), do not seem to me to coincide with those previously expressed (page 484). The following extract gives the ipsissima verba used by Greef in the latter place referred to:—"Ich meinerseits habe keine bestimmte Anzeichen finden können, die mit Sicherheit eine besonders abgegrenzte und erhärtete Rindenschicht, oder was doch wohl dasselbe sagen will, eine Membran bekunden, wohl aber mehrere die auf die Abwesenheit einer solchen schliessen lassen." And he adduces as evidence of the foregoing view the fact (not unfrequently to be seen), that large finger-like sarcode projections are capable of being extruded through the outer boundary, these withdrawn, and the place of their exit effaced, and the spines again all normally *in sitû*. He also refers as additional evidence to the extrusion of green granules by a separation of the spines, and by intervening opening which again disappears. All this is just and according to fact, but I think the foregoing extract does not seem to accord with the following, in alluding, in the eneysted state (fig. 15), to the very same outward boundary where the bases of the spines touch and form a hollow globe, now standing off leaving a vacant margin around the encysted body:—"Die Oberfläche dieses Saumes wird vielmehr durch eine zarte und glashelle aber starre und undurchdringliche Kieselhülle gebildet, in oder unter welcher die Fussplättchen der Stacheln festsitzen." It does not appear that they can truly be held imbedded in a rigid pellucid siliceous coat, from the strong reasons previously adduced. The intervening bond of union, whatever it may be, has then no connection with the central sarcode body, for the spines can independently maintain their mutual position, and yet this with greater or less readiness gives way to forces sometimes from within, and most likely sometimes from without, and in the living animal recovering position and coherence, as, for instance, pressing on the covering glass, or the action of the little parasitic rotatorian previously described by me. In addition to the fact of finger-like protrusions of colourless sarcode being projected and withdrawn, one sometimes sees likewise two individuals of *A. turfacea* united by an isthmus ("conjugated"?) (showing that the spines can recede from one another and come back to position) the isthmus gradually stretched, till it becomes a mere connecting thread of sarcode, which eventually snaps, and the two examples pass away from one another, no trace of the place from which the mutual band of sarcode emanated being left. Can such examples represent the state of "fission" (Zweitheilung), adverted to by Greef, as one of the reproductive conditions of *A. turfacea?* Under ordinary circumstances, the circular discoid bases of the spines must leave triangular spaces between, quite large enough to allow of the passage forth of the ordinary pseudopodia through the medium which we must assume causes the mutual coherence of the spines, and this without any displacement of the latter.

In this genus Acanthocystis Greef brings to notice three other forms; two he names as new, and the other he leaves in abeyance. One, which he calls *Acanthocystis pallida,* seems to me to differ in no respect from *A. turfacea* further than in the absence of the green colouring granules, except that he attributes to it the existence of the bacillar or linear spicula said to occur along with the radial spines, and as to which, indeed, he is silent as regards Carter's form. But, as I have mentioned, I now am quite disposed to hold these as very problematical. There remain only then the colourless granules to distinguish this form from *A. turfacea.* Now, quite colourless examples of what I have always thought could be none else than *A. turfacea* often occur here; nay, examples present themselves with a great proportion of the body mass bearing the green (chlorophyll) granules, and the remainder colourless, and these regions marked by a sudden transition. I think the same green granules sometimes become colourless. We see sometimes Difflugiæ loaded with chlorophyll granules. *Raphidiophrys viridis* sometimes, but it is rarely, shows some colourless granules. I cannot but think, therefore, that *A. pallida* is only a colourless, not at all uncommon state, of *A. turfacea*.

But as regards another form (not named) referred to by Greef I would, indeed, very deferentially think he had far better ground to establish it as a distinct species. I mean the form he gives in Pl. XXVII, fig. 18. This form seems to be distinguished from *A. turfacea* by the want of the shorter series of radial spines—by the longer series, less copiously present, being, according to Greef's account, immersed to a certain extent within the periphery of the inner body—and, further, by the outer sarcode region, here strongly pronounced, being subdivided at the outer margin into a great number of exceedingly delicate, linear, acute processes, the pseudopodia passing, just as in *A. turfacea,* right out through the outer region afar into the water. This fringe-like subdivision of the marginal or circumferential portion of the outer sarcode-region strongly resembles that characteristic of my *Heterophrys myriopoda* (see my Pl. XVII, fig. 4). In that form, however, there are *no* radial spicula like Acanthocystis. Further, judging from Greef's figure his form is of a more olive-coloured green than *A. turfacea.* I do, therefore, venture to think this must really be quite distinct.

Greef, moreover, names a further new form *Acanthocystis spinifera,* of all the described forms, coming nearest to my *A. Pertyana,* but, as will be seen in the remarks following the diagnosis of that species, under the head of Affinities and Differences, seemingly, indeed, quite a different thing therefrom.

The following will, I think, serve as a diagnosis of the new form occurring in this country :—

Acanthocystis Pertyana (Arch.).

(Pl. XVI, fig. 1).

Specific characters.—*Radial spicula very short, shaft comparatively thick, tapering, pointed at the apex; pseudopodia very slender, in length about equal to the diameter of the body, bearing minute granules passing up and down; body mostly colourless, but sometimes green when more or less loaded with chlorophyll-granules.*

Measurements, variable; diameter of body from $\frac{1}{800}$ to $\frac{1}{600}$, length of spine $\frac{1}{5000}$ to $\frac{1}{3300}$ of an inch.

Localities.—In pools near Carrig mountain, county Wicklow, and in county Westmeath; rare.

Affinities and Differences.—This form is readily and at once distinguishable from *A. turfacea* (Carter) by its considerably smaller size, short tapering and pointed spines, of one kind only (not elongate, hardly tapering, bifurcate, and of two dimensions). It is, likewise, far more usually devoid of chlorophyll-granules than *A. turfacea,* nor are they ever seemingly present in the dense quantity which ordinarily characterises that far more striking species. The pseudopodia do not seem to be so delicate in my form, at least they are usually more readily noticed than in *A. turfacea.*

Its points of difference from Greef's new species *A. spinifera*[1] are not so manifest at first sight, but on closer comparison the two forms seem abundantly distinct. That species seems pretty nearly to agree with the present in dimensions (though apparently averaging a little larger), as well as in the radial spines being slender and pointed, but in it the spines are much longer, far finer, and much more elongate, and more acute. In *A. Pertyana* the spines are in length not more than a fifth or sixth part of the diameter of the body, whilst in *A. spinifera* they seem to be in length about three fourths of the diameter of the body. They agree in having a discoid base (Fussplättchen), as do, so far as we know, indeed all the species referable here. In my form there never appear any colouring granules except green, never yellow bodies ("yellow cells"?) like those often occurring in Greef's. Further, I have never seen in the former any indication of the central vesicular body (representative of "central capsule"?) which forms a distinguishing feature in the latter. I have never, indeed, myself seen Greef's form, but I cannot entertain a doubt as to its complete distinctness from mine.

As regards other described or figured forms I have already (p. 255) referred to its resemblance to *Actinophrys brevicirrhis*[2] (Perty); if, indeed, we conceive for a moment the pseudopodia absent from my figure, and assuming that the rays bordering the figures given by Perty, may be actually spines, not pseudopodia, the resemblance to Perty's is certainly greater than to any other published figure I know of; but the very great uncertainty due to the insufficient account given by him of his animal fully justifies my appropriating to this species a distinct name.

But in contrasting my form with Greef's *A. spinifera,* and on reading over his very interesting account of that species,

[1] Loc. cit., p. 493, t. xxvii, figs. 20—23.
[2] L. c., p. 493, t. xxvii, figs. 20—23.

I am led on to advert to his further suppositions regarding the assumed developmental or transition states thereof drawn by him.[1] Beyond any question Greef's figures 26, 27, and 28 represent nothing else than Barker's *Diplophrys Archeri*,[2] and nearly equally certainly Greef's figure 29 represents my own *Cystophrys oculea*.[3] The nearly orbicular (Fig. 26) or broadly elliptic (Figs. 27 and 28) figure of Diplophrys is there, the large characteristic conspicuous oil-like, amber coloured, refractive body, with the same little granular bodies, are there, and the two pencils of delicate pseudopodia emanating from opposite ends, but set slightly obliquely to one another, are there,—all just as they occur in this very marked little form, as it has presented itself in gatherings made from the east, south, west, and centre of Ireland. But although this wide distribution must be attributed to it, it is always seemingly scanty, and rarely encountered; this may indeed be, in part, due to its great minuteness. Perhaps Greef's otherwise excellent representation of this form would have been improved if he had indicated that sometimes the pseudopodia slightly subdivide dichotomously, and occasionally show more or less of a changeable dilatation at the point of ramification or along the length of a pseudopodium. I myself have never seen anything like Greef's figure 25.

(*To be continued.*)

On a New Polyzoon, "Victorella pavida," *from the* Victoria Docks. By Wm. S. Kent, F.Z.S., F.R.M.S. of the Geological Department, British Museum. With Plate IV.

In November, 1868, I briefly referred, in the pages of 'Science Gossip,' to a representative of the Ctenostomatous Polyzoa, taken by myself in the brackish waters of the Victoria Docks.

Though at the time possessing strong reasons for premising the species to be new to science, no name was conferred upon it, and it was rather brought forward with the view of attracting attention and possibly of recognition.

Having been fortunate enough this last autumn to secure fresh samples from the same locality, and feeling now convinced that the form represents not only a new species, but moreover, serves as the type of a new genus, and even family,

[1] L. c. t. xxvii, figs. 25, 26, 27, 28, 29.

[2] Proceedings Dublin Microscopical Club, in 'Quart. Journ. Micr. Science,' loc. cit.

[3] Ante in this paper, p. 265.

I proceed to describe it at greater length, and to bestow upon it a name which shall distinguish it from its congeners.

The sub-order of the *Ctenostomata*, in accordance with the system of classification most recently accepted, is subdivided into the two families of the *Alcyonidiadæ* and the *Vesiculariadæ*. The first of these is distinguished by the polypidom being sponge-like, fleshy, and irregular in shape, and in which the cells furnished with a contractile orifice are immersed. In the second the polypidom is plant-like, horny, and tubular, having free deciduous cells, whose extremities are flexible and invertile.[1] The species to which I would now direct attention, though possessing the circlet of setæ characteristic of the order, secretes a polypidom referable to neither of the two forms just indicated.

Its affinities with the *Vesiculariadæ* are the most marked, but, as will be seen on reference to the accompanying plate, the contour of the polypidom is entirely irregular, and wholly wanting in that uniformity and complexity of structure so characteristic of that family, possessing neither the main rachis nor the distinct deciduous cells by which all the members of the *Vesiculariadæ* are so readily distinguishable.

Hence, it appears essential that another family should be specially constructed for its reception, and, taking into consideration the uniform structure of the polypidom, I propose the acceptance of the term *Homodiætidæ* as a family name expressive of that same structure. The diagnosis of this third family may be briefly summed up as follows:

Polypidom horny, tubular; cells not deciduous nor separately distinguishable, but throughout freely communicating, their terminations flexible and invertile. The generic name I propose for this new polyzoon is *Victorella*, a somewhat lame acknowledgment of the great variety of animal life which the locality from whence it was procured has afforded me. The same characters above given serve also for its generic distinction, to which must be added that the animal has no gizzard, and is provided with eight ciliated tentacles.

[1] In Gosse's 'Marine Zoology,' and in the 'Micrographic Dictionary,' the genus *Pedicellina* is admitted as the representative of a third family of the *Ctenostomata*, but the error of such an arrangement becomes apparent on considering that the individual animals in the various species of this genus simply roll up their tentacles when at rest, and are unable to withdraw them within their polypidom; and hence the crown of protective setæ surrounding the orifice of the cell and characteristic of all the true Ctenostomes would be to them a simple superfluity, and the absence of these characteristic setæ, in conjunction with other well-marked structural peculiarities, has furnished ample grounds for setting this genus apart as the type of a separate sub-order known as the *Pedicellineæ*.

It now only remains to furnish the specific title, and for reasons which I shall explain hereafter, I have chosen that of *pavida*, and, as *Victorella pavida*, its characters may thus be cited:

Polypidom minute, confervoid, adherent, or semi-erect, irregularly branching. Tentacles eight in number. No gizzard. Inhabiting brackish water, parasitic on the polypary of *Cordylophora lacustris*.

As is above intimated, I have so far only met with this minute and very elegant little polyzoon attached to the polypary of *Cordylophora lacustris*, and, to the unassisted eye, the only visible indication of its presence is afforded by the appearance of, as it were, a more or less entangled mass of confervoid filaments adherent to the protective sheath of its more robust though less highly organised supporter. The assistance of the microscope thus becomes essential for the elucidation of the true nature of the organism, and on the removal and examination of a fragment of the filamentous mass with the aid of that instrument, its true affinities are immediately made apparent. Each slender filament now proves to be a tiny tube containing a living organism of the most delicate and complex structure, and which, on expansion, is at once recognisable as a true representative of the Infundibulate Polyzoa; the presence, moreover, of the circlet of protecting setæ aiding still further to refer it to the Ctenostomatous section of that order.

One of the most striking features connected with the life-history of this little Polyzoon is its remarkably shy and retiring habits, and hence my choice of the specific name adopted. On first transferring a fragment from the aggregate mass to a hollow slide or zoophyte-trough, for more convenient microscopic examination, it at once retreats to the remotest corner of its domicile, and many minutes, and sometimes even hours, pass away before it again ventures to display itself; on its doing so, however, the patient observer is amply rewarded for the brief delay. A forward motion of the tentacles first takes place, then the circlet of setæ, in the form of a fascia, appears beyond the orifice of the cell, extended to their utmost, the tentacles push onward through and beyond them, and in far less time than has been occupied to record it, the little creature is expanded in its fullest glory. A crown of eight long, delicate, and remarkably flexible tentacula now surmounts the transparent stalk, and a tiny yet rapid maelstrom responds to the rhythmical vibration of the many thousand scarce visible cilia which clothe them, engulphing and hurrying away to a living tomb such unfortunate infusorians or monads as may happen to wander within the precincts of its eddying vortex.

Of specimens of *Cordylophora* detached from the timber-baulks in the Victoria Docks in search of this Polyzoon, by far the richer supply has been found adherent to that which was taken some five or six feet below the surface of the water, little, if any, being met with on such as from its proximity to the surface was easily attainable with the hand alone; and this apparent partiality for deep water readily accounts for the difficulty that has hitherto been experienced in keeping it alive for any length of time in shallow vessels.

The morphological affinities of *Victorella pavida* are somewhat remarkable. Overlooking, for a brief interval, the structure of the polypidom, its relationship to various representatives of the Vesiculariadæ are at once palpable and striking. Its possession of eight tentacula point out its affinity to the genera *Serialaria, Vesicularia, Valkeria, Mimosella,* and *Bowerbankia,* but the absence of a masticatory organ or gizzard restricts the comparison to the first and third alone of those five genera; and happening to have at hand an admirably prepared slide of *Valkeria pustulosa,* with the tentacles in a state of full expansion, I found that, except in size, the latter being much the larger, any histological difference between the two was difficult to determine.

Of as high importance, however, as the various modifications of the endoskeletal system of ossification in the different groups of the Vertebrate division of the animal kingdom, must be ranked the structure of the polypidom or exoskeletal system of support which obtains in that invertebrate section of the same kingdom to which we are now referring; and here, as has been already demonstrated, there is a most essential and important difference, and one which very few words will suffice to show carries with it a peculiar and no less important significance.

The infundibulate arrangement of the tentacula and, above all, the coronet of protecting setæ at once suggest the necessity of referring *Victorella* to a sub-order,—the *Ctenostomata,* which has hitherto been known as having none other but marine representatives; while, on the other hand, the irregular and homogeneous structure of its polypidom is precisely what we meet with in *Plumatella* and other members of an order,—the *Hippocrepia,* of which not a single representative has yet been discovered inhabiting pure sea-water, and hence the sum total of the structural peculiarities of this minute denizen of brackish water seems to go far towards furnishing us with one link of a series which future investigation and comparison may demonstrate to unite the two.

In other papers[1] I have alluded to the extraordinary variety of animal life that is to be met with in the waters of these docks, some being of undoubted marine origin, and the remainder representing forms commonly frequenting our inland rivers and ponds; while again, as has just been shown, it is not wanting in a type of structure peculiar to itself,—or rather one possessing characters shared in by the two, though common to neither. On the whole, however, and if it is right to judge from the higher forms represented, the balance is greatly in favour of the freshwater species, since among these all the fish must be included, those which are taken there in considerable abundance being such fluviatile forms as perch, roach, dace, rudd, and bream, &c.

Among the Polyzoa *Plumatella repens* is frequently met with, and this last autumn I have had the satisfaction of taking from the same locality a no less truly marine representative of the class than *Bowerbankia imbricata,* though only in that immature and creeping condition which was at first discribed as *B. densa* by Dr. Farre, but which was subsequently demonstrated by Mr. G. Johnston to be simply the early stage of the first named species.

The idea may possibly suggest itself to some that the form just described as *Victorella pavida* is identical with the species last referred to. A glance, however, suffices to dispel any such illusion. *Bowerbankia* was at once recognised by its conspicuous and well developed gizzard, by its shorter and more rigid tentacula, and, lastly and most essentially, by the structure of its polypidom, which was in entire accordance with what obtains in and is characteristic of every representative of the *Vesiculariadæ,* having a main rachis bearing distinct cells constricted towards their point of attachment, and whose deciduous nature was made apparent by the facility with which they became detached from that point of juncture.

Another interesting form, which has not been previously referred to, as inhabiting these waters is that low type of the Annelida *Ælosoma quaternarium* Ehr. whose morphology has been so ably figured and described by Mr. E. Ray Lankester in vol. xxvi, part iii, 1869, of the Transactions of the Linnæan Society, and a recent excursion to the 'Docks' in company with the above-named gentleman resulted in the capture of other representatives of the same sub-kingdom, which will probably be shortly introduced and described by him as new to science.

[1] "On a New British Nudibranch" (*Embletonia Grayi*), 'Proc. Zool. Soc.,' Jan., 1869, and "On some New Infusoria from the Victoria Docks," in the 'Monthly Microscopical Journal' for May, 1869.

In the accompanying plate (Plate IV) the small figure in the left hand corner indicates a small detached fragment of *Victorella pavida* of the natural size (fig. 1); the centre group (fig. 2) is the same considerably enlarged, and beneath that again is a piece of *Cordylophora lacustris* magnified a few diameters, and showing the mode in which the Polyzoon attaches itself to it (fig. 3).

On a CRUSTACEAN PARASITE *of* NEREIS CULTRIFERA, Grube. By W. C. McINTOSH, M.D., F.R.S.E., F.L.S. With Plate V.

CRUSTACEAN parasites (Ectozoa) of the Annelids would not appear to be very common; but this may arise in some measure from their having been overlooked, rather than from their actual rarity. So far as I at present am aware, Dr. H. Kröyer is the only author who has described such a parasite upon an Annelid ('Naturhistorisk Tidsskrift,' 1864, p. 403, Tab. 18, f. 6, *a—g*). This species, *Silenium polynoes*, Kröyer, he found on an example of *Polynoë cirrata*, O. Fabricius, (the well-known and widely distributed *Harmothoë imbricata*, Lin.). In *Silenium polynoes* the body of the female is in the form of a simple saccate mass, with a petiolate process for attachment, and furnished with two flask-shaped ovaries at the posterior extremity. There are neither antennæ, rostrum, nor feet. The male, again, is much more minute, though more complex in organisation, for the produced anterior region of the cephalothorax is supplied with four pairs of feet, and the caudal process is triarticulate and setose. Dr. Baird also mentioned to me that he had seen a small crustacean parasite attached to a foreign *Lepidonotus*, while examining the collections in the British Museum.

On the rich shores of the Channel Islands several examples of *Nereis cultrifera*, Grube, a plentiful species under stones and in other places between tide-marks, were infested by a crustacean parasite, which, from its size, and the colour of the ovisacs, was very easily observed. They generally occurred towards the posterior end of the worm, and the largest specimen of the worm had about fifteen examples of the parasite, most of them, however, being small. The usual number was from three to five. They either adhered to the groove between two feet, or to the sides of the feet. In the living state they are of a dull whitish colour, and cling most securely to the worms, but they do not seem to incommode them to any extent; and it may be supposed that the Nereis can at any time easily rid itself of its visitors if so inclined. The worm, however, is by no means an active species.

Male.—In the examples of the parasite which appear to be males (Plate V, fig. 1), the cephalothorax is somewhat ovate, narrowed behind, and furnished in front with minute chitinous processes like hairs. The head has a pair of rather stout antennæ, composed of four regularly diminishing segments. The edges of the last three under pressure are not smooth, but have minute projections from which the setæ spring. The terminal segment is tipped with several setæ, which are longer than itself. The next two limbs arise from the region between the former and the mouth, are more slender, and appear to have a similar number of articulations, but, from their softness, it is difficult to determine the exact number. The exposed surfaces of the two terminal segments are marked by certain microscopic serrations, which have a definite arrangement. The tip of the organ has about five long setæ, one of which much surpasses the other in length. The sides of the mouth are furnished with two pairs of folded limbs, the posterior of which has its terminal articulations furnished with a single strong claw.

The cephalothorax behind the foregoing has on each side three processes. The anterior consists of a short and rounded basal portion, with two slender terminal processes. The latter differ from each other in shape, the outer having the form of a sabre-sheath, with numerous microscopic spikes along its convex surface, and a tip furnished with about three setæ, two of which are very long. The other process is generally straight, has also the microscopic spines along its outer border, but there is only a single long seta on its tip. The same description applies nearly in all respects to the succeeding pair of limbs. The third pair is represented by a simple conical papilla with a terminal bristle.

The cephalothorax generally is filled with a minutely granular substance, except when the well-defined longitudinal and transverse muscles of the body appear. Posteriorly there are also two elongated translucent bodies, probably the testicles (?). Behind the posterior narrowed region is a bifid caudal process with two long lateral styles. Each of the latter consists of a strong basal segment, and an elongated filiform process with traces of segmentation.

The male is very much smaller than the female.

Female.—In the young condition the body has a somewhat ovoid figure, with a produced posterior and a blunter anterior extremity, and both antennæ and caudal styles can be observed from the dorsal surface (fig. 4). In the fully developed condition, however (fig. 2), the thoracic region becomes greatly enlarged, so that when viewed from the dorsum, the animal has a rounded-ovate form, with a narrower region in front, and it is only in a lateral or ventral view that the snout and caudal portions are fully seen (fig. 3). In many specimens there is a series of somewhat regularly arranged dimples on the dorsum.

The head is supplied wlth two antennæ composed of four segments, the sides and terminal articulation of which have numerous chitinous spines. The next pair of appendages is similar to the corresponding pair in the male, being more slender than the antennæ, and tipped with strong spines, one of which is longer than the others. There next follows a series of foot-jaws, but they are not easily differentiated. Two pairs of thoracic feet at least are visible behind these.

Posteriorly the abdominal region is bifid, and furnished with two long styles similar in structure to those first described. In the adult, the ovisacs are attached by a slender pedicle to a slight prominence on each side near the posterior end of the abdomen, and form cylindrical pouches filled with a vast number of whitish ova.

The microscopic spines occur on the anterior part of the cephalothorax, the snout, and other parts of the animal.

NOTE *on the* DISTRIBUTION *of* NERVES *to the* VESSELS *of the* CONNECTIVE TISSUE *in the* HILUS *of the* PIG'S KIDNEY, *and on the* GANGLIA *found in Connection with these* NERVES. By JAMES TYSON, M.D., Lecturer on Microscopy at the University of Pennsylvania. (With three woodcuts.) Read before the Biological and Microscopical Section of the Academy of Natural Sciences of Philadelphia.

PREMISING that these observations are incomplete, though, I believe, accurate as far as they go, my excuse for bringing them under the notice of the Section at this stage is, that its members may have the opportunity of examining the preparations before they become altered. Having been mounted quite two months, they are still in good keeping, though it is impossible to state how long they will remain thus. The observations, when complete, will also have for their object to determine whether these nerves and ganglia are also distributed to the vessels and tubules in the secreting structure of the kidney, as the labours of Beale appear to have shown; failure to demonstrate this latter point being attributed to want of success in the injecting process—a matter of no inconsiderable difficulty, as the entire organ (that of the young pig) was but an inch in length, and the artery, therefore, exceedingly small. It is hoped that further attempts will accomplish this, though Dr. Beale himself, of whose observations these are a repetition, makes no allusion to having found the nerves upon the tubules and vessels of the pig's kidney, but has demonstrated them most satisfactorily in this portion of the kidney of the frog and newt, as will be learned later.

FIG 1.

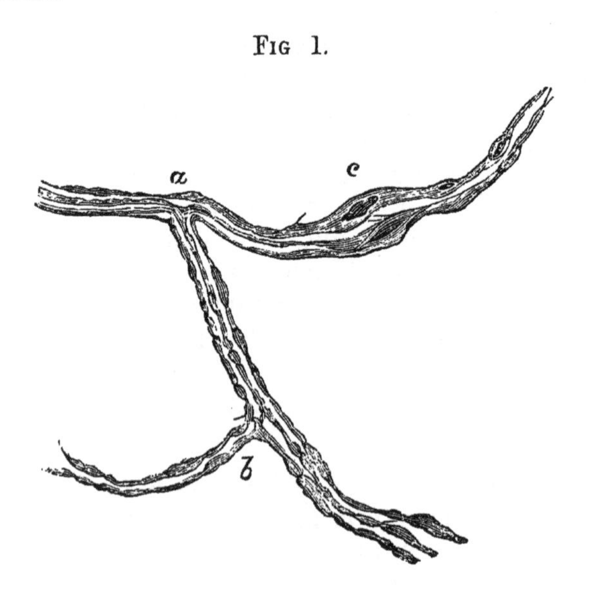

The preparations were made from the kidney of the newborn pig, which was treated as nearly as possible after the method of Dr. Beale—staining with carmine, washing with dilute glycerine (2 glycerine, 1 water), and finally preserving in acetic acid glycerine (5 drops to oz.). In this latter solution, also, have the preparations been mounted.

The appearances can be well traced in the diagrams which I have had prepared (Figs. 1 and 2). In Fig. 1 is shown a bundle of fine nerve-fibres, characterized by the presence of germinal matter at different parts, and longitudinally throughout their course, dividing at the points *a* and *b*, and expanding into large ganglion-cells at *c*, with each extremity of which a nerve-fibre is continuous. Fig. 2 exhibits a finer set of nerve fibres, constituting a network in the vicinity and immediately about the arteriole, easily distinguished, though uninjected, by its transverse nuclear markings. The filaments here correspond each to a single one of the bundle of

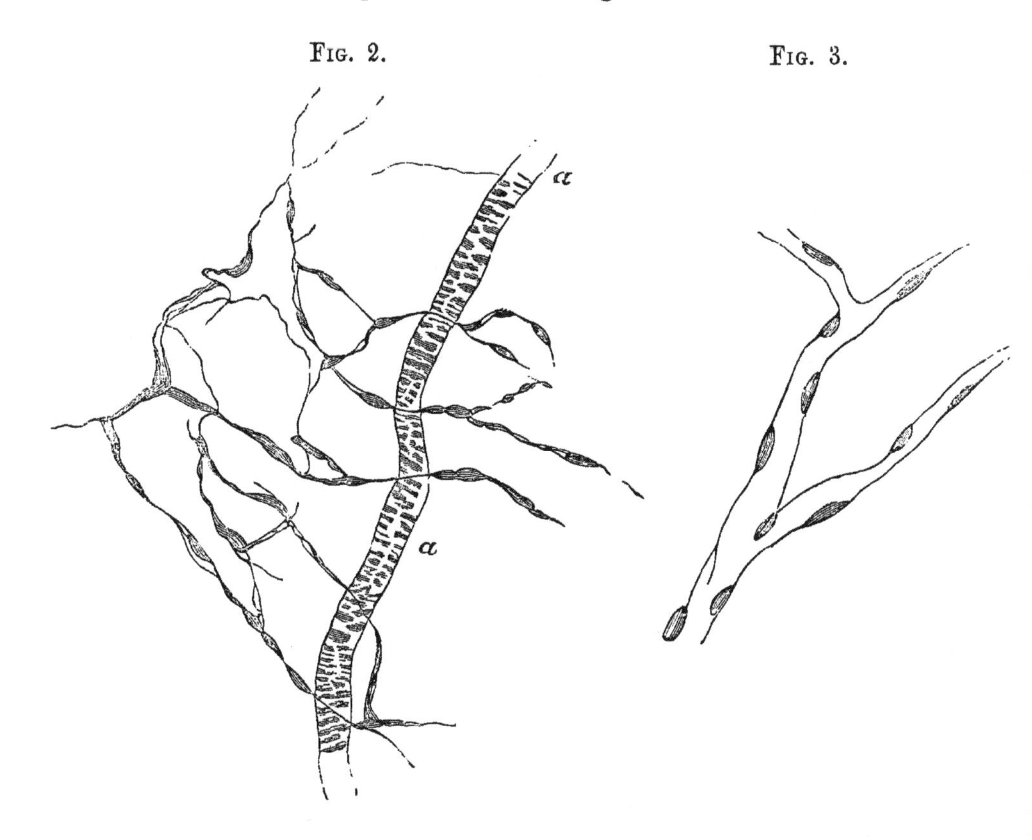

Fig. 2. Fig. 3.

three represented in Fig. 1, and exhibit the same subdivisions now accorded as characteristic of the finer nerve fibres in their ultimate distribution, when they are deprived of the adventitious membranous sheath and white substance of Schwann.

No difficulty will be encountered in distinguishing these bundles of fine nerve-fibres as drawn in Fig. 1, from the arteriole and capillary with which it is barely possible they may be confounded. The transverse nuclear markings of the arteriole are at once characteristic, and although capillaries are also characterized by a more or less longitudinal arrangement of their nuclear markings, yet by comparison of Fig. 1 with Fig. 3, it will be seen that they are still different from those of the nerve fibres, being distinguished by an alternate arrangement of the nuclei on the sides of the vessel, while the germinal matter in the nerve-fibre occupies its entire thickness, bellying it out, as it were, while the immediate continuation is often even more slender than the body of the nerve fibril at a point midway between two nuclei.

Further evidence that these delicate fibres are fine nerve-fibres is seen in the fact of their continuation with the large ganglion cells, and most conclusive proof in the fact that they have been traced by Dr. Beale in continuation with dark bordered nerve-fibres, and resulting from the subdivisions of the axis cylinder of these.

No ganglia are represented showing more than two cells, for this reason, that, although cellular collections revealed themselves under low powers resembling those indicated by Dr. Beale as masses of ganglion cells, yet, as they could not be resolved by high powers into individual bodies exhibiting the characters of ganglion cells, I was unwilling to present them as such. This difficulty may have been due to the intensity of the staining, which was too vivid in the case of many cells, all portions being alike deeply tinged, so that the different parts could not be easily resolved; nor, in aggregations, could the cells themselves be separated. Yet we believe the fact of the existence of the ganglia is no less established, at least until some more satisfactory interpretation is put upon them.

These ganglia and nerves, apparently quite abundant in the connective tissue of the notch of the young pig's kidney, have also been found by Dr. Beale in the kidneys of children. They are, however, more difficult to obtain in the adult human kidney, probably in consequence of the increased amount of formed material which rapidly grows at the expense of the germinal matter, rendering the latter more indistinct and difficult of demonstration with increasing age. This change is readily demonstrable by microscopic examination of the tendon of a child or young animal, in comparison with a corresponding structure of an adult. The former will be found crowded with nuclear or germinal matter, while, in the latter, it will be placed at longer intervals, separated by larger quantities of the fibrous formed material now in excess. Thus is produced a so-called connective tissue in which uninjected capillaries and shrivelled nerve fibre, which have performed their function, together with the more indistinct ganglia and fibres still active, form a part. Hence it is believed by Dr. Beale, that the oval masses of germinal matter seen in the intervals between the tubes, and upon the surface of the vessels of the mammalian kidney, belong to nerve-fibres, the structure of which is so delicate, or so much obscured, that it is extremely difficult to trace their course.[1]

[1] Beale, 'Kidney Diseases and Urinary Deposits,' 3rd edition, 1869, p. 17.

In the frog and newt, however, he has demonstrated the ganglia, and traced from them the nerves which he has traced to their ultimate distribution around the tubes and capillary vessels of the kidney.[1]

These nerves and ganglia, mainly derived from the sympathetic, in connection with those referred to as described by Dr. Beale in the secreting portion of the kidney, also from the same source, are believed by the latter physiologist to be a part of a system of afferent and efferent nerves, distributed to the capillaries and tubules of the kidney, by means of which so-called nervous influences, as emotions, and it might be added also, in some instances at least, remedies, produce their effect upon the secretion of this organ. The ganglia are considered as bearing the same relation to the kidney that the ganglia, believed by many to be in connection with the cardiac nerves, bear to the heart. The nerves distributed to the walls of the uriniferous tubes, are believed to be *afferent* or sensory, conveying to the centres an impression, the response to which is conveyed by the *efferent* nerves, or those distributed to the capillaries, whereby the latter dilate or contract their calibre. Thus, these nerves regulate the supply of blood to the secreting structure of the kidney, producing on the one hand an abundant and rapid supply, accompanied by a distended vessel and proportionate secretion of watery urine; on the other, a diminished and slower current, and corresponding secretions of urine, probably containing a larger amount of solid ingredients.

If these views are correct, and they certainly involve no dislocation of the principles of deductive and analogical reasoning, in connection with the recent accepted views of the function of the sympathetic, they become most important not only in explaining physiological and pathological phenomena of the kidney, but also in their relation to therapeutics, while they account most satisfactorily for certain recognized phenomena in the secretion of urine. It would seem that their correctness must rest entirely upon that of the observations. If the facts be correctly observed, it would appear to me that the view is at least a legitimate conclusion, though, in the uncertainty of our science, it may not be the only one.

If, then, so much depends upon the accuracy of observation, it is exceedingly important that other observers should go over them, to confirm if possible, or to deny if need be. To this end, it is hoped, these efforts contribute.

It can scarcely be objected to the conclusions drawn that the facts have not all been demonstrated in a single species;

[1] Ibid., p. 16.

that while the nerves and ganglia have been found in the connective and submucous tissue in the notch of the pig's kidney, and in the same situation in the child, they have not been found in connection with the tubular structure of these kidneys, but in those of the frog and newt. The elements of general physiology, however, show most conclusively that except there be some fundamental difference in function, structural arrangements which hold good in the organs of one animal hold good in the corresponding organ of another. Now, we not only have no such fundamental difference in the function of the kidney of man, the frog, and newt, but the structure, so far as the acknowledged secreting elements are concerned, is almost identical; and the arrangement of the tubules and capillaries in the organs of these lower animals has been of invaluable service in determining the true anatomy of the human kidney. With equal propriety, then, may the arrangement and the distribution of the nerves in the same organs be used to determine more delicate points in the anatomy of the human kidney.

The Minute Structure *of the* Human Umbilical Cord. By Dr. K. Koester.[1]

(*Abstract.*)[2]

In a short historical introduction Dr. Köster recapitulates some of the views which have been taken of connective tissue in general, and of tissue such as that of the umbilical cord in particular. The peculiar interest of this structure lies in the fact that it is nourished without any proper vascular system. Virchow, who first showed the identity of bone-, cartilage-, and connective-tissue-corpuscles, was also the first to suggest that these bodies, with their prolongations, formed an anastomosing system of channels by which nutriment is conveyed. These views were energetically opposed by Henle and, to some extent, by Kölliker, and a long controversy arose, to which a new direction was given by the researches of Recklinghausen.[3]

This observer regarded the anastomosing system of channels not as formed by prolongations of the connective tissue cor-

[1] 'Die feinere Structur der Menschlichen Nabel Schnur. Inaugural Dissertation von Dr. K. Köster.' Würzburg, 1868.

[2] Prepared by Dr. Frank Payne, Lecturer on Pathology, St. Mary's Medical School, London.

[3] "Die Lymphgefässe und ihre Beziehung zum Bindegewebe," 'Virchow's Archiv,' Bd. 28. Berlin, 1862.

puscles, but as spaces without distinct walls merely excavated in the fundamental substance of the tissue, and standing in direct communication with the lymphatics; the connective tissue cells properly so called being contained within these cavities, and taking their shape from the shape of the particular part of the space in which they might happen to be contained. The two methods of investigation used by Recklinghausen, viz. the method of injection, and that of silver impregnation, have been much assailed, but, as Dr. Köster thinks, wrongly. It is by the aid of the same methods that he has succeeded in demonstrating a 'plasmatic system' in the umbilical cord; and his views on the construction of this system are, up to a certain point, the same as those in Recklinghausen. As early as 1832, Fohmann[1] described a system of vessels in the umbilical cord, and regarded them as lymphatics; but his observations seem to have referred chiefly to extravasations.

The investigation is divided into two parts, one relating to the mucous tissue itself, the other to the epithelium. In the investigation of the former, three methods were adopted: (1) injection; (2) silver impregnation; (3) simple examination of perfectly fresh specimens.

1. The method of injection was thus carried out. By means of a "Pravaz" syringe, a somewhat concentrated solution of Prussian blue was simply forced into the denser parts of the cord, not into any vessels. In this way patches an inch long were injected in fresh specimens with a delicate blue network, and still better results obtained by using specimens hardened for some hours or days in 20 per cent. alcohol. The microscopical examination of parts thus injected showed in the deeper layers a network of intercommunication by channels varying much in width. They were for the most part spaces unequally distended in a varicose manner, sometimes lying close to one another with hardly any intervening substance, and in other cases connected by finer branches. No regular arrangement could be made out, but the whole gave the unmistakable impression that the blue injection had found its way into previously existing channels and not into mere arbitrary extravasations produced by violence. This impression was strengthened by the number of smaller channels which connected the larger spaces. In the more superficial parts a similar blue network was seen (fig. 1), but the channels were here narrower, and also more uniform in size; though still with sufficient varicosity of outline to show that they were very greatly dilated.

[1] 'Tiedemann u. Treviranus Zeitschrift,' Bd. iv, 1832, p. 276.

2. The method of silver impregnation was first resorted to

Fig 1.

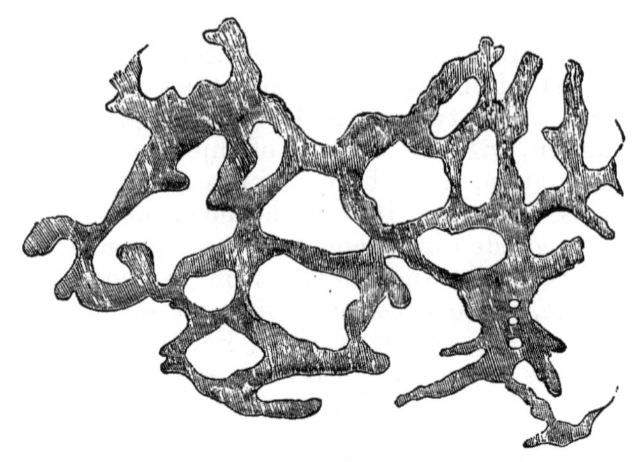

Injected system of channels from the cortical portion of the cord. Magnified 90 diameters.

to decide the question whether the system of channels thus injected might not possibly be lymphatics.

The injection, however, of silver solutions, entirely failed to produce the characteristic epithelial markings; though this method, as well as the simple staining of superficial sections with nitrate of silver, brought into view the well-known communicating stripes and spots which have been regarded, in accordance with the views of Recklinghausen, as indicating a system of plasmatic channels. They were, however, especially in the deeper layers, enormously large and very much wider than those seen in other forms of connective tissue or in the cornea. The resemblance of this network in size and arrangement with that demonstrated by injection was so close as to be unmistakable; it remained, however, to show its connection with the proper cells of the mucous tissue.

3. The examination of unaltered sections of the cord brought into view a mucous substance, traversed by delicate fibrillar outlines. The fibrils were not uniformly distributed, but collected into bundles, which by their divergence and decussation enclosed alveolar spaces of unfibrillated mucous substance. The network thus formed was not, however, composed of closed alveoli, but rather of spaces communicating freely with one another, and containing the fundamental mucous substance.

The cells of the umbilical cord are of two kinds, one constant, the other extremely variable in size and shape. The variable cells appear in the unfibrillated mucous spaces in the form of roundish cells with a round nucleus, and a more or less distinctly granulated protoplasm without any limiting membrane. In the fibrillar bundles, on the other hand, are situated stellate and caudate cells which precisely resemble these in their nucleus and protoplasma. That these two forms are really identical is shown by the phenomena of contractility and migration which they exhibit. A round cell may be seen to stretch out in the direction of a fibrous partition towards which it slowly moves, and at length becoming fixed, there assume the spindle-shaped or stellate form; while, on the other hand, a spindle-shaped cell may be seen to move into a mucous space and become round; so that the form of these cells depends entirely upon the nature of their temporary resting-place. Quite different from these are the fixed cells of the cord which Virchow has described and figured (in the 'Cellular Pathology') as branched connective tissue cells. These when examined fresh are ramified bodies pale or granulated, the prolongations of which form numerous anastomoses; sometimes tapering off into mere threads, sometimes remaining uniformly wide. The system thus formed is provided with nuclei which require special mention. They are more frequently contained in the fixed corpuscles than in the prolongations, though the larger of these may possess several. Many at least of these nuclei were evidently situated in the walls of the bodies, looking as if they adhered to the outside. The whole system thus formed is found to traverse the unfibrillated mucous space, no less than the fibrillated portions. It evidently corresponds to the connective tissue network of Virchow; only that it is in the author's view not solid, but a hollow system of channels. Where the outline looks dark it is filled up with granular protoplasma; and this is owing to the presence within the cavity of the above-mentioned variable cells, which are caudate or stellate when lodged in the angular intersections of the hollow network, and become round when the channels in which they are contained are susceptible of easy dilatation, as when passing through the mucous alveolar spaces. The identity of this network with the system of channels demonstrated by injection and silver impregnation is shown first by the unequal dilatation of both, and especially by the appearance of partly injected portions, where spaces into which the blue injection had found its way were plainly continuous with the stellate corpuscles and thread-like processes of the connective tissue network (see fig. 2).

The last and most difficult question is the nature of the walls of the plasmatic channels. The notion that they have no distinct walls, but are merely excavations in the funda-

FIG. 2.

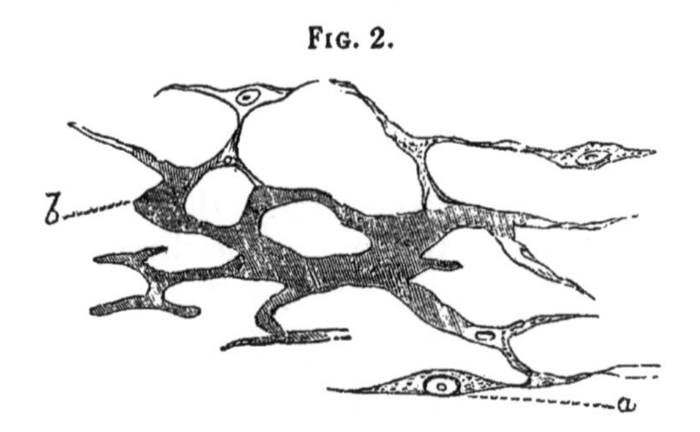

Partly injected portion of the cord.
(*a.*) Network of connective tissue cells, plainly continuous with (*b*) darkly shaded portions, representing channels filled with the injection.
Magnified 270 diameters.

mental substance, seems especially inapplicable in the case of a tissue where this substance is semifluid. Considering the arrangement of the nuclei (mentioned above), which are evidently imbedded in the walls, the author thinks no other hypothesis tenable than that the connective tissue cells themselves constitute the walls of the hollow spaces, which usually go by that name; and since there is no distinction between the walls of these spaces and those of the channels which connect them, the whole system must be enclosed in the same way. The connective tissue cells would then be membranous or epithelial nucleated structures which roll themselves up to form a tube, just as some observers (Aeby and Eberth) have supposed that the blood-capillaries are formed by adherent epithelial cells, only that for the short channels of the connective tissue network a single cell suffices. The whole system is thus, strictly speaking, *intercellular*, not *intracellular*, as supposed by Virchow.

Dr. Köster's memoir concludes with observations on the epithelium of the umbilical cord. These coincide generally with those of other observers, but he has observed some peculiar structures among the cells of the lowest epithelial layer. These are round or oval, sharply defined openings (*stomata*) which are unequally distributed. They generally contain a cell with granular protoplasma resembling the contractile (migratory) cells contained in the plasmatic channels of the cord. Small drops of mucus are also met with in the *stomata*, and after injection they are often choked with plugs of the same material. A still more complete injection forces out the coloured mass in little drops on the surface of the cord, and it seems probable that these make their exit by the *stomata* here described; but this point could not be distinctly seen. Were this clearly made out, it might at once be concluded that the stomata are the openings of the plasmatic channels; as it is, the author does not venture to pronounce a positive opinion on this point.

On a NEW SPECIES *of* GREGARINA *to be called* GREGARINA GIGANTEA. By EDOUARD VAN BENEDEN, Doctor of Natural Science, Louvain.[1] With Plate VI.

THE little organisms known as Gregarinæ have much attracted the attention of naturalists during late years. Their organisation is exceedingly simple; they consist of merely a single cell, but the history of their development, of their reproduction, and of their metamorphoses, has exercised the sagacity of a great number of eminent observers, and the profusion with which they are spread through nature has caused them to come under the notice of many naturalists. The literature relating to these inferior organisms is very rich, but, nevertheless, several points relating to their organisation and to their development are still completely unknown.

The Gregarinæ live as parasites either in the intestine or in the perivisceral cavity, or in the reproductive organs of animals belonging to various classes. They have been pointed out in the various groups forming the class of worms—from the Turbellaria (P. J. Van Beneden and Claparède) and Sagitta (Diesing), as also from the Nematodes (Walter), and the Annelids, where they are extremely abundant. The Gregarinæ of the earthworm have chiefly served as the subject for the numerous works of which these animals have been the subject. Schneider has described a species peculiar to the Holothuriæ; some are known among Molluscs, even among the Tunicata (*Salpa*, Diesing). They are very abundant in a great number of insects; they have been described as infesting Myriapods; both Scolopendra, and Julus, and the Crustacea also are sometimes infested with these minute organisms. Cavolini, who was the first to find Gregarinæ, observed them in the appendicular organs of the stomach of *Cancer depressus*; Von Siebold a long time ago made known the fine Gregarina of *Gammarus Pulex*; Lach-

[1] Read before the Academy of Brussels, and communicated by the author.

mann has described a Gregarina from *Gammarus putaneus;* and Kölliker made us acquainted with the *Gregarina Balani* which he had observed in *Balanus pusillus.*

I have the honour, on the present occasion, to present a description of a new species of Gregarina which lives in the intestine of the Lobster. It is chiefly remarkable for its great size, and it is on this account that I propose to call it *Gregarina gigantea.* Thanks to its size I have been able to ascertain some new facts relative to the organisation of these animals, and some of these observations are not without importance in relation to the theory of the cell. I have found besides some facts relative to the development of this Gregarina which will contribute to complete our knowledge relative to the history of the development of these singular parasites.

Description.—The Gregarina of the Lobster has a very much elongated form, which gives it a superficial resemblance to a Nematode. The largest individuals I have had under my observation did not measure less than sixteen millimètres in length, with a breadth of ·15 of a millimètre. The body presents nearly the same breadth throughout, excepting in its posterior part, where it diminishes progressively. The anterior extremity, on the contrary, rounded in front, is slightly swollen out into a globular form: the *Gregarina gigantea* belongs to the division which M. Kölliker characterised by these words—" Eingeschnürte Gregarinen, mit einem einfachen abgerundeten Vorderende."

The structure of these Gregarinæ is very simple; I consider them with M. Kölliker as monocellular animals; in them are found all the constituent elements of a cell, and it is impossible to find in their structure a fact which is of such a nature as to throw doubt on their monocellular constitution. Here we have, therefore, an animal cell which can reach the length of sixteen millimetres. I believe that, as far as simple cells are concerned, there are only the eggs of birds and of some other animals which surpass in their dimensions those of the cell which concerns us here. And there is this difference between the egg of these animals and our monocellular organism, that in the egg of the bird one has to distinguish a living, active part (the cicatricula), and an inert passive part (the yellow), which forms almost the whole of the egg. For this inert part of the egg-cell does not form an integral part of the cell. It is found within the cellular membrane truly (vitelline membrane), but the only living part of the cell is the cicatricula; it alone divides and gives rise to the cells of the embryo. In the Gregarina, on the contrary, all the parts are living, active, and contractile, and I believe that, from this point of view, one may say that the *Gregarina gigantea* is the largest simple cell which is known.

In our Gregarina a membrane with a double contour can be very clearly distinguished, perfectly transparent, and without structure. It presents no opening, and its thickness is throughout the same, excepting, however, at the anterior part of the rounded bulb which terminates the body of the animal in front. This structureless membrane represents the "cell membrane" of this monocellular animal.

Under this membrane can be distinguished clearly a layer of some resistance, formed of a substance which is perfectly transparent and devoid of granulations. It is this structure which Leidy and Ray Lankester first made known, and I have been able fully to establish in this species that it is truly in this layer that the parallel striæ arise, whence results that fibrillar aspect of the animal which one remarks on using high powers. It is when the Gregarina contracts that these striæ appear, and they disappear when it is in repose. I believe, with Leidy and Ray Lankester, that the substance which constitutes this layer possesses essentially contractility, and that it is this which is concerned in the production of the movements of which the animal is capable. It represents, physiologically, the muscular subcutaneous layer of many animals more elevated in organisation, and of Nematodes especially; but instead of being formed of distinct muscular fibres it is a continuous layer of contractile,—if you like, muscular—substance.

The consistence of this layer is much greater than that of the granular matter which occupies the centre of the cellular tube. The central granular matter is very mobile; it moves about in the interior of the cavity of the cell. This is by no means the case with the transparent and contractile substance of which we were speaking; this is fixed and intimately blended with the membrane of the cell. The limits between this layer and the granular matter of the centre are, however, not sharply marked. I imagine that the density of the layer decreases from the periphery towards the central axis of the tube.

The thickness of this layer is nearly the same throughout the whole length of the animal. It augments, however, a little at its anterior extremity, principally at the point of union of the globular enlargement with the rest of the body. There, this layer sends a prolongation inwards in the form of a transverse septum to the interior of the tube, in such a way as to divide the central granular mass into two parts, of which the one, very small, occupies the cavity of the anterior globular enlargement, and the other fills all the rest of the body of the animal. The whole cavity of the body of the Gregarina is filled with a granular matter, formed by a viscid liquid which is perfectly transparent. This holds in suspension fine granulations of a rounded form, which are formed by a highly refractive and slightly yellow matter. The quantity of granules with which the liquid matrix is charged increases with the size of the Gregarinæ, and, moreover, the opacity of the animal is greater in proportion as its dimensions are larger. This granular liquid which occupies all the cavity of the cell is very mobile, and may be seen shifting about in the interior of the membrane whilst the animal executes its movements.

It is always easy to distinguish the nucleus of the cell in suspension in the granular liquid which occupies all its cavity. It has, normally, a regular ellipsoid form, and its dimensions vary with those of the Gregarina. In the largest individuals which I have found it measured ·13 of a millimètre across its major axis, ·08 to ·10 of a millimètre across its minor axis. This nucleus presents a membrane which is perfectly distinguishable, and the cavity of the vesicle is filled by a homogeneous, colourless, and transparent liquid. This nucleus is not a solid body destitute of a membrane, as M. von Frantzius thought it. It is easy to assure oneself of this by isolating the nucleus, and subjecting it to pressure. One may then see, at a given moment, a bursting of the membrane take place, and the liquid contents of the nucleus escape by the aperture produced. The membrane of the nucleus is, however, very thin, and it is this which explains the modifications of form which the vesicle undergoes when external pressure is brought to bear on it. I have seen a nucleus affect successively the forms represented in fig. 6, *a*, *b*, *c*, *d*, in a Gregarina which glided on the glass slip placed on the microscope-stage, picking its way among the various solid contents of the intestine of a lobster; but the form of the nucleus never changed itself spontaneously, and apart from the influence of external causes.

The most important fact of all that are put out in this notice concerns the spontaneous apparition and the disappearance of the nucleoli of the nucleus in a very short space of time. If one of these Gregarinæ of moderate size is observed, the nucleus is seen at first provided with a single nucleolus, presenting, some seconds later, a great number of little refracting corpuscles, of very variable dimensions, which are also nucleoli. Some of these enlarge considerably, whilst the primitive nucleolus diminishes in volume little by little, finally disappearing. The number of nucleoli varies at every instant. Some disappear whilst others are forming; they commence in the form of a minute point scarcely perceptible. This point grows to a certain limit; it becomes a veritable corpuscle, formed of a homogeneous, highly refractive substance, then the corpuscle diminishes in volume; it refracts the light less and less; finally it disappears. It even happens that all trace of nucleolus disappears in the nucleus, and some instants later one or several nucleoli can be distinguished there, which undergo afresh all the variations which I have just described. This fact of the successive apparition and disappearance of the nucleoli in a nucleus of a cell, and the modifications which occur in the nucleus as to number, dimension, and character of the nucleoli, has never yet been pointed out to my knowledge. It appears to me to have a great importance in relation to the cell theory. The idea of the existence of a membrane round these little bodies (the nucleoli), and of their vesicular nature, is by no means reconcilable with the rapidity of their formation, and the modification which they undergo in the course of some minutes; and if it is demonstrated that the membrane is not an essential part of the cell, and that the *nucleolus* is sometimes, if not always, devoid of membrane, may it not be presumed that the *nucleus* of a cell is not necessarily a vesicle, and that, contrary to the generally received opinion,[1] a nucleus of a cell may be equally devoid of membrane?

The Gregarinæ move, and three kinds of movements may be distinguished in them:

1st. They present a very slow movement of translation, in a straight line, and without the possibility of distinguishing any contraction of the walls of the body which could be considered as the cause of the movement. Further, it is very difficult to account to oneself for the cause of this movement of translation, at least to admit—what is difficult to demonstrate—that the Gregarina acquires an adherence to the surface on which it moves. Invariably the appearance of this movement recalls completely that of the Turbellaria; but it is impossible to distinguish the least trace of vibratile cilia on the surface of the body of these animals, even with the highest magnifying powers.

2nd. The Gregarina of the Lobster presents another kind of movement, consisting in the lateral displacement of every part, taking place suddenly, and often very violently, from a more or less considerable part of its body. Thus the posterior part of the body may be often seen to throw itself out laterally by a brusque and instantaneous movement, forming an angle with the anterior part. At the vertex of the angle the body then presents a regular fold, and the animal forms a broken line. Folds can be thus formed in a great number of points, more or less approximated; and it results from this that the animal can describe a spiral, if all the folds occur on the same side, or twist itself about in various ways. It is probably due to the contractility of the transparent subcuticular layer that the Gregarina has the power of executing these movements.

3rd. In consequence of the different contractions which are produced, and by the action of which the Gregarina folds itself so as to form broken lines, the granular liquid which occupies the cavity of the cell is seen to move, and the granulations to shift about in the interior of the body of the animal.

I have found as many as twenty-five Gregarinæ in the intestine of a single Lobster, and at certain times every Lobster presents this parasite. I have observed them in the months of May, of June, and of August, in lobsters coming from the coasts of Norway. It is probable that they will be found equally in those of the coast of Brittanny. I have found no traces of these parasites on Lobsters kept for a long time in captivity in the piscicultural parks at Ostend. Is it the same with the lobster confined in these parks as with the animals of our zoological gardens and the fishes of our aquariums? May the loss of their parasites be due to their captivity?

At the end of the month of last September I examined a great number of lobsters freshly arrived from Norway, with the object of refinding this beautiful Gregarina. Not a single one contained in its intestine the parasite I was seeking; but I perceived that all presented, in the walls of the rectum, little white grains of the size of the head of a small pin. These were the *cysts* of Gregarinæ, situated beneath the epithelium; and, what is remarkable, the cysts were disposed one by the side of another, forming little rectilinear series of 3, 4, 6, and even of 7 cysts.

By the beautiful researches of Von Siebold, Henle, Kölliker, Bruch, Stein, Lieberkuhn, and other eminent naturalists, the evolution of the Gregarinæ has been in great part elucidated. We now know that a single Gregarina may become encysted, and that the frequent fact of the existence of two granular masses in the same cyst is explained by the division of the contents of the encysted Gregarina, and not by the conjunction of two Gregarinæ in one and the same cyst, as Stein and other naturalists had supposed. It is chiefly to Bruch, Lieberkuhn, and A. Schmidt that the honour of demonstrating this fact belongs. It is known also that after a kind of cleavage of the granular masses of the cysts these masses become transformed into little vesicles, which in turn give rise to the psorosperms or pseudo-navicells. Lieberkuhn has shown that the psorosperms produce amœboid forms, and he thinks these amœbæ themselves are developed into Gregarinæ, or give origin to Gregarinæ. But this last phase of the evolution of these little beings is still problematical, and will require further serious investigation. But what is perfectly established now, thanks to the labours of Stein, Kölliker, Lieberkuhn, and several other naturalists, is that there exists no relation of filiation whatever between the Gregarinæ and the Filariæ, and that the opinion maintained on this question by Henle, Bruch, and Leydig, must be finally abandoned.

I have not been able to observe these different phases of the evolution of the Gregarinæ of the lobster; I have not even succeeded in establishing the transformation of the granular masses of the cysts into psorosperms; but I have fully recognised, in confirmation of the observations of Bruch, Lieberkuhn, and A. Schmidt, that the contained granular matter of the cysts is at first a simple sphere, always devoid of nucleus, and that the two rounded masses which one observes frequently in the cysts come from the first in consequence of a sort of cleavage, in fact from a division. A groove appears at first at the surface of the granular sphere, into which the wall of the cyst immediately is applied. This fissure advances progressively towards the centre of the sphere, and eventually divides it into two parts. Each of them has the form of a hemisphere, and they are applied to one another by their plane surface; but soon the diameter of the cyst increases, a space which is filled with a limpid, colourless liquid as fast as it forms appears between the wall of the cyst and the surface of the two granular masses, which lose little by little their hemispherical form, becoming gradually rounded. The diameter of the cyst continues to increase, and the two masses at last become each a perfectly spherical globe. I have seen all these changes take place on the stage of the microscope. But what has not yet been observed is, that after this division of the primitive sphere into two spheres, the wall of the cyst formed by several concentric layers of a diaphanous material decomposes into a soft granular matter, whilst each of the two globes surrounds itself with a new membrane. Soon the traces of the envelope

[1] Kölliker's 'Elements of Human Histology.'

of the primitive cyst can be observed, but very obscurely, and the two globes of the second generation only are distinguishable, surrounded by a common granular substance. The globes grow little by little, at the same time that their envelope thickens. Henceforward each of them is to be regarded as a new cyst, whose contents will divide in their turn, to give rise to two new globes or spheres, which will become cysts of the third generation after the absorption of their walls. The upshot of this is that the cysts of the Gregarinæ can multiply by division before giving rise to psorosperms, and the manner in which this phenomenon presents itself recalls completely the multiplication of the cells of cartilage. There, too, the cells multiply by division, and the capsules of the cartilage change little by little their character, and are transformed into intercellular substance. The granular globes of the Gregarina-cysts may be compared to cartilage cells, and the granular matter which surrounds them to the intercellular substance of the cartilaginous tissue. In this mode of multiplication of the cysts the reason of their rectilinear arrangement in the walls of the rectum of the lobster is at once obvious. It is hardly necessary to add that these observations on the multiplication of the cysts give the explanation of the fact, so often observed but not yet interpreted, of the existence of two granular masses in one and the same Gregarina-cyst.

It appears from what precedes that in certain circumstances—perhaps at a fixed period of the year—the Gregarinæ, which were living freely in the intestine of the lobster, migrate into the rectum. There they become encysted, after having penetrated under the epithelium of the intestinal walls, and these cysts multiply by division. It is not possible to doubt that, after a certain time, the cysts are resolved into psorosperms; but it has yet to be found out what becomes of these psorosperms, as also how the Gregarinæ arrive in the intestine again, and under what form they first appear there.

Note.—I may take this opportunity of remarking that the view entertained in the very interesting paper of my friend Dr. Van Beneden, as to the encystment of the Gregarinæ, can hardly be considered as yet demonstrated. It may be fully admitted that single Gregarinæ do become encysted; but there are facts which lead to the supposition that two are *usually* thus encased, and that the formation of pseudo-navicells thus presents a remarkable approach to the conjugation of Algæ, as remarked by Huxley. Not only are two masses most commonly seen in a cyst, but in a spherical cyst of *Gregarina Blattarum* I found two *nucleated* Gregarinæ ('Quart, Journ. Microsc. Sci.,' vol. iii, new ser., pl. vii, fig. 17). The habit of attachment of two individuals head to tail, as it were, common in many bilocular Gregarinæ (loc. cit., figs. 18, 19), and in *Monocystis* also (fig. 6), and *Zygocystis cometa* of Stein, indicates a tendency to conjunction which would favour the enclosure of two individuals in a single cyst. At the same time it is only right to state that this head-and-tail-attachment is commoner in small than in large specimens of Gregarina, and may be due possibly to some process of division occurring without encystment, though this is quite hypothetical.

E. RAY LANKESTER, Ed. 'Q.J.M.S.'

The KINSHIP *of* ASCIDIANS *and* VERTEBRATES.

IN 1867 Kowalevsky published, in the transactions of the Imperial Academy of St. Petersburg, some observations which are more profoundly interesting than anything of the kind which has appeared of late years, since, if correct, they indicate distinctly a bridge over the chasm, supposed to separate Vertebrates from all other animals. They prove the existence of *a bicavitary structure,* of a neural tube and a visceral tube, separated by an axial cartilaginous rod, in the early stages of development of Ascidians, an arrangement previously believed to be essentially characteristic of Vertebrates.

We are indebted to Professor Michael Foster for the following account of Kowalevsky's paper :—

The observations of Kowalevsky were made on several Ascidians; but in all the processes are remarkably similar. The earlier stages were chiefly studied in *Phallusia mammillata,* the metamorphosis of the larva into the sessile form in *Ascidia intestinalis.*

The ova of Ascidians possess, on leaving the parent, a somewhat complex structure. Each consists of a *vitellus,* devoid of any proper vitelline membrane, and surrounded by a layer of gelatinous material, in which are strewed yellow nuclei or cells. These nuclei or cells are small homogeneous vesicles, very similar in appearance to the blood-corpuscles of the higher vertebrata, and are probably purely material elements derived from the follicles in which the ova are developed. Their subsequent history is remarkable, inasmuch as they become transformed into the "white cells" of the mantle. Outside this gelatinous layer is a hard membranous capsule, which in turn is studded with peculiar structures, varying exceedingly in the various species. These, however, are of no importance and speedily disappear.

The vitellus varies a good deal. In *Phall. mammillata* it is highly transparent and refractive; in *Asc. intestinalis, Cynthia,* &c., it is opaque, brown or dark yellow. No nucleus can be detected in the mature ova, though it is readily seen in the immature.

The Ascidians are true hermaphrodites, and artificial impregnation can readily be conducted with ova and spermatozoa taken from the same animal. The spermatozoa attach themselves to the ova not by their heads but by their tails. Their entrance into the ovum was not observed.

Cleavage commences within an hour of impregnation, and advances very rapidly. It is confined entirely to the vitellus, the gelatinous layer with its vesicles taking no share in the process. Two meridian furrows divide the ovum first into

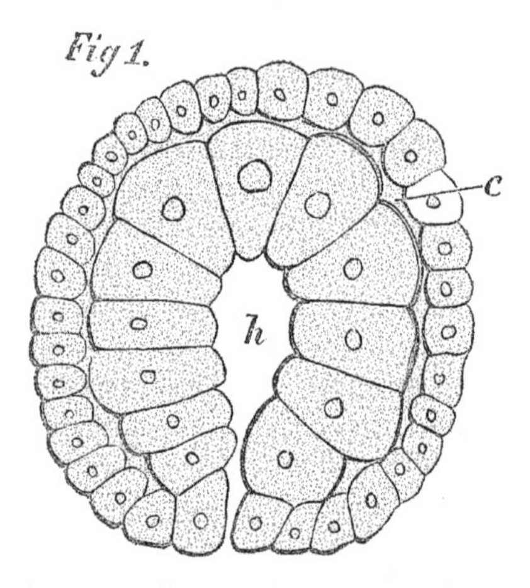

two then into four masses, which are further divided into eight by an equatorial furrow. Even in the two first masses a nucleus may be detected, and the cleavage masses arrange themselves around a central cleavage cavity (cavity of Baer).

As soon as the number of cleavage masses amounts to about thirty-two the ovum begins to be flattened below, and then is doubled in on itself. Through this involution the ovum becomes gradually transformed into a double sac, the cells from the upper set of cleavage masses, numerous and small, forming the outer sac, while those from the lower set, fewer and larger, forms the inner or lining sac, the original cleavage cavity being reduced to a narrow space between the two, as is shown in fig. **1** *c*. The new cavity thus formed corresponds to the alimentary cavity of the future animal (*h*); at first it is of the form of a shallow cup, but it gradually becomes deeper and its orifice narrower, until at last if the external opening does not close up all together it becomes so small as to escape notice.

The cells constituting the inner sac or walls of the alimentary cavity now increase more rapidly, and the ovum at the same time lengthening and becoming more elliptical the central cavity is much narrowed.

At the same time it is observed that opposite to the above-mentioned external opening, or to the spot where it was, the surface of the ovum is raised up into two folds running lengthways. At one end (the posterior extremity that is to be) these folds curve round and join each other; at the other (anterior extremity) they gradually fade away. They are best seen in transverse (optical) section, fig. 2, which shows the digestive cavity *h*, with its walls, still composed of large cells, separated by the remains of the original cleavage cavity *c* (now the general cavity of the body) from the body wall composed of smaller cells, and raised. At *w* the body wall is raised up into the above-mentioned folds which thus give rise to the groove *f*. New cells may also be seen to have made their appearance between the outer and inner sacs beneath the folds.

Although Kowalevsky was unable to trace out step by step the arching over and coalescence of the (medullary) folds, he believes that they do thus behave; at all events an hour later we meet with the form shown in fig. 3, in which *h* is, as before the alimentary cavity, surrounded by its wall of cells *g*, and *c* is the general cavity of the body. Outside the body is now a new cavity, *n*, as yet opening externally at its front by the orifice *d*, and roofed in by a double layer of cells originat-

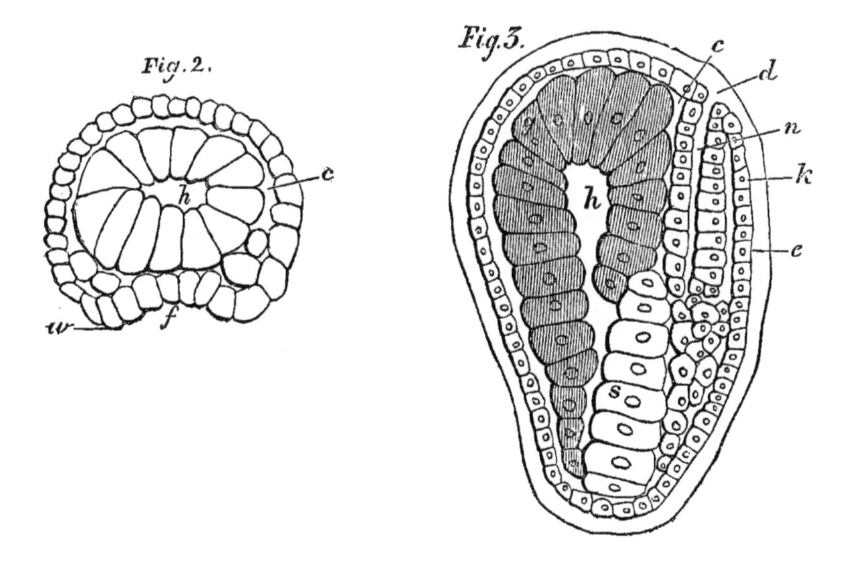

ing apparently from the folds just spoken of. One of the layers, *k*, forms the proper roof of the new cavity, while the other is continuous, and forms a part of the cells constituting the walls of the whole body. This new cavity or incomplete tube is the rudiment of the embryonic nervous system, and may, therefore, be called the neural tube. Between it and the alimentary tube at the hind part of the body a new structure has now made its appearance—the row of large cells *s*, the immediate origin of which is uncertain. This row of cells is the rudiment of the axis band (or notochord) of the tail.

The embryo now becomes bent at its hind part, the bending takes place more on the left side than the right; and the bent portion grows more rapidly than the rest of the body, thus giving rise to the tail.

The stage represented in fig. 4 is now soon reached. The digestive tube *h*, with its thick walls, occupies the fore part of the body; immediately above it is the neural cavity *n*, wider and more capacious in front than it was, but filled up with cells behind *m*. It is still open at *d* (thus presenting a very close correspondence with *Amphioxus*, in which the front opening of the neural tube is long patent). The cells previously forming the general body wall have now become differentiated into an epidermis, underneath which are, at certain spots, numerous loose rounded cells. The most conspicuous feature, however, is the row of cells, *x*, each possessing a nucleus stretching from behind the neural cavity to the tip of the tail. (The figure also shows the gelatinous envelope omitted in the previous figures.) A transverse section taken

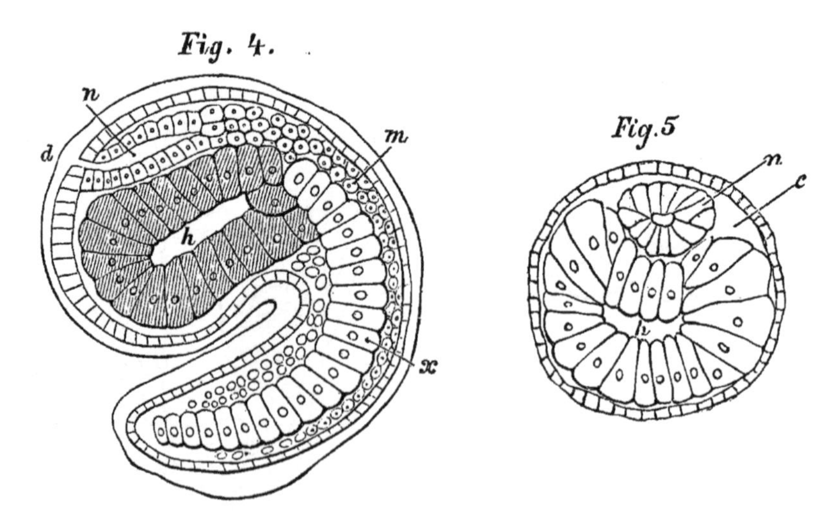

at this stage through the forepart of embryo present, as fig. 5, with a neural tube *n* placed above a visceral or alimentary tube *h*, both enveloped by the wall of the body. Did the row of cells *x* (or notochord) but reach a little farther forward we should have an exact diagram of the section of a vertebrate body.

The neural cavity now becomes completely closed in, and at the same time takes on more of a spherical shape. Within, especially behind and underneath, a proliferation of cells takes place, to a certain extent filling up the cavity. Pigment is deposited in the form of a round patch which in time becomes almost completely covered up by small cells. This pigment patch is undoubtedly an organ of vision, and the cells surrounding it constitute a ganglion. The upper and anterior wall becomes very thin, and from the lower and anterior wall there grows out a peculiar body not unlike a wine-glass covered at its top by pigment, with a stem of transparent, highly refractive, material. This body, believed by Kowalevsky to be an auditory organ, projects freely into the neural cavity.

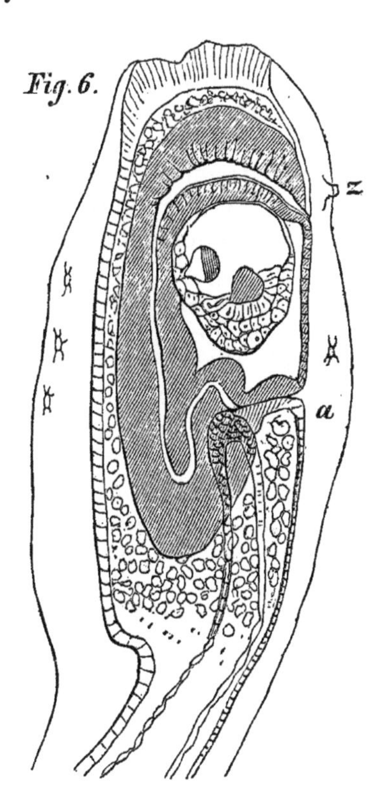

No trace of nerves leading from the neural cavity can be detected.

The digestive cavity lengthens, its walls remaining thick. It curves forwards and upwards, and reaching the upper surface immediately in front of the neural cavity opens out into the *mouth* (see fig. 6); behind it lengthens out with one or more windings, and the anus (*a*) (generally somewhat on one side) is now distinctly visible. It is highly probable that the anus is identical with the primary orifice of the digestive cavity, but its identity cannot be traced throughout satisfactorily.

The epidermis becomes thin and flattened out over the whole of the body except at the extreme front, where it is developed into three prominences, varying in shape in different species, and serving for the purposes of attachment. The cells beneath the epidermis multiply rapidly, being in some places very numerous.

The axis band of the tail undergoes remarkable changes. The roof of cells seen in fig. 4 increases very much in length, the tail becoming in consequence very much curved. Each cell possesses a well-marked nucleus. Very soon, along the middle line of the axis band, particles of some highly refractive material make their appearance at the junction of the cells. These increase very rapidly in size, pushing the cells on one side and gradually taking their place, the new refractive body appearing between every two of the old cells. When two enlarging refractive bodies meet they coalesce, and continue to do this repeatedly, the whole of the axis band thus becoming made up of a mass of homogeneous refractive material forming a case, the original cells being left in a reduced condition at the outside only, where they constitute a sheath. In this sheath elongated nuclei are visible, but whether they are the transformed nuclei of the original cells is not quite clear.

This mode of origin of the axis band quite corresponds with the development of the notochord of Amphioxus. There is in both cases the same primary strand of nucleate cells, and the same appearance of refractive particles which invade and eventually replace the primary cells. The chief difference is that, whereas in the ascidian the new material finally forms an unbroken mass, in Amphioxus it exists in the form of plates applied closely to each other.

The rounded cells lying in the tail between the notochord and the epidermis are gathered round the former in a mass which ultimately becomes transformed into a tubular muscular coat. This strong likeness supports the view that the axis band of the tail of Ascidians is really a notochord, and thus we have in the Ascidian larva the three great vertebrate features:—1st The double tube, neural and visceral; 2ndly, the formation of the neural tube by means of medullary folds; 3rdly, the notochord.

While these changes have been going on in the material derived directly from the cleavage masses of the vitellus, the yellow vesicles found in the gelatinous envelope have also suffered transformation. Losing their colour they acquire a nucleus, their substance becomes distinctly protoplasmic, and they send out processes in the gelatinous substance; in fact they take their place as constituent cells of the mantle.

Very soon after the larva has left the egg it seeks to fasten itself to some object. This it accomplishes by means of the three processes above described. Forthwith the tail is retracted, the homogenous, highly refractive, core becoming broken up, and the elongated cells of the sheath being drawn together and appearing as conspicuous rounded cells. The muscular coat at the same time is resolved into round cells which lie in the cavity of the body. The gelatinous envelope of the tail is also retracted and very much wrinkled. A little later on the remnant of the tail appears as a confused heap of fatty tissue in which veritable cells can hardly be distinguished, and which is gradually absorbed as Krohn describes.

The epidermic processes by which the larva attaches itself in great measure disappear, the constituent cells "wandering" into the gelatinous sheath, and the portion of the body which serves for attachment widening out into a foot.

The neural vesicle collapses, the cavity becoming less, and the cells constituting the ganglion becoming round, small, and more alike. Later on the organ is reduced to a heap of cells gathered irregularly, without distinct outline, round the pigment spots, and thus dwindles into the permanent ganglion of the adult animal. Very many of the cells originally forming part of the nervous system are transformed into blood-corpuscles.

The fore part of the alimentary canal, by appropriation of the cells lying round it, acquires thickened walls, and is dilated into a wide branchial sac which is marked off by a constriction from the rest of the digestive tube, the latter also growing stouter and its convolutions becoming more obvious.

At the front part of the branchial sac the wall of the sac and the wall of the body (*i.e.* the mucous and epidemic layers) coalesce in two oblong patches on either side. The centre of each patch is absorbed, and the holes thus formed widen transversely into the branchial slits. The succeeding slits are formed in a similar way.

Close to fore and under corner of the branchial sac there

appears a patch of cells, the source of which was not clearly made out. This is the rudiment of the heart and pericardium, but the exact steps of its subsequent development were not satisfactorily observed.

Water enters the branchial sac from without through the orifice which was the primary mouth, and flows through the branchial slits into the space which is now found between the epidermis and the gelatinous envelope, and which extends far back to an orifice, which appears close to the anus, and through which excrementitious matter is ejected.

As development proceeds peculiar organs, forming apparently a network of blind tubes, arise round the intestines; they possess a complicated structure, to the study of which the author hopes to return.

Professor Kupffer, of Kiel, has addressed the following letter to Professor Max Schultze, of Bonn, on this matter. The letter is published in Schultze's 'Archiv,' Part iv, 1869.

"You, of course, know Kowalevsky's work on the development of the simple Ascidians, which brought to light facts such as nothing else has done before, to span the gulf between vertebrates and invertebrates, and has given positive foundation to the doctrine of a phylogenetic connection between apparently entirely different circles of life. As far as I know, no one but Haeckel has paid any particular attention to the work, and it does not seem to have been generally taken as trustworthy. I own I did not myself belong to those who believed in it. It behoves me the more to acknowledge this, as through continuous observation during this summer of the *Phallusia canina* which is found in the Bay of Kiel, my opinion has been entirely changed. The first phase of development, the formation of the free swimming larva from the egg, shows in such elementary clearness the chief features of the development of a vertebrate animal, that the observation is quite convincing. The animal which I examined does not seem to have been any one of those on which Kowalevsky worked. Apart from his first stages of development having been described from *Phall. mammillata*, his *Ph. intestinalis*, Linn., is not to be identified with the *canina* of O. F. Müller. I leave out of consideration, among other things, that he speaks of the peculiar appendages of the egg-membrane in the forms used by him, as *soon* falling off, whilst the analogous structures in our species present characteristic cellular tufts which remain fastened in regular order to the egg-shell until the larva is hatched, so that the emptied egg membrane is still recognisable by these appendages. The chief or fundamental features of the development are however the same, as will be gathered from the following description:—The ripe egg consists inside the egg canal of the reddish-brown yelk mass, the soft egg membrane, and lastly these accessary structures, namely, on the inner surface, a simple layer of small yellow cells, and outside a regular layer of long, blunt oval-shaped tufts joined one to another at the base. Both structures exist already in the ovarium. The yellow cells during development approach nearer the embryo, and form, with a gelatinous layer which arises between it and the embryo, the mantel. The mantel is, therefore, a persistent egg envelope, which is formed in the ovarium and derives no elements from the yelk. The fertilisation of the yelk takes place when free, after the egg has been laid. A *cleavage cavity* (Furchunghöhle) is observable as soon as thirty-two cleavage masses have been formed; towards the end of the cleavage process I cannot detect it any longer; it appears at last to be filled up by the increased number of cleavage masses, so that its persistence as the *body cavity*, that is, as a narrow interspace between the outer skin and the alimentary layer, cannot be considered as proved. The formation of the alimentary canal I saw exactly as Kowalevsky describes it; the spherical cleft-yelk-mass bends in like a cup, the cavity of the cup becomes the alimentary tract; the uppermost simple layer of cells are marked off on its walls from the deeper ones by a division, and become the foundation of the outer skin; the inner layer becomes the alimentary wall. There lies between this and the outer skin, when the cup is crescent-shaped, yet a third layer, as I can distinctly distinguish in my specimens.

The hemispherical cup tends now again to take the shape of a ball, narrowing towards the mouth, which later on closes itself entirely. Before this closure is effected the nervous system has become shaped out as a fusiform cavity, and the foundation of the tail separates itself from the main body, which contains the alimentary cavity and nervous cavity. I must here explain the singularities of the beginning of the formation of the nervous system; it is not quite clear to me whether the outer skin plays a part in it or not. This part of the examination presents extreme difficulties, because the transparent space formed by the cavity is still very narrow, whilst the rudiments of other organs are packed closely together, and the connection of the cells is still too loose to bear much pressure from the covering glass. Everything appears much clearer at a period soon after this, when the embryo has assumed a pear shape. It lies then in a bent-up position in the egg. On the convex side of the body-portion the fusiform nervous cavity lies immediately under the upper skin, surrounded by a wall of its own; beneath this lies the alimentary canal, distinguishable by the reddish colouring of its cells, at present without an opening. The *chorda*, consisting of a double row of square cells surrounded by the as yet round muscle cells, reaches a little into the body-portion, so that its lengthened axis would penetrate between the nervous cavity and the alimentary canal. It is impossible to think of a more beautiful model of a vertebrate embryo. On the convex side, above the axis, we have the nerve tube; on the concave side under the axis, the visceral tube; the contrast of "dorsal" and "ventral" is so clearly represented that I can scarcely be said to exaggerate when I say that the sight of it must work overpoweringly on those who enter upon the examination doubtingly. Without giving you in detail the progressive history of development in such a hasty letter as this, I will only add a little more concerning the nervous system. I differ herein somewhat from Kowalevsky, although in a sense which strengthens his assertions. In my Tunicate the nerve-tube does not form itself around the spherical bubble which contains the rising tufts which have the peculiar tactile apparatus which he describes and figures. What takes place is very different from what he describes in *Phal. mammillata*, for from the bubble towards the tail, a thickish nerve-cord stretches forward which contains a fine central canal which opens into the cavity of the bubble. This cord extends its posterior end between the muscles of the tail, so that its outline is there lost. The foremost end of the Chorda is, therefore, underneath the central nerve-system in *reality*, and not only when lengthened ideally. A vesicular front part and a cord-like hind part of this nerve-system are distinguishable.

So, too, one may observe the degrees of progressive development of the escaped free-moving larva. When this period is reached both openings of the alimentary canal have been formed by the outer skin rapidly increasing inwards (in a conical form) until it becomes imperceptibly connected with the alimentary wall; a canal then appears in the axis of the newly-grown part. The mantle still closes both openings for some time. With the exception of the dorsal side, where the nervous system lies between the skin and the alimentary canal, the outer skin is separated entirely from the intestinal wall, so that a roomy body cavity is formed which contains a continually increasing number of small round cells.

Here the highest point of development in this direction is reached, and it is now that the second phase begins, which, measured from the stand-point of the original development, is retrogressive. This condition is introduced by a period of rest after the free-swimming larva has attached itself in one of the manifold ways which have been described. Whilst the chorda, the original muscles, and the upper skin of the tail shrink together and degenerate into a single heap, the nervous system takes part in the retrograde change, but continues longer of the same form which it before possessed, although continually decreasing in volume. As long as a remnant of the tail is to be seen a thread-like cord of the nervous system is to be seen stretching into it, which shows the former relationship between this and the tail muscles. The respiratory pharynx does not make rapid progress, for the gill sac was already separated from the intestine in the free-swimming larva, and it had besides already developed the foundations of the ciliated ridges without, however, showing the cilia.

During this condition of rest, for it is characterised by the absence of any movement, the heart is formed, which consists of a group of those small round cells spoken of above as occurring inside the cavity of the body, whilst the rest, decreasing in numbers, become amœboid cells, which are set in motion by the first pulsations of the heart. They are the blood-globules or rather lymph-globules, the cavity of the body being a lymph sinus. With the commencing action of the heart the condition changes afresh; a true chrysalis stage is terminated, the gill sac widens, the slits of its walls become edged with cilia, the inhalent and exhalent orifices break through the mantle, and with these changes the outward form of the Ascidia appears distinctly.

So much for the present concerning these things. I hope during the vacation to send more minute particulars to the 'Archiv.'

REVIEWS.

A History of the British Hydroid Zoophytes.
By Thomas Hincks, B.A.[1]

The appearance of Mr. Hincks' long-expected work has supplied a desideratum that must have been felt by nearly all microscopic observers, at any rate when on a visit to the seaside.

The abundance, beauty, and variety of the objects supplied by the class of animals which constitutes the subject of his work, must at all times have attracted the attentive admiration of numerous observers with the microscope; but as the next step beyond mere vague admiration and curiosity is, or ought to be, a desire to become acquainted with the nature and life-history of the objects of this admiration, and to be able to classify and name them, it must long have been felt by numerous students of microscopic nature that they were without any satisfactory guide, and their gratitude to one who has supplied such an efficient one as the "History of the British Hydroid Zoophytes," should be equally deep and lasting.

Up to the present time the only systematic introductions of a satisfactory kind to a knowledge of these curious organisms since the time of Ellis have been Dr. Johnston's "History of British Zoophytes," and the late Dr. Landsborough's "Popular History of British Zoophytes;" both excellent works in their day, but whose scientific value has long been much diminished by the subsequent advance of knowledge, which has led to an entire, or almost entire, change, not only in the principles of classification, but in our conceptions of many of the forms upon which they treat. But this cannot be better shown than in the words of Mr. Hincks' preface:—

> The appearance of this work cannot certainly be accounted premature. Twenty-one years have elapsed since the second edition of Dr. Johnston's 'History of British Zoophytes' was published, and during that period the whole aspect of his favourite science has changed. His classification of

[1] Two vols. Van Voorst, London, 1868.

the Hydroida has long been in great part obsolete, while the numbers of known species has been almost trebled since he wrote. Like his predecessor, Ellis, he rendered in his day invaluable service to Zoophytology, and gave an impulse to the study of it of which we are now reaping the fruits. It should be added that there is a charm in his work which does not become obsolete with its science; it will always rank with the 'Corallines' of Ellis, amongst the classics of Natural History literature. As a manual for the use of the student, however, it has long ceased to be of any value; nor is there any work in existence that contains a complete account of the British Hydroida. The place, therefore, is vacant which the present work aspires to fill. I have endeavoured to make it a full and faithful exposition of our present knowledge, and to do for students of this day what Johnston's history accomplished for those of his own generation. It is certainly time that the remarkable results attained since he wrote, and now widely scattered, should be presented in a connected form, and made available for general use; and that the difficulties should be removed which interfere with the cultivation of one of the most delightful branches of Natural History.

The enormous advance that has been made in our knowledge of the structure and life-history of the Hydrozoa generally within the last few years, has necessarily led to wide and sweeping changes in their classification, and in no division of the class has this advance been greater, nor the necessity for an improved nomenclature and classification been more marked, than in that of the Hydroida.

Nothing can more plainly exemplify the nature and extent of the changes thus brought about than a comparison of the contents of Mr. Hincks' work with that portion of Dr. Johnston's classical "History of British Zoophytes," the second edition of which dates to no more than twenty years back, which relates to the *Hydroida*, or, as they are there termed, the *Anthozoa hydroida*.

The entire subject of these animals is comprised in 137 pages. They are classified in five families and fifteen genera, under which we find enumerated about fifty-five species; whilst Mr. Hincks has found it impossible to confine his account of the same order to less than a volume of 300 pages, in which he discriminates not less than fifty-seven genera, and no less than 170 species.

It is obvious, therefore, that such a work was imperatively called for, and it only remains to see how it has been executed.

To those who are aware of the assiduous attention that Mr. Hincks, in the midst of other important labours, has for many years past paid to the subject, and are acquainted with his numerous contributions upon it in various periodicals, it is needless to remark that there is no one among British naturalists better fitted for the task he has undertaken. And to those who upon the appearance of his book may, for the first time, be induced to direct their attention to the study of the Hydroida or of any of the subdivisions of the order, no better or clearer guide can be recommended.

The introduction conveys an excellent "general sketch of the structure of the Hydroida, and the history of their reproduction and development," which, as the author observes, is a fitting prelude to the study of the British species.

This sketch commences with a terminology, or definition of the terms employed, with description of the various parts or organs, which, in the Hydroida, though neither numerous nor complex, yet, in many respects, are sometimes so obscured by the different modes in which they are developed, as to cause the uninformed or inattentive observer entirely to overlook their true homologies. By simple adherence, however, to a strict nomenclature, this difficulty is at once almost entirely obviated. As the terms employed by Mr. Hincks are used pretty nearly in the same sense in which they are understood by most naturalists, mainly from the labours of Professor Allman, it is to be hoped that future writers will consent, so far as is possible, to adhere to them, and avoid the temptation which is too often yielded to, of manufacturing new ones.

Mr. Hincks, however, does not always adhere to this list of terms; as, for instance, in several places when speaking of gonozooids he employs the term *spadix* for the prolongation of the cænosarc into the gonozooid, which term is not to be found in his list of definitions.

The various, and some of them very curious, modes in which reproduction is effected in the Hydroida, have long afforded to naturalists and physiologists a oopious field of observation. Although, since the appearance of Steenstrup's work on Alternation of Generations, the main principle, as it may be termed, upon which all these various modifications of the reproductive process are based, has been more or less distinctly perceived, it is only of late that the entire subject can be said to have been fully elucidated, towards which few, if any, have contributed more largely than Professor Allman. Of this part of the subject Mr. Hincks gives a good condensed account, in which, however, we are surprised to find that he has said so little on the histology of the tissues, except as regards the thread-cells and nematophores. Other portions of the introduction contain remarks on the geographical distribution, mode of collecting, including the habitats of different forms of Hydroida, together with dichotomous tables of arrangement which "are added to enable the student at once to refer any species to its place." Tables of this kind would be very advantageously given in all systematic works, for although, as it is said, "they do not [necessarily] represent natural affinities and relationships, but are a purely artificial contrivance," yet they are undoubtedly calculated "to save time and somewhat wearisome labour to the inexperienced student."

The introduction concludes with an account of the system of natural classification adopted by the author, in which a very fair and interesting summary of the history of this department of knowledge will be found. As in the case of all truly natural systems, the mere perusal of these observations will serve to convey a very good knowledge, not only of the mutual relations of the various families and genera, but also of their structure. Upon this part of his task Mr. Hincks has bestowed great pains, and has, as we think, arrived at very satisfactory conclusions, or at least at a classification as satisfactory as the present state of our knowledge will allow to be formed.

In comparing the extent of the present work with that of Dr. Johnston's, it should have been added, that whilst the latter is content with twenty-nine plates, the former presents us with no less than sixty-seven, besides innumerable woodcuts. Of the execution of these plates, and their fidelity to nature, it is impossible to speak too highly; but as regards them there is one point upon which we are compelled to remark, with much regret, that in the small-paper copies, which, we presume, constitute the bulk of the issue, the impressions, for some reason or another, are not at all equal to those in the large-paper copies. For the credit of the publisher this should have been avoided, both in justice to the artist and the purchaser.

G. B.

QUARTERLY CHRONICLE OF MICROSCOPICAL SCIENCE.

Histology.—*New Handbook of Histology.*—Dr. Stricker, of Vienna, has secured the co-operation of a number of the most distinguished histologists of Germany in the preparation of a work on the tissues of man, which will be the most complete and trustworthy statement of the views of the leading German school that can be obtained. Two parts of the work have already been received in England, containing chapters by Max Schultze on nerve-tissue, by Stricke on microscopical methods, by Kuhne on nerve and muscle, by Pflüger on the salivary glands, by Rollet on blood-corpuscles, by Waldeyer on teeth, &c. &c. Numerous well-executed woodcuts illustrate the work, which is by no means an expensive one.

NERVE.—*Stricker's Handbuch der Lehre von den Geweben*, 1868, Cap. III. Dr. Arndt, Schwalbe, and Koschennikof, in recent numbers of *Max Schultze's Archiv.* Grandry in the *Journal de l'Anatomie et de la Physiologie*, 1869, p. 219, &c.—A number of recent papers on the structure of nerve-elements, of which the above are some, and whose appearance, together with others we have duly chronicled, is reviewed by M. Claparède in No. 141 of the 'Archives des Sciences.' M. Claparède observes that this remarkable series of works tends profoundly to modify our ideas on the intimate structure of the elements of the nervous system.

The results of these new researches are set forth in the most decided manner in the contributions furnished by M. Max Schultze to the treatise of histology which is now being published under the direction of M. Stricker. According to the concordant results of these divers researches, the constituent elements of nervous fibres are *primitive fibrillæ*, very fine filaments, recognisable only by the aid of at least 800-diameter magnifying power. By their aggregation these primary fibrillæ form the bundles known till now under the name of naked axial cylinders.

These strands or bundles can become surrounded with a medullary tunic and with a tunic of Schwann, so as to constitute the nervous fibres known to everybody.

Attached to this important discovery are names of mark in histology, as, besides that of M. Max Schultze, we must mention MM. Frommann, Babouchine, Arndt, Schwalbe, and others. This structure, moreover, is not at all restricted to the nervous fibres only, but extends, besides, to the ganglion-cells.

The form of ganglion-cells had not been sufficiently studied till the present time. In the cerebral cortical substance MM. Meynert, Arndt, Koschennikoff, and others, find that the ganglionary cells have constantly a pyramidal shape. The base of the pyramids is always directed towards the centre, the summit towards the circumference. From this summit proceeds a prolongation larger than all the others, but of what sort histologists are not agreed.

Researches on the histologic development of the nervous system will be able, perhaps, to throw light on the nature of the constituent elements of this system. Histologists are far from being agreed on the nature of the granular framework of the central nervous system—a framework to which different names have been given, such as "reticulum," "neuroglia," "glia," &c. Some regard it as a nervous substance, others as a variety of connective tissue. M. Besser ('Archiv für Patholog. Anat.,' xxxvi, p. 307) was the first to endeavour to sift the question by the study of the nervous cerebral elements in the newly born infant. He found that at the moment of birth the development of the neuroglia is enormous, and that this tissue, composed of nuclei and of a network of granular fibrils, becomes by this transformation the grafting stock, as it were, of all the elements found in the brain later on. The differentiation of the nervous elements is preceded, above all, by the formation of capillary vessels on a large scale. The nuclei of the nervous cellules are the result of a transformation of the nuclei of the neuroglia; the substance itself of the cells results from a transformation of the fibrillar network of this same neuroglia, which is also the origin of the axial cylinders of the nervous fibres. Such a method of formation leaves no place for a membrana propria of nerve-cells. Besides, the greater part of modern histologists do not hesitate to proscribe it. This method of generation accounts for the fibrillar structure that so many authors, especially since the works of Frommann and of Beale, have recognised in the ganglion-cells of the most diverse regions of the nervous system. This structure is far from being the result of a coagulation produced by the reagents employed, as it is easier to recognise in fresh preparations than in those which have become hardened by the different processes employed. The prolongations springing from these cells, which are generally multipolar, are nothing more than bundles of primitive fibrils. Do certain of these bundles (or fascicles) unite directly with the nucleus of the ganglion-cell or not? Histologists are now much occupied by this question. Harless has answered it in the affirmative. More than twenty years have passed since his researches on the electric organ of the Torpedo were made. Then came MM. Axmann, Lieberkuhn, Guido, Wagner, and Owsjannikow, who also maintained, in the most positive manner, the union of nervous fibres with the nuclei of ganglion-cells. They found, it is true, ardent contradictors in Rud. Wagner and MM. Valentin and Stilling, who have made the majority of histologists accept the absence of all direct relation between the nuclei of cells and the fibres or prolongations springing from the bodies of these cells. Nevertheless, in 1864, M. Frommann revived the discussion by his researches on the motor cells of the lumbar region of the spinal cord. He described a complete system of fibres, some finer and others coarser, which not only envelope the nucleus, but actually penetrate into it to the nucleolus. According to this author the very fine filaments that he calls nucleolar filaments (Kern Körperchenfäden) cross the nucleus to spread out in the body of the cell; a part of them are enveloped with a sort of sheath furnished by a prolongation of the nucleus; in consequence he calls this sheath the nuclear tube (Kernröhre). A short time after, Arnold published some very similar observations relative to the sympathetic ganglion-cells in frogs. MM. Hensen, Kollmann, Arnstein, Courvoisier, Guy, and Bidder have likewise observed these filaments in the sympathetic cells. On the other side, M. Arndt, who has also seen a sort of thread in communication with the nucleus, maintains that it does not arise from the nucleus itself, but from a clearer zone which envelops this structure. The union of this filament with the nucleolus would then be a simple appearance, resulting from an optical illusion. It is this, it appears, which Harless understood so far back as the year 1846. "The nervous fibre," says he, "never penetrates into the ganglion-cell in the plane of its largest circle, but it always describes a little arc, so as to arrive at the nucleus; also, when you bring this latter strueture in focus, you see the section of the fibre projecting itself like a nucleolus on the nucleus." Whatever may be said about these observations, the penetration of fibres into the interior of nerve-cells seems less inexplicable now that the body itself of these cells is ascertained to be entirely formed by extremely thin fibrils blended and interwoven in different directions. The question promises, besides, not to be settled so quickly, for we see M. Grandry describing recently, both in the cylinders of the axis and in the nervous cells themselves, certain transverse striæ that can be rendered very distinct by the action of nitrate of silver. There would appear to be here a regular alternation of two sorts of discs, different both by their refractive properties and by their chemical characters. As M. Grandry does not deny the longitudinal character of the fibrillæ, he is obliged to admit for the nerves a structure very like that of the muscular fibrillæ.

The ganglion-cells are, then, veritable knots of fibrils. Involuntarily one asks oneself with M. Arndt if this structure can be reconciled with the function that is ordinarily attributed to these organs. Until now these cells have passed for the central points of all nervous irritation, for the centres which give birth to all nervous phenomena. The discoveries of which we have just rendered account no longer permit us to regard them as aught else but the points of concentration of irritations coming in different directions, destined, perhaps, to transmit them to other spots. In the actual state of science the purely fibrillar and granular substance of the cortical layer of the cerebrum, and without doubt of all grey substance, ought, it seems, to be considered as a tissue essentially irritable. The irritations of this tissue are transmitted by the fibrils to the ganglion-cells, which, after having concentrated them, transmit them to the cylinders of the axis to which they give birth. It is thus, at least, that things should happen in response to centrifugal excitation of nervous activity. But it is clear that the movement ought to take place in an inverse direction for centripetal irritations. These irritations, accumulated in the various cells, are distributed by the prolongations of these bodies to the fibrillar and finely granular matter of the grey substance. This matter, it appears, ought to play henceforward an important part in physiological theories, and it is urgent to give it a name. That of "terminal fibrillar network," proposed by M. Stephany, is perhaps the best. Although placed in the nervous centre, this network is truly terminal in this sense, that all the nervous elements finish by terminating there.

Without doubt we must not diminish too much the importance of ganglion-cells, but it must be agreed that it has, perhaps, been exaggerated. In the first days that follow birth there does not exist in the brain any nervous fibre nor any ganglion-cell, but only elements of the fibrillary and granular terminal substance with its nuclei, and yet there is already at this time transmission of orders from the centre to the periphery. M. Arndt, the principal advocate of the importance of the fibrillar substance in the grey cortical layer of the cerebrum, does not extend this importance to all the reticulum of the base of the encephalon and of the spinal cord. He continues to regard this reticulum as of a connective character, and even in that of the brain he distinguishes both nervous and connective elements. M. Besser, as we have already said, describes the neuroglia thus developed in the child at birth as producing, in consequence of its evolution, not only the nervous elements, but also the vessels of the encephalon. This view of things is energetically repudiated by M. Arndt. According to this observer the vessels of the grey cortical substance, like those of the whole hemispheres, owe their formation to fusiform cells which have nothing to do with the real nervous element. The cerebral *pia mater* is a late formation of membrane, since it does not exist as such until the fifth month of fœtal life. Its formation is in direct relation with that of the vessels.

On the Termination of Nerves in the Retina of Men and Animals. By Max Schultze, with one plate. 'Archiv f. Mikrosk. Anat.,' 4th part, 1869.—Certain fibrillæ ensheathing the rods and cones are described as the true nerve-endings.

Recent Researches on the Retina.—W. Krause has published a very complete *résumé* of the vast amount of work which has been done on this matter during the past three years, in which he combats Max Schultze's views as to the function and structure of the rods and cones. The article has been translated in the last three numbers of the year of Robin's 'Journal de l'Anatomie.'

On the Termination of Nerves in the Epithelial Layer of the Skin. By Dr. Podcopaew. 'Schultze's Archiv,' 4th part, 1869.—Dr. Podcopaew states that he has been able to trace the nerves into the epithelial layer of the skin of the rabbit and other animals, by means of solutions of chloride of gold. Branched lines come into view lying between the cells of the rete, continuous with easily demonstrable nets lying beneath the rete. From the former, very delicate darkly tinted lines may be traced, which run up between the epithelial cells, and near the surface again form fine plexuses. The subepithelial plexus of nerves consists of non-medullated fibres, on the sides of which a few nuclei are attached. It thus appears that a distinct nervous plexus exists between the rete mucosum and the proper laminate epithelium.

On the Relation of the Nerves to the Smooth Muscular Fibres of the Frog's Bladder. By Dr. Tolotschinoff.

On the Sense-organs of the Lateral Line of Fishes and Amphibia. By Franz E. Schulze, with two plates. 'Schultze's Archiv,' 1st part, 1870.—On this subject Leydig has lately published a most elaborate work.

Some Remarks on the Nerves of the Salivary Glands. By Dr. Sigmund Mayer. 'Schultze's Archiv,' 1st part, 1870.

On Sensory Cells carrying Hairs in the Skin of Molluscs. By Dr. Flemming, with one plate. 'Schultze's Archiv,' 4th part, 1869.

LYMPHATICS.—*Plasmatic Circulation in Connective Tissue.*—L. Ranvier, in the 'Archives de Physiologie,' No. 4, 1869, describes the structure of tendons and areolar tissue. He believes in the existence of a plasmatic circulation in connective tissue, and judges from the presence of cells, like the white globules of blood, that it is a true lymph circulation. A canalicular network is stated by him to exist, enclosing the cells of tendon and areolar tissue. It is suggested also that a great space exists between the bundles of subcutaneous areolar tissue, analogous to a serous cavity.

Communication of the Arachnoid Space with the Lymphatics.—R. Bochn states ('Virchow's Archiv,' xlvii, 218) that he has convinced himself that there exist pores or stomata on the free surface of the dura mater, which open into the arachnoid space, and place it in communication with plasmatic canals in the connective tissue of the dura mater, just as the openings which Recklinghausen found in the peritoneum and pleura place those serous cavities in continuity with the lymphatic system.

On the Distal Communication of the Blood-vessels with the Lymphatics, and on a Diaplasmatic System of Vessels, is the title of a very remarkable paper by Dr. T. Albert Carter, of Leamington, published in No. V of the 'Journal of Anatomy and Physiology,' 1869. It appears that this paper was written and presented to the Royal Society of London in 1864, and that that discriminating body did not grant it a place in the 'Philosophical Transactions,' but allowed a short abstract of it to appear in the 'Proceedings,' and preserved the drawings "among the archives of the society." Dr. Carter has done well to publish his paper, for it contains a statement of careful researches leading to the most interesting results. Dr. Carter is known to many microscopists as a most skilful manipulator of the injecting syringe, having been one of the first to use a fine carmine injecting fluid. In the course of his researches, by the aid of this process, he discovered minute vessels surrounding cells in the mucous membrane of the frog's palate. These vessels were too minute to admit of the passage of blood-corpuscles, yet were filled by the injecting fluid. Dr. Carter is inclined to suppose that they are in communication with lymphatic terminal branches, but in any case they form a series of plasmatic canals similar to those conceded as existing in bone and dentine, but in direct continuity with the capillaries. In other tissues, as well in other animals, *e. g.* tongue of man, hair-follicles of mouse, &c., the author has detected this network.

Dr. Carter is not, however, content with this demonstration, but endeavours to prove that the capillaries communicate distally with the lymphatics through the intervention of such fine canals as those described. Numerous instances are cited in which injection was seen to pass from the capillaries through a fine system of canals, such as the canaliculi of bone, into lymph-spaces. Dr. Carter attaches considerable importance to what Goodsir termed "germinal spots" or centres of nutrition, in connection with this matter, and he brings forward a large amount of evidence to show that fine canals place the capillaries in actual communication with the nuclei of various tissues, *e. g.* connective tissue, muscular tissue, cartilage and bone. He points out that the existence of this fine set of canals facilitates the explanation of the rapid formation of capillaries in inflammation, and their subsequent disappearance; he would attribute inflammation to a loss of tonicity in the walls of these finer canals, as also in the capillary walls. Dr. Carter entertains the opinion that every living cell or fibre of the higher organisms is in direct connection at some part of its surface with a channel conveying nutritive fluid. With regard to the offices performed by these fine tubular networks, he considers all those found in the epidermal or mucous tissues to be especially and peculiarly connected with the function of secretion, and probably also, but in a minor degree, with that of absorption; while those situated in the deeper parts of the organism, such as muscle, fibrous tissue, &c., are employed in conveying blood-plasma to, and effete matters from, the tissues through which they run or with which they may be in contact. See also on this matter, the abstract of Dr. Köster's paper, p. 46.

General.—*Histology of the Earthworm.* Six plates. By Dr. Edouard Claparède. 'Kölliker's u. Siebold's Zeitschrift f. Wiss. Zoologie,' 4th part, 1869.—We must defer a more extended notice of this memoir until our April number. Prof. Claparède points out that the histology of the earthworm had yet to be described, since the papers of preceding authors were incomplete. In particular he mentions that of Ray Lankester, which, he says, fails in this respect, while containing much that is interesting on some points. The histology of the worm is very carefully treated by the author, who has had the advantage of starting from the point to which Leydig's admirable memoirs on the allied worm Phreoryctes, (see this Chronicle, vol. vi, new ser., p. 37), and on the nervous system of Annulosa, &c., have brought our knowledge on these matters during the last few years. The minute structure of the integument and of the muscular system, as also of the nervous system, are shown to have been misunderstood by Lankester, who merely epitomised Lockhart Clarke's observations as regards the nervous system. It should be understood that Ray Lankester professed in his memoir, published in this journal more than five years since, to describe in the first place the œsophageal glands which had not been previously well observed; secondly, to introduce to English readers and to confirm D'Udekem's and Hering's observations on the generative organs; and thirdly, to give a brief sketch of the general anatomy of the other structures of the worm. In this latter portion of the paper there are omissions and misinterpretations, which are corrected in the magnificent essay of the illustrious Swiss naturalist.

Embryology.—*Mémoire sur la Formation du Blastoderme chez les Amphipodes, les Lernéens, et les Copepodes.* Par M. Edouard Van Beneden et Dr. Emile Bessels ('Mémoires couronnées de l'Acad. Royale de Bruxelles,' tom. xxxiv). 2. *Recherches sur l'Embryogénie des Crustacés: Asellus aquaticus.* 3. *Mysis ferruginea.* By the same (Bulletins of the same, tom. xxviii, 1869). 4. *Sur la mode de formation de l'œuf et le developpement embryonnaire des Sacculines.* By the same ('Comptes Rendus,' Paris, November, 1869).

Dr. Edouard Van Beneden, son of the illustrious Professor P. J. Van Beneden, of the University of Louvain, is engaged in a most valuable series of researches on the early stages of development. Last year he gained the prize of the Belgian Academy for a really magnificent work on the formation of the egg, and the significance of its various parts in different classes of the animal kingdom. This essay, of which we have had the good fortune to examine the illustrations, has not yet been published, but will shortly appear. In the list above are the titles of the papers which form the natural sequel of Dr. Van Beneden's elaborate investigations, and which he is continually extending.

In the first of these memoirs it is pointed out that in these crustacea the egg consists, 1st, of a germinal vesicle enclosing one or more nucleoli; and, 2nd, of a vitellus, in which two distinct parts must be recognised, viz. the protoplasm of the egg-cell, and what is termed the deutoplasm (nahrungsdotter of La Valette), consisting of the highly refracting vesicles which form the greater part of the vitellus, and sometimes called "vitellus of nutrition." These two parts separate when the blastoderm is in process of formation. The young egg has no cell-membrane, as proved by its amœboid movements and aspect. In the oviduct it gains a covering, which is really a chorion, and not a vitelline membrane. An exochorion forms in some crustacea. In *Chondracanthus* there is a micropyle in the chorion for the passage of spermatozoids, whilst in the Amphipods there is none. In the various groups of crustacea the cells of the blastoderm give rise to a structureless membrane formed by secretion, which is the primary embryonic membrane—the Larvenhaut of German writers. Some crustaceans present yelk cleavage, others do not; and even in the same genus (Gammarus) some species show the phenomenon, others offer no trace of it. The blastoderm does not result, as generally asserted, from a change of the yelk-masses at the periphery of the ovum after cleavage into blastodermic cells, whilst others fuse together at the centre to form a mass of nutritive matter; but this phenomenon results from the accumulation of the deutoplasm at the centre of the egg, whilst the protoplasm separating from this, and taking with it the nuclei of the cleavage masses, appears at the periphery to form the blastodermic cells. *Gammarus* (marine forms), *Dermophilus, Chondracanthus*, and *Copepoda* present the separation of deutoplasm and protoplasm immediately after complete cleavage of the yelk. *Anchorella, Clavella, Caligus*, &c., present no yelk cleavage, but the deuto- and proto-plasm separate directly after fecundation. The blastodermic cells develop by division from the original egg-cell thus separated from the deutoplasm, and enclose and spread over the whole surface of the yelk. In a third type, realised in the *Gammarus* of fresh waters, whilst the deutoplasm is not included in the multiplication of the egg-cells, so as to constitute true yelk cleavage, yet the multiplication does not take place from a point at the periphery, but a number of cells are produced simultaneously, which pass from within to various parts of the periphery. In this memoir many interesting facts are detailed, such as the effect of heat and light in promoting cleavage, and the existence of parasitic Amphipods forming the new genus Dermophilus, living on the *Lophius piscatorius*.

In Memoir No. 2 Dr. Van Beneden gives the results of inquiry into the development of the Isopod *Asellus aquaticus*, so common in ponds. He criticises the papers lately published by G. O. Sars and by Dohrn. The egg on quitting the ovary is surrounded by a true chorion. Sars and Dohrn have described a second membrane as existing when the egg is laid; but this is really, according to Dr. Van Beneden, an embryonic structure—" the blastodermic cuticle." In *Asellus* the first phases of yelk cleavage are wanting, the cleavage being merely superficial, and the formation of the blastodermic cells results from the multiplication of the original egg-cell, a gradual separation between the protoplasmic and deutoplasmic elements of the yelk taking place. The blastoderm is *not* formed, as Dr. Dohrn believed, at the expense of a blastema in which the cell-nuclei develop by free formation. The enigmatical embryonic organs known as "Blättformige Anbange," "micropyle apparatus," "trefoil-like appendages," which Dr. Van Beneden terms "appendices foliacés," arise from the posterior part of the cephalic lobes in the course of embryonic growth. They burst through the blastodermic cuticle and the chorion in the course of their development. As soon as these transient appendages have attained their full growth the permanent appendages commence to make their appearance, and Dr. Van Beneden maintains that the first and second pair of antennæ are the first to appear, contrary to Dr. Dohrn's statement. In this state the embryo represents the well-known four-armed Nauplius form of crustacean development. A cuticle now forms on the surface of the embryo, which is the larvenhaut or nauplian cuticle, and is shed after the blastodermic cuticle has been moulted. The chorion, blastodermic cuticle, and nauplian cuticle having been successively shed, the embryo comes forth from the maternal pouch in its adult form.

No. 3 treats of the aberrant Decapod *Mysis ferruginea.* In this case, also, there is a partial yelk cleavage, resulting in a complete envelopment of the egg by the blastoderm. The existence of a primordial cellular ridge is pointed out, which is exactly homologous with the cellular column (Keimhugel) of Hemiptera, Orthoptera, and Lepidoptera. By the study of the changes of this part the homologies of the parts and appendages of the cephalic lobes, &c., in these two groups of Arthropods may be fixed with certainty. The mandibles and antennæ give their first indications at the same time, and after the caudal lobe, which is bent back under the ventral region, as in all Decapods. The nauplian cuticle is the first which *Mysis* develops. There is no blastodermic moult. The trefoil appendages of *Asellus, Peneus,* &c., are represented on the flanks of the embryo *Mysis.*

No. 4 relates to the eggs of the crustaceans of the genus

Sacculina, and is in reply to and correction of M. Gerbe. Dr. Van Beneden shows that M. Gerbe had mistaken the whole egg for the body which produces the yelk, and that the egg of Sacculina cannot be compared to that of Birds at all; nor is there anything in this egg representing the curious "vitelline body" of the egg of spiders and some Myriapods, which is very variable and even accidental in occurrence.

Dr. Van Beneden's series of researches, as well as his prize essay on the ovum in various classes of animals, will repay careful study, and place the whole subject, which is now of such vast interest and importance, within the reading of those who have the use of the French but not of the German language. Their chief recommendation lies, however, in their great originality, and the important new considerations they contain.

Miscellaneous.—Dr. Greef's paper on 'New Fresh-water Radiolaria' is noticed by Mr. Archer in another part of the journal, whilst Prof. Kupffer's letter to Prof. Max Schultze on the 'Kinship of Ascidians and Vertebrates' also appears amongst the memoirs. Both these papers were published in 'Schultze's Archiv,' 4th part, 1869.

New Coffee Fungus.—The Rev. M. J. Berkeley forwards to the 'Gardener's Chronicle' a letter from the well-known botanist, Mr. Thwaites, of Ceylon, in which he speaks of the consternation caused among the coffee-planters of that island in consequence of the rapid increase of a parasitic fungus in the coffee-plantations, causing the leaves to fall off before their proper time, and endangering the safety of the crop. It is a singular fact that among more than one thousand species of fungus which have been received in this country from Ceylon this particular one does not occur; not only is it an entirely new species, but it is with difficulty referable to any recognised section, being intermediate between the true moulds and the *Uredos.* Mr. Berkeley establishes for it a new genus *Hermileia.—Nature.*

PROCEEDINGS OF SOCIETIES.

DUBLIN MICROSCOPICAL CLUB.

15*th July*, 1869.

REV. E. O'MEARA exhibited a specimen of *Donkinia recta* new to Ireland; this form had been found as British as yet only on the coast of Northumberland.

Dr. John Barker wished to record the occurrence of a campanulate-cased freshwater Vaginicola, in which the case was furnished with a *valve* similar to that of the form designated *Vaginicola valvata.* This would bear out Mr. Archer's note of a valve being existent in the large Vaginicola from Victoria Docks kindly sent by Mr. Reeves, and would, perhaps, go to indicate that the valvular structure may pervade even all the forms in the genus.

Mr. Archer had again to thank Mr. Reeves for kindly sending from London for exhibition a slide showing, in fine fertile condition, two species of Bulbochæte—that form restricted by Pringsheim as *Bulbochæte setigera* (Ag.) Prings., along with *Bulbochæte pygmæa*, major (Prings.). The latter is very much the smaller, and in every respect a different plant from the former, upon which, however, it grew epiphytically. Indeed, if within a single genus two species might with any propriety be called the antithesis of each other, these might. In *B. setigera* (Prings.), which is a large species, the oospores are large and depressed; the septum of the supporting cell somewhat above the middle point; dwarf-male straight, with foot and "inner" antheridium; in *B. pygmæa* (Prings.) the plant is very minute, oospores elliptic, no septum in supporting cell; dwarf-male with foot and "outer" antheridium.—These fine specimens were accompanied by a very minute little filament, seemingly agreeing with that named *Œdogonium turfosum* by Kützing; but on the best examination of the present specimens there was no evidence of cap- or sheath-cells, and it might be hence problematical whether this could be rightly considered as belonging to Œdogonium at all. It, in fact, consisted of a very slender confervoid filament, cells three to five times longer than broad, with an elliptic fruit-like cell, a little longer than an ordinary joint, here and there interposed. The contents of this oogonium-*like* cell had become somewhat contracted, but had evidently, when fresh, completely filled the cavity, and was of a rather densely granular nature. The extremely fine filaments of this form might, under a moderate power, be overlooked readily, occurring as it did along with the Bulbochæte, as the diameter did not greatly exceed that of the bristles of the *Bulbochæte setigera;* but their distinct appearance and jointed nature, with the apparent fruit-cells, were quite evident on applying a power sufficiently high.

Mr. Archer, thanks also to Mr. Reeves, exhibited examples from England of the "water-net" (*Hydrodictyon utriculatum*). This remarkable alga has never yet been found in Ireland.

Dr. John Barker exhibited a remarkable little Infusorium, clearly a new species, if not, indeed, the type of a new genus. This was minute, pear-shaped, mostly green, covered by scattered vibratile cilia, those at the extremities being longer and more prominent than the rest, at one side furnished with an almost semicircular, vibratile, constantly undulating flap. The green colour was due to the presence (imbedded in the substance of the body) of numerous chlorophyll-granules, but this form has been occasionally met with colourless. This little animal inhabited an elongate, barrel-shaped, hyaline glassy test, open at both extremities, in which it could freely move and completely turn round and round. This test appears, at first at all events, to be attached by the side to various filaments, but is often met free. A part of the exercise of this creature appeared to be performed in spinning itself round and round on its longitudinal axis, like a top, and another part in efforts to push itself through the opening at one end of the test, speedily turning round and driving itself strongly so as to partially emerge at the other end. It even sometimes succeeded in swimming away, and the empty "barrels" were sometimes met with in the water. The animal seemed to divide by transverse fission. A more detailed account of this remarkable, though very minute little creature, will be prepared by Dr. Barker for a future occasion.

Dr. Barker exhibited a Staurastrum, taken by him in Co. Westmeath, presenting the peculiarity of having but two arms in end view, but which had all the aspect in front view of *Staurastrum gracile*, from which it thus differed in being plane, not triangular. This circumstance rendered this form a very pretty object, because, being so nearly flat, all its superficies, arms and all, could be simultaneously brought into focus, and thus seen to perfection. Dr. Barker was inclined to regard this compressed form as a species distinct from any other, even *St. gracile.*

From the latter view, expressed by Dr. Barker, Mr. Archer observed that he as yet could not but differ; be regarded this as a singular and interesting form of *Staurastrum gracile*, and would even venture to foretell that if the pools whence these specimens had come were searched the triangular form might disclose itself. These examples were just an additional proof of the view he had long entertained, that the forms generally referred to the second section of the genus Arthrodesmus (Ehr.) were, strictly speaking and naturally, but plane or two- (not three-, four-, or more) sided Staurastra. He was well acquainted with a very minute and slender triradiate Staurastrum, which he regarded as nothing but a triple-rayed form of *Staurastrum tetracerum* (Ralfs), ordinarily, however, occurring compressed. The number of rays or angles in the end view of a Staurastrum, as is well known, is of no importance; the same form occurs with sometimes three, sometimes four or more angles (nay, sometimes with a different number at either end of the same specimen), therefore whether with two only or three would seem to be equally of little real importance, except that the reduction to two only is greatly more rare. It is the character or kind of arms, not their number, which is of importance. Mr. Archer would not by any means, however, go so far as Reinsch does in some cases. For instance, that author regards *Arthrodesmus convergens* (Meneghini), Ehr., as but a variety of *Staurastrum Dickiei* (Ralfs), and he merges the former into the latter, suppressing *St. Dickiei.* Altogether Mr. Archer would reiterate the view he had before now expressed (largely confirmed, he thought, by Dr. Barker's specimens in another section of the genus), that *Arthrodesmus convergens* was truly a two-sided Staurastrum, and should be placed in that genus, so far coinciding with Reinsch; but he could not acquiesce, on the other hand, that it could be considered as but making up with *Staurastrum Dickiei* one species only, for these two forms are not only distinct in size and contour, but absolutely different in their zygospores. Reinsch, indeed, takes no note of this latter circumstance in this instance, though he considers it of importance in other similar cases, so that, perhaps, he may hereafter modify his view. In the form called *Arthrodesmus convergens* the zygospore is large and absolutely destitute of spines, whilst in *Staurastrum Dickiei* it is densely beset with subulate acute spines. With regard to Dr. Barker's form, and the typical *St. gracile,* there did not appear to be those material differences in size and contour that exist between the two forms alluded to; Dr. Barker's seemed, indeed, to have a greater proportional distance from extreme to extreme of the projecting processes than the triangular form, but this would appear to arise from its being plane, the observer thus seeing the whole distance nearly in one line, whilst in the triangular form the two (of the three) processes in view would be turned a little up towards the observer, thus rendering the distance between their extremities apparently, not really, shorter; the absolute distance, if it could be measured by traversing along the upper outer corner or edge of the top or end of the Staurastrum, would be pretty much alike. There is no doubt, however, so long as we are unacquainted with the zygospores of those forms, as regards either view, we have not exhausted the argument.

19*th August*, 1869.

Dr. John Barker showed two pretty forms of Spirulina, and, as they writhed and twisted under the microscope, very singular looking objects. Without venturing to say very certainly, these Mr. Archer had some time ago identified as *Spirulina Zanardinii* and *S. Jenneri*. The former is a slender form, with large coils or spirals, and, the filaments being intertwined or doubled on themselves, large, continually changing openings or intervals occur, giving, as they mutually and somewhat quickly twist and untwist, a very graceful appearance. The latter is a thicker and stouter filament of short coils, and when intertwined they lie in closely, curve to curve, leaving no interspaces, the result being a rope or plait-like appearance. These occur in greater or less strata, at the bottoms of pools, and sometimes rise to the surface in little masses, like Oscillatoriæ buoyed up by bubbles.

Mr. Crowe showed sections of fossil tooth of shark and fossil palm.

Dr. John Barker likewise showed the conjugated state of a minute Cymbella, the four valves of the original pair of conjugating frustules lying closely applied to the pair of young frustules, these lying parallel and surrounded by the former, the whole involved in a mucous envelope.

Dr. Moore showed examples of the alga *Botryococcus Braunii*, which occurred in long sheets, of some yards in length, floating on the surface of Lough Bray, so as to become a conspicuous object. In this plant the clusters of green cells are imbedded in a colourless mucous matrix, and sometimes such clusters remain united by strings of this mucous investment.

Mr. Archer mentioned that *Botryococcus Braunii* was not seemingly an uncommon alga in moor pools, sometimes coating submerged sedges and the like with a greyish-green stratum, sometimes, however, suspended in the water in streaks, and often isolated. It passes through a red condition. He had, however, never seen it in anything like the masses described by Dr. Moore. More than once, when a single group or family of this alga, from gatherings kept for some time in the house, had turned up under a low power of the microscope, he had been to some extent deceived by the way in which it resembles some radiolarian rhizopod, strange as it may seem. The mucous matrix containing the families of cells seems not unfrequently to give off rather long, filiform, prolongations, which stand out more or less radiantly, looking not unlike pseudopodia, and these are undoubted rhizopoda containing chlorophyll. It might, indeed, be a good example of two objects, with no affinity in any respect to each other, still superficially simulating one another.

Mr. B. Wills Richardson exhibited some very beautiful sections of hempseed calculi, cut for him by Mr. Charles Baker, of Holborn, London. About 600 specimens of this description of calculus were passed by the patient—a gentleman advanced in life—in the space of a few months. Several being facetted, led him at first to suppose that they came from the prostate gland; but analysis proved their composition to be oxalate of lime, with traces of lithic acid.

Rev. E. O'Meara referred to two packages of material kindly supplied to him by Sir Leopold McClintock. One was raised from 800 fathoms, twenty-five miles north-east of Vera Cruz, Gulf of Mexico, in which no trace of animal or vegetable life was found. The other, taken from the bottom of the sea in 750 fathoms, fifteen miles north-west from Cape Antonio, west extreme of Cuba, July, 1865, was full of organisms. This latter material having been treated with hydrochloric acid, was greatly reduced in bulk, and the residuum having been submitted to examination, was found to consist of some sand, with a considerable number of Polycistina, sponge-spicules, and diatoms, for the most part in a fragmentary condition. Among the diatoms were found several small species of Coscinodiscus. *Pinnularia pandura* and *Cocconeis punctatissima* occurred frequently, and often in a perfect state. Among the rarer species were found some forms of Glyphodesmis, and also a specimen of Campylodiscus, which, though not quite perfect, he could identify with *Campylodiscus ecclesianus* of Greville ('Quart. Jour. Mic. Science,' Jan., 1857, Plate III, fig. 6).

Mr. Archer showed fertile examples from near Multyfarnham of *Bulbochæte setigera* (Prings.), showing both forms of fructification, same as those sent from England by Mr. Reeves, with a much more minute form likewise growing thereon, not, however, *B. pygmæa* (Prings.), as in Mr. Reeves' specimens, but *B. gracilis* (Prings.). It was to be regretted, however, that the latter did not show the antheridium, and the more so as Pringsheim himself had not seen it in this species. The oogonia in the examples now shown were fully matured, and had acquired the amber colour characteristic of that state; hence it might perhaps be assumed that the antheridia had fallen away, and, as Pringsheim's figure shows the fully ripe condition of the oogonia, perhaps the same may have happened in the examples he had under observation.

Mr. Archer exhibited conjugated examples of *Spirogyra orthospira* (from Co. Tipperary), showing the zygospores formed, as is specifically characteristic in the uninflated cells.

Mr. Archer showed likewise the odd-looking rotatorian, *Actinurus neptunius*, remarkable for its extreme length and great comparative slenderness. This gaunt example of rotatorian life probably measured $\cdot\frac{1}{13}$th of an inch in length.

Mr. Archer showed examples of a minute little Staurastrum taken from a bog near Mullingar, which he thought undescribed; this he had not found at all near Dublin, but it was identical with the minute form he had met with in one or two gatherings made in the spring near Glengariff. It most resembled *Staurastrum læve*, but is distinguished therefrom by the entire ends of its rays or arms, which bear one or two little knob-like elevations about half way down their length. The description of this form Mr. Archer would reserve for a future occasion.

23rd September, 1869.

Rev. E. O'Meara showed *Pleurosigma arcuatum* from a gathering made by him recently at Bannow, in the County Wexford.

Also a species of Nitzschia from the same place, which form Grunow has described doubtfully as a variety of *N. reversa* (W. Smith), ('Verhandlungen der Kaiserlich-Königlichen Zoolog. Botanischen Gesellschaft in Wien.,' Band xii, 1862, tab. xviii, fig. 4). This distinguished microscopist had not seen specimens of *N. reversa;* if he had, his doubts as to the identity would have been confirmed. This form differs from *N. reversa* both in outline and in the character of its sculpture, so as to entitle it to be regarded as a distinct species. Mr. O'Meara suggested it should be named *N. Grunovii.*

Mr. B. Wills Richardson exhibited one of M. Nachet's cameras for making drawings of objects with the microscope in the upright position, the image being apparently projected in front of the stand. He (Mr. Richardson) spoke highly in favour of the use of this camera for drawing objects in cells containing fluid; for, as the slide lies "on the flat," there is but little risk of an object moving, which is so liable to occur during the use of cameras that require the compound body to be at a right angle to the uprights or pillars of the stand.

Mr. Archer exhibited, new to Ireland, the plane form (var. β) of *Aptogonum desmidium*, or, better, *Desmidium aptogonum* (Bréb.). This was taken, very sparingly, from a bog close to the town of Mullingar. It is very rare. The triangular form had been recorded from Connemara last year by Dr. Barker. Nothing could surpass, as a pretty object, a portion of a filament of this plane form, for its flatness admits of all coming into focus at once under a quarter-inch.—Mr. Archer likewise presented, new to Britain, *Arthrodesmus bifidus* (Bréb.), and in the same gathering. This is a very minute form, but one which cannot be confounded with any other. *Arthrodesmus tenuissimus* (Arch.) is somewhat like it in front view; but a side or end view of that form, showing the *pairs* of minute divergent spines, at once settles the matter; and though these two resemble each other in front view more than either seems to resemble any other form, the bidentate lobes of *A. bifidus* presents something quite distinct from *A. tenuissimus,* which latter is a species even still more minute.

Professor E. Perceval Wright exhibited *Dehitella atrorubens* of Gray. A small portion of this remarkable organism had been very kindly given to him by Dr. J. E. Gray, who had described it as follows:—"Sponge or coral, dichotomously branched, expanded, growing as a large tuft from a broad, tortuous, creeping base, of a dark brown colour, and uniform hard, rigid substance. Stem hard, cylindrical, opaque, smooth; branches and branchlets tapering to a point, cylindrical, covered with tufts of projecting horny spines on every side, those on the branches often placed in sharp-edged, narrow, transverse ridges; those of the upper branches and branchlets close, but isolated, and divergent from the surface at nearly right angles." ('Proc. Zool. Soc. London,' 1868, p. 579, fig. 1, p. 578.) This genus has been placed by Dr. Gray with a second new genus, *Ceratella*, in a family called Ceratellidæ; and while strongly inclining to locate this family among the true horny sponges, Dr. Gray at the same time calls attention to the fact that many of the characteristics of the keratose sponges are not to be met with in the dry horny skeletons of the two species described.

On treating a small portion for some (thirty-six) hours with caustic potash, the only effect observed was a greater transparency of the keratose fibres, and perhaps a greater flexibility of the entire mass. There was no tendency, however, of the frame or network to break up into detached pieces, such as might occur if it were formed of a series of horny spicules united the one to the other; indeed, the skeleton must be looked upon as continuous.

On placing another portion in some weak nitric acid, effervescence at once occurred, and the coloured horny material pretty speedily disappeared, leaving, however, behind, a semi-transparent basis, which in great measure preserved the form of the original little twig, and which had all the appearance at first sight of being siliceous. On being placed on a glass slide, and covered with a piece of thin glass, it yielded to a slight pressure, and when examined under the microscope showed a gelatinous basis, in which were entangled a few biacerate siliceous sponge spicules, and a few diatoms (*Navicula*, *Pinnularia*, *Coscinodiscus*, *Amphitetras*), the former, without doubt, just as much foreign to the *Dehitella* as the latter.

On examining the structure after it has been gently boiled in distilled water for a few moments, it will be found to have absorbed a certain quantity of the water, so that even after it has been tightly pressed and flattened it will, on the pressure being removed, soon recover its shape. It would appear, however, that the so-called tufts of horny spines met with on the sides are really not so much spines as sharp-ending prolongations of the common skeleton; indeed, they cannot be called tufts of "spicules" in the ordinary meaning of this word, and the arrangement of the network will be best learned from the accompanying woodcut (fig. 1). On making a transverse section of the stem it is apparent that the main fibres are continuous, and that thus there is a series of canals permeating the entire mass (fig. 2). The skeleton is to a certain extent regular; that is, it is made up of a series

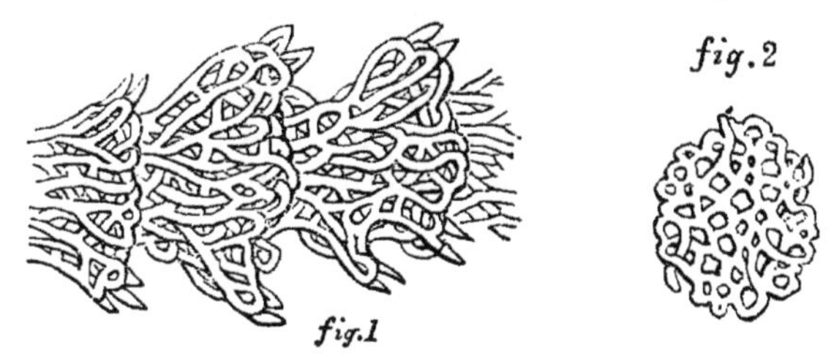

of long horny strands, which run almost parallel to one another, either terminating in a loop or in an obtuse point. The horny fibres are obscurely striated, and, when young, are not only much lighter in colour, but also appear to be hollow. It will thus be evident that but little additional light can be thrown on this curious form, beyond that already thrown on it by Dr. J. E. Gray, until its rediscovery in a recent state; but Professor Wright trusted that, in exhibiting this specimen and detaining the Club with these remarks, he would not be considered as altogether wasting the time of the members, and he would simply now express his opinion that Ceratellidæ was a family of arborescent keratose sponges.

Mr. Archer showed *Cosmarium annulatum* (Näg.), seen by him for the second time only. The first occasion was in a gathering made in the "Rocky Valley," near Bray, and the present was made from pools near Mullingar; but as the species is very minute, it may have been overlooked in other localities.

Mr. Archer wished to record that he had since had an opportunity to make some collections from the same locality in County Westmeath from which Dr. Barker had obtained the remarkable plane form exhibited at the July meeting, which he (Mr. Archer) would refer to *Staurastrum gracile*, and that he had met with the triangular form, as well as the plane, in one of the gatherings, thus, he thought, fully bearing out the forecast he had made, and the views he had expressed; at least, so far as could be done pending the discovery of the zygospores, but which, indeed, might just possibly tend to decide the question in the other direction.

Mr. Archer likewise showed *Characium tenue* (Hermann).

Mr. Archer further drew attention to some extremely minute bodies of a crystalline appearance, occurring inside the cells of *Spirogyra nitida*. These floated just close under the spiral bands, presented a greenish hue, and were of a general cruciform or X figure, the arms very slender, and sometimes branched or feathered more or less; sometimes an H figure, with the horizontal connecting line produced beyond the vertical ones at both sides; the extremities, as in the cruciform ones, likewise somewhat branched. These were clearly not chlorophyll-granules, though showing a greenish tint.

Dr. Moore showed a gathering of the little minute unicellular clustered alga which sometimes forms a scum on the surface of the waters in the houses at the Botanic Gardens, drawn attention to by him at the Club meeting of July, 1866. The identity or naming of this production would, however, be a matter of great difficulty, but it was interesting to note its periodic recurrence.

Mr. Archer drew attention to a puzzling production he had lately noticed in some quantity in the gatherings made from pools near Mullingar. This consisted of a tabular or foliaceous, very variously shaped, frond (so to call it), composed seemingly of bacillar greenish bodies, held together in close approximation and in variously arranged positions, but so as to leave no irregular intervals, by a reddish-coloured, common agglutinating substance. This, therefore, had some resemblance to a reddish-coloured, orbicular-celled alga, forming roundish clusters not infrequently found in these situations; but whether any state of the latter, occurring in the same water, it would be hard to say, but *à priori* they looked distinct. But a curious circumstance, as regards the production now specially drawn attention to, was that the young "fronds," which, unlike the very irregularly lobed further advanced ones, were of a globular figure, with, of course, a cavity within, possessed the power of rotating hither and thither with some energy, but by what agency defied all efforts to perceive. Sometimes these hollow globular examples seemed to possess some kind of granular core or body in the centre. This very unattractive looking object—algal apparently—would not deserve a much closer examination, and Mr. Archer would be obliged to content himself with this imperfect and crude note for the present; perhaps, indeed, he should apologise for having on several late occasions brought forward some very puzzling little nondescripts so very crudely, but he had done so in the hope (it might be faint, indeed) that, even roughly as he was able to record them, the possibility might be that observers elsewhere might be able to recognise having met them, and be, perhaps, in a position to throw some light upon the obscurity.

Royal Microscopical Society.

October 13*th*, 1869.—The President in the chair.—A list of donations was read, and the thanks of the meeting presented to the respective donors; a special vote being accorded to Mr. Ross, who, as the President announced, had given to the Society new immersion front lenses for the $\frac{1}{8}$th and $\frac{1}{12}$th object-glasses which he had already presented to the Society.

Mr. Hogg exhibited a phial containing a quantity of *dichroic* fluid [1] which had been found by Mr. Allbon in a ditch, between Mortlake and Kew, and contained Batrachospermum atrum in a decomposed state. The fluid obtained by Mr. Sheppard, of Canterbury, who first discovered and described it, contained a great deal of animal life, while that exhibited by Mr. Hogg was almost entirely composed of a confervoid growth, the sides of which were covered with cells filled with pseudo-naviculæ. When examined by transmitted light the fluid gave a delicate bluish pink colour, and by reflected light a reddish hue. Under the micro-spectroscope, its spectrum is just that described by Mr. Browning in vol. vii, 1867, of the Society's 'Transactions.' A few pieces of camphor serve to preserve the fluid; and although the specimen exhibited had been corked up for several months, the colour is

[1] This fluid, like that described by Mr. Shephard in Vol. VII of this Journal, is clearly, as I pointed out in opposition to Mr. Reade, identical with the Phycocyan of Cohn, and is derived from dead Oscillariæ.—E. R. L.

nearly as good as when it was fresh gathered, and the spectrum reaction quite perfect.

Mr. Carruthers read a paper on the "Plants of the Coal-measures."

Mr. Slack wished to call the attention of Fellows conversant with crystallography to the curious instance mentioned by Mr. Carruthers, in which, after the charring of the vegetable structure, although the particles of carbon preserved the exact form of the vegetable cells, they had opposed no obstacle to the crystallization of the carbonate of lime, which had gone on through their interstices as though no obstacles had intervened.

Mr. C. Brooke stated that structure is much interfered with by foreign matter—the sandstone of Fontainebleau, for instance, assumes the form of rhombohedral crystals of calcite. The stone does not contain more than 5 to 7 per cent. of carbonate of lime; but the 95 per cent. of silex seems to be dragged into form by the 5 per cent. of carbonate of lime which controlled the character of the crystallization.

A paper was read from Brevet Lieut.-Colonel Woodward, Assistant Surgeon of the United States army, on "Immersion Objectives and Nobert's Test-plate."

The 'Quarterly Journal of Microscopical Science' for October, 1868, contains (p. 225) a short article in which I record the results of certain experiments made by me with the new nineteen-band test-plate of Nobert. In that paper I stated that I had obtained the best results with the $\frac{1}{25}$th objective of Messrs. Powell and Lealand, of London. The $\frac{1}{50}$th of these makers, which in my hands had excelled their $\frac{1}{25}$th on Podura and other test-objects, proved inferior on this plate, apparently because the cover of the object was too thick to allow the lens to do its best. With the $\frac{1}{25}$th I satisfactorily resolved the true lines of the fifteenth band of the plate; and subsequently my friend and assistant, Brevet-Major E. Curtis, Assistant-Surgeon, U.S.A., prepared a series of photographs of the several bands, showing the true lines in each, from the first to the fifteenth inclusive. I was, however, unable to make out the true lines in the last four bands with any lens then in my possession. I conceived the idea, nevertheless, that if I could procure a test-plate ruled on a thinner cover, so as to give the $\frac{1}{50}$th full play, I might go farther. I therefore wrote to Nobert, who, after long delay, furnished me with a new test-plate, which reached me during March of the present year. This test-plate cannot be too highly praised for its delicacy and beauty. The lines are ruled on the under-surface of a square of thin glass the $\frac{1}{275}$th of an inch thick, which is cemented to a glass circle the $\frac{1}{110}$th of an inch thick. This circle is mounted over a round aperture in a strip of burnished brass 3 inches by 1, on which is inscribed the usual memoranda placed by Nobert on his nineteen-band plates.

The results of using this test-plate are as follows:

A careful count of the lines in each band, with Messrs. Powell and Lealand's new $\frac{1}{16}$th immersion objective, gave the following results:—

15th band ...	45 lines	18th band ...	54 lines
16th " ...	48 "	19th " ...	57 "
17th " ...	51 "		

In obtaining the above results I illuminated the microscope, as in my former work on the Nobert's plate, with a pencil of monochromatic light obtained by reflecting the direct rays of the sun from a heliostat upon a mirror, by which they were thrown through a cell filled with a solution of the ammonia sulphate of copper, upon the achromatic condenser. As an achromatic condenser I substituted for that belonging to the large Powell and Lealand stand of the Museum a $\frac{1}{5}$th of an inch objective of 148° angle of aperture, and used it without a diaphragm; obliquity of light was obtained by moving the centering screws of the secondary stage.

I also obtained satisfactory resolution of the nineteenth band, with the same lens, by using for the illumination violet light, obtained by throwing the violet end of the solar spectrum produced by a large prism upon the achromatic condenser used as above, and, subsequently, by shifting the prism, got successful resolution of the nineteenth band with blue, green, yellow, orange, and red light, These results I had the pleasure of exhibiting to Dr. Barnard and several others.

As for other lenses, carefully tried on the same plate, I obtained the following results:—

The $\frac{1}{8}$th of Wales and the $\frac{1}{25}$th and $\frac{1}{50}$th of Powell and Lealand, all dry lenses, resolved the fifteenth band, but not the sixteenth.

An immersion $\frac{1}{15}$th by Wales resolved the sixteenth band, but failed to go farther. An immersion $\frac{1}{20}$th by Wales resolved the seventeenth band, but failed to go farther. A Hartnack immersion No. "11," belonging to President Barnard, also resolved the seventeenth band, and failed to go farther.

A Tolles' immersion $\frac{1}{5}$th, just constructed for Dr. J. C. Rives, of this city, resolved the fourteenth band, but failed to show the true lines on the fifteenth. This result with the Tolles' immersion $\frac{1}{5}$th corresponds with the results very recently obtained with a Tolles' immersion $\frac{1}{6}$th, just received by my distinguished friend, Mr. W. S. Sullivant, of Columbus, Ohio, who wrote me May 25th of the present year:—"The immersion lens you inquired about, which Tolles sent me, was marked $\frac{1}{6}$th, but was a strong $\frac{1}{5}$th English standard. The utmost it could do was to show true lines on the fourteenth band."

These results confirm the opinion expressed in my former article, that the lines claimed to have been seen, but not counted, in the nineteenth by a Tolles' immersion $\frac{1}{6}$th were spurious lines, an opinion to which still greater weight is added by the following result:—A Tolles' immersion $\frac{1}{10}$th of 175° angle of aperture was received at the Museum, May 26th, from Mr. Charles Stodder, who stated in his accompanying letter that it might be regarded as a fair sample of Mr. Tolles' work. With this lens, after numerous careful trials, I was unable to see the true lines beyond the sixteenth band.

It will be seen, then, that in my hands the best definition was obtained by the immersion $\frac{1}{16}$th of Messrs. Powell and Lealand; and I may here say that, on a thorough comparison of this objective with the dry $\frac{1}{25}$th and $\frac{1}{50}$th of the same makers, I found that not merely did their new lens resolve higher bands on the Nobert's plate than could be made out with the $\frac{1}{25}$th and $\frac{1}{50}$th, but that it would bear the use of eye-pieces and amplifiers so as to give higher powers than can be obtained with the $\frac{1}{50}$th, with much better illumination, with better definition, as well as with a practical working distance. The lens may therefore be especially commended for anatomical work when the highest powers are desirable.

In conclusion the author referred to some remarks by Mr. Mayall, Junr., in a paper on "Immersion Objectives and Test Objects." [1]

Mr. Mayall says:—"Dr. Woodward seems not to have been sure of the accuracy of the count he made on his photograph: for although in one part of this paper in the current (October) number of the Journal of this Society, he says the photograph shows the twelfth band as resolved into thirty-seven lines, farther on he says that forty is the real number in that band." This misapprehension on the part of Mr. Mayall arose from a misprint in the Journal. On p. 231, fourteenth line, "12th band" reads in my original MS. "13th band;" on the thirtieth line of the same page, I find "12th band" printed instead of "19th band," which is the reading of the original.[2] The same article contains some other singular misprints, most conspicuous among which may be mentioned, "Starting's work on the microscope," p. 225, instead of Hartings; and "Greenhap," p. 288, instead of Greenleaf. At the time my article was prepared I had no doubt whatever of the true number of lines in all the bands resolved, except the fifteenth, about which, as I stated, I was uncertain whether the true number of lines was forty-five or forty-six. At present, additional work has satisfied me that forty-five is the number, and I am also well assured of the correct number as given above for the remaining bands. I freely admit that the difficulty of determining which is the last real, and which the first spectral line is very great even on glass positives; nevertheless, a comparison of several photographs with each other, and with the bands as seen in the microscopes, has satisfied me that my count is correct.

[1] See this Journal, July, 1869.

[2] Dr. Woodward was, unfortunately, not able to correct his own proofs, on account of distance. The errors pointed out read in his original MS. as they were printed. We are glad to have the opportunity of correcting them.

A paper was read from G. W. Royston-Pigott, M.A., M.D. Cantab., "On High Power Definition with illustrative examples."

The researches of the author with regard to the defining power of object-glasses of diminutive focus have led him to the conclusion that many objects are falsely represented under object-glasses of from $\frac{1}{16}$th to $\frac{1}{20}$th inch focus. The misrepresentation he attributes to a certain "residuary aberration (chiefly spherical)," which he thinks he detects on all glasses of high power. He applies his views more especially to the most ordinary microscopic test-object, the Podura scale. He states that he has been able to resolve what have hitherto been regarded as minute spines into rows of beads.

It is well known that under a low power, as 80 or 100, the Podura is remarkable for its wavy markings (these are a safe guide in selecting the scale), aptly compared to "watered silk." It is here that the *secret* of their cause and nature is to be sought for: hitherto one which has baffled the most famous glasses of modern times. As a simple fact sometimes leads to a suggestion, view carefully against the light two pieces of silk woven with the finest weft and warp placed one over the other: accordingly as one is lightly *stretched* more than the other or as the weft of one is inclined more or less to the weft of the other, instantly an endless series of waves are developed by the lines of optical interference, mesh intersecting mesh with infinitely varied effect; but always *waves*. Can the waves of the Podura be similarly caused?

Raising the power to 200 or 250 and using a side light upon our scale athwart its length, all waviness disappears, and in its place is seen a longitudinal *ribbing*, shaded very darkly; with a less oblique side light, lucid rhomboid chequers glitter brightly: the rhomboidal sides, crossing at acute angles, may be seen with a low power of 500. With 1200 these ribs have divided themselves into a string of longitudinal beads. But with 2300 they appear to lie in the same plane, and terminate abruptly on the basic membrane: upon focussing for the strings of beads attached to the lower sides the beading appears in the intercostal spaces. The upper beads are best seen either green upon a pink ground or pink upon a greenish ground, which phenomena may possibly arise from the different dispersive powers or refraction of the various structures or the correction of the glasses; or even more recondite causes.

When the light is much more oblique, yet achromatic, the beads appear shaded as roughly represented in the diagram, the intervening spaces showing fine traces of intersecting lines.

Using now an adjusting $\frac{1}{2}$-inch at 250, and rotating the scales, some of the most favorable positions, with oblique light, inclined about 15 degrees to the axis of the scale, show a double set of longitudinal lines forming a lattice-work. These lines are the markings existing on the other side of the scale.

With 300 to 500 the celebrated "spines" appear, according to the size of the scale, as very dark short tapering marks (like

" notes of admiration " without the dots!!!). To see these clearly with 2500 has been considered the *ne plus ultra* of microscopical triumphs, and it is consequently with no little diffidence that the writer ventures to traverse the belief of twenty-five years. The object of this paper is to show that definition can be further improved under the use of high powers, and if he should succeed in accomplishing this, the leisure of some years will not have been spent in vain.

A paper was also read by Mr. S. McIntyre, on "The Scales of certain Insects of the Order Thysanura."

A discussion ensued, in which Mr. J. Beck, Mr. Browning, Mr. Hogg, Mr. Slack, and the President, took part. They all stated their opinion that Dr. Pigott was in error with regard to his observations.

December 8th.—The President in the chair.—A paper was read by Professor Rymer Jones on " Deep Sea Soundings in the neighbourhood of Sandalwood Island." The paper gave an account of the microscopic examination of the contents of a phial given to the author by Lieutenant Ross, R.N., grandson of Mr. J. Ross. The matter was obtained from the bed of the sea at a depth of 1080 fathoms. It contained the *spicula* of twelve genera of Sponges, and of a large number of species of *Foraminifera*, *Polycystinæ* and *Diatomaceæ*, many of which the author believed to be new.

The following papers were taken as read—on the " Stylet region of the Ommatoplæan Proboscis," by Dr. McIntosh, and on " Organisms in Mineral Infusions," by C. Staniland Wake, Esq.

Brighton and Sussex Natural History Society.

December 9th.—The President, Mr. T. H. Hennah, in the chair. —A paper was read by Mr. C. P. Smith on the Gemmæ of Mosses. In flowering plants the seed is an embryo plant provided with stem, root, and leaves, only requiring developing to produce a perfect plant. In mosses the spore is but a simple cell, without any germ or embryo, which gives rise to an intermediate state, so that mosses are plants of two or rather alternating generations. In addition to this mode of generation there is another by means of gemmæ or sprouts, which have been defined as loose granular bodies, capable of becoming plants. In all known British mosses none of the Pleurocarpie, or side-fruiting, have been found producing gemmæ, whose situation varies in different mosses; thus in Tortula papillus, which grows on trees in Sussex and elsewhere, the gemmæ are found in the upper part of the inside of the leaf; the fruit of this moss is unknown, except in Australia; *Didymodon gemmascens*, having the nerve excurrent, has the tip covered with gemmæ. *Tetraphis pellucida* produces them in pedicellate clusters at the ends of separate stems. In *Webera annotina* they assume the form of birds in the axils of barren branches. *Bryum atropurpureum* has tubercles or bulbs in the axils of leaves. On the leaves of *Orthotrichum Lyelli* grow little strings of cells, which, presenting a confervoid appearance, were named Conferva castanea. It has since been demonstrated that these confervæ are but an intermediate stage of mosses. *Oncophorus glaucus* has a great number of cells forming a dense mass at the tip of the leaf; these, in the damp season, give rise to numbers of young plants; hence this plant is common in countries where it is not known to fruit. The subject of the growth from gemmæ had not been thoroughly investigated; he purposed studying the phenomena, when he hoped to lay before the Society. some new facts. The paper was illustrated by drawings and microscopic objects prepared by Mr. Smith, which were exhibited afterwards by the following gentlemen, the most striking being by Mr. Hennah:—*Innium cuspidatum*, hermaphrodite flowers, showing archegonia, antheridia, and paraphytes; *Innium Hornum* and *Polytrichum commune*, showing ♂ flowers; *Neckera oligocarpa* ♀ flowers, consisting of Archegonid and Paraphytes.

Mr. Sewell.—*Pottia carifolia* section of leaf, exhibiting layers; *Orthotrichum Lyelli*, with confervoid gemmæ with leaves, this is the *Conferva castanea* of the early botanists.

Mr. Wonfor.—*Aulacomnion androdyum*, showing gemmæ or pseudospodia; *Ullota phyllantha*, with gemmæ on the tips of the leaves, and forming aggregate cells; and *Tetraphis pellucida*, in which the gemmæ were enclosed in a lenticular bud.

JOURNAL OF MICROSOPICAL SCIENCE.

DESCRIPTION OF PLATE I,

Illustrating Professor Perceval Wright's Notes on Sponges.

PLATE I.

Aphrocallistes Bocagei, sp. nov.

From the specimen in the British Museum. The lower portion of the figure represents the basal portion of a second and older specimen. The network lid is slightly imperfect. Nat. size.

MEMOIRS.

On some Freshwater Rhizopoda, New *or* Little-Known. By William Archer.

(*Continued from vol.* ix, *N.S., p.* 34.)

In Diplophrys the body seems bounded by a definite integument very like the appearance in that respect of the form met with in our waters, which I would, as yet, identify as *Plagiophrys spherica* (Clap. et Lachm.); and the places for the passage forth of the tufts of pseudopodia are indicated, in examples met with in which they are not projected, by a little rounded depression. The internal amber-coloured body sometimes appears as if fractured into a number of portions, but these still clustered in pretty nearly the same spot. The region beyond the amber-coloured globule is pellucid and of somewhat bluish tint, with a few colourless granules.

Diplophrys (Barker) might then, to a certain extent, be said to bear a parallel relationship to Plagiophrys (Clap. et Lachm.) (*Pl. spherica*), somewhat similar to that of Amphitrema (mihi) to Pleurophrys (Clap. et Lachm.).

But Greef is disposed to suggest that the yellow bodies of his *Acanthocystis spinifera*, escaping forth therefrom, each surrounded by a hyaline halo or rather hyaline vesicle, have the power to put forth, and indeed actually do put forth, at opposite poles, those radiating tufts of delicate pseudopodia depicted in his figures 26, 27, and 28—in a word assume the characteristics of *Diplophrys Archeri* (Barker). Greef seems further inclined to regard the form figured in his Fig. 25 as but a modification of the same, and his Fig. 29 as but great consociated groups of the former. But as before mentioned, I cannot but regard his Fig. 29 as representing one and the same thing as my *Cystophrys oculea*.

Now, one or two considerations appear to me to be opposed to Greef's view. The yellow oil-like granules, with the surrounding definitely bounded body, as if by an outer wall, of Figs. 26-28, are in dimensions considerably larger than the yellow granules with their surrounding cell-like structures of Fig. 29—or, as I would in other words be disposed to express it, the individuals of Diplophrys are notably larger in size than the contained cell-like structures of *Cystophrys oculea*—therefore, supposing the latter to be only a group of the former bound by some common matrix into a colony, the individuals of Diplophrys must in some way have suffered a great dwindling down in size in the process. In fact the average diameter of the body in Diplophrys may be taken at $\frac{1}{2000}$ of an inch, whilst the average diameter of the contained cell-like structures in *Cystophrys oculea* may be taken at $\frac{1}{5000}$ of an inch.

Again, the view that Diplophrys springs from a development of the yellow bodies of *Acanthocystis spinifera* seems to be controverted by the fact, *quantum valeat*, that the latter species has not been ever found in this country. Further, as I have mentioned in a previous part of this paper, the cell-like structures in *C. oculea* are invested by a common matrix of sarcode, from the rather definitely bounded surface of which emanate the pseudopodia; and this enclosing sarcode body has the power to tear itself in two, new pseudopodia emanating from the just disjoined surfaces—that is to say, they are in the sarcode contained, not themselves the active sarcode-containing bodies. Of course I am going on the assumption that Greef's figure 29 actually does represent my *C. oculea*, which I can hardly doubt; nor can the yellow colour of the central bodies in this figure, as opposed to the red colour in mine, much militate against this view, for the tint expressed to the eye seems to me to depend a good deal on the focussing and on the illumination. But it may be premature to judge of Greef's opinions on the forms as figured by him as he promises to return to them in the next forth-coming section of his memoir. For my own humble share, in the meantime, I cannot but think it impossible that Diplophrys can be a phase of the yellow bodies of *Acanthocystis spinifera* at all, whilst I should have as little doubt that it can be either any preceding or subsequent state of Fig. 29, here supposed by me to be one and the same thing as *C. oculea.* I am myself a good deal puzzled by the occurrence of the curious little organism drawn attention to by me at one of our Club meetings,[1] which

[1] Proceedings of Dublin Microscopical Club, March, 1869, in 'Quarterly Journal of Miscroscopical Science,' vol. ix, N. S., pp. 323-4.

presents such a perplexing *resemblance* to a Diplophrys, pseudopodia drawn in, and the body surrounded by an aggregation of small diatom frustules, fragments, and miscellaneous *splinters*. I need not, however, further advert to this little organism here, as in the place alluded to I have recorded the little I have ever been able to make out about it. Had I not overlooked it, I would have given a figure of this queer little thing on my previous plates—perhaps I may take a future opportunity of doing so.

I hope I may not be thought to have made a too wide and unnecessary digression from the affinities and differences of my *Acanthocystis Pertyana*, but in contrasting it with Greef's *A. spinifera*, the remainder became unavoidable. I need hardly say I await with interest Greef's further communication hereupon.

Genus, *Raphidiophrys* (Arch.).

Generic characters.—Rhizopod composed of two distinct sarcode regions—the inner forming one or several rounded individualised definitely bounded hyaline sarcode masses, each containing a subperipheral stratum of colouring granules—the outer more or less coloured, soft, and mobile, bearing numerous elongate irregularly scattered siliceous spicula, acute at both ends, and forming a common investment to the inner globular masses, which latter give off long slender non-coalescing pseudopodia.

The main distinction of this genus from Acanthocystis is the spicula being solid, no difference in the extremities and scattered every way. The absence of a central capsule excludes it from marine Radiolaria to which it otherwise possesses affinity.

Raphidiophrys viridis (Arch.).
Pl. XVI, fig. 2.

Specific characters.—Inner rounded masses hyaline, globular, with numerous large chlorophyll granules, arranged in a hollow globular manner within the periphery; outer region slightly buff-coloured, containing densely numerous elongate very slender more or less curved, acicular spicula, acutely pointed at each end and lying in every possible direction; pseudopodia numerous, extremely slender, very long, hyaline, comparatively rigid, never coalescing. No evident nucleus nor pulsating vacuoles.

Measurements.—Diameter of inner globes ranging about $\frac{1}{300}$th of an inch, the size of the compound clusters varying according to the number of contained globes, sometimes so large as to be seen by the unassisted eye, poised in the water like specimens of Actinosphærium, but greenish, not white.

Localities.—Pools at Ballylusk, and one or two other situations near Carrig mountain, near Tinnehely, county Wicklow, and near Multyfarnham, county Westmeath, very sparingly at Glengariff, county Cork; rare and local, and sometimes seemingly confined to some very restricted area of the few pools which have produced it, but in those same spots found, by careful search, at various seasons.

Affinities and Differences.—If this fine form possessed a central capsule there would be, so far as I see, no necessity to form a new genus for it, for in that case it would be simply a new fresh-water species of Spherozoum (Meyen) Haeckel. There is not seemingly any other fresh-water rhizopod for which it could be mistaken. The presence of the spicula would alone quite decidedly separate it from *Heterophrys myriopoda.* Greef, indeed, in his paper already cited[1] accuses me, by reason of hasty observation and of faulty comparison with Carter's description, of having misapprehended the true characters of *Acanthocystis turfacea*, and suggests that I must, therefore, have only applied a new name to that already-known form, and he cites my brief reference to it at our Microscopical Club meeting.[2] But I may be here forgiven for venturing to observe that if Greef had more closely looked over the record of that meeting, he would have seen that as well as Raphidiophrys, I likewise exhibited at the same meeting, in contradistinction, examples of *Acanthocystis turfacea,* then for the first time identified and exhibited in Ireland. Further, even in the cursory record there made of Raphidiophrys, it was described as possessing "immersed and entangled in the outer region, beyond all computation densely numerous, very slender, elongate spicules, *acute at both ends lying in every possible direction*"—thus showing characters which could in no way, even most superficially examined, be mistaken for the radiant vertical spicules of *Acanthocystis turfacea*, discoid at one end and furcate at the other. But, after all, in the eyes of some, our *Raphidiophrys viridis* may be, perhaps, only specifically, not generically, distinct from the recognised members of the genus Acanthocystis—both mu-

[1] L. c., p. 482.
[2] 'Quart. Journal of Micr. Science,' vol. vii, 1867.

tually agree, as well as with others, in my opinion, in having two distinct sarcode regions, an inner more dense and an outer more fluid; but all the acknowledged species of the latter—*A. turfacea* (Carter), *A. viridis* (Grenacher, if distinct from the former), Greef's form (fig. 18) loc. cit., *A. spinifera* (Greef), *A. Pertyana* (mihi)—have radial spicula with discoid bases mutually approximated around the periphery of a single globular sarcode body, whilst in *Raphidiophrys viridis* (mihi), the spicules are crowded without order, and are not discoid at one extremity.

Genus, *Pompholyxophrys* (Arch.).

Synonym, *Hyalolampe*, Greef.[1]

Generic characters.—Rhizopod composed of two distinct sarcode regions—the inner a dense, coloured, globular sarcode mass—the outer colourless, bearing a number of separate hyaline globular structures, these disposed in a more or less thick layer around the inner globe, which latter gives off more or less elongate slender non-coalescing pseudopodia.

Pompholyxophrys punicea (Arch.).

Pl. XVI, fig. 4.

Syn. *Hyalolampe fenestrata*, Greef.[1]

Specific characters.—Inner mass loaded with numerous scattered, reddish or garnet-coloured pigment granules, accompanied by others, colourless or sometimes green; outer region of great tenuity, colourless, the spherical vesicle-like bodies, colourless, hyaline, numerous, somewhat varying in size, arranged therein so as to form a more or less thick stratum two or three deep all over the periphery of the inner globe, leaving a hyaline minute interval all round the latter, which gives off not very numerous nor very long but very delicate pellucid linear pseudopodia, some of which sometimes appear thicker above than below, that is as if suddenly attenuated near the base; no evident nucleus nor pulsating vacuoles.

Measurement.—Diameter variable; that of inner body averaging about $\frac{1}{700}$th; total, including outer region, averaging about $\frac{1}{550}$th of an inch.

[1] L. c., p. 501, pl. xxvii, fig. 37.

Localities.—In various heath pools in Co. Wicklow, Cork, Kerry, Westmeath; scanty, but not unfrequently encountered.

Affinities and Differences.—So far as I can see this little rhizopod appears to be unique, for the outer stratum bearing the problematical hyaline vesicles or globules has no parallel. They are, as I have mentioned, readily cast off, and appear to me sometimes as if they became collapsed in specimens some time kept. In my previous account of it (ante) I doubtfully referred to its possession of an outer sarcode border or region containing the hyaline globules; but, by a renewed examination, I have now little doubt but that such actually exists, though of considerable tenuity and very pale in colour. I had likewise supposed that, amongst freshwater rhizopoda, it was without a parallel in the red pigment granules, but that statement must be modified inasmuch as some of the forms belonging to Greef's new genus Astrodisculus also show red granules in their interior, but our form by no means belongs to that genus. I doubt not, however, that the present is actually identical with another of Greef's forms as I have indicated above—I mean that which he has named *Hylolampe fenestrata.* But as my description and nomenclature preceded his, mine naturally takes the priority, and his name falls to the rank of a synonym. But whilst, indeed, I have no doubt of the correctness of this assumption, as will be seen I am necessarily at variance with Greef in my interpretation of the structure of the form in question. I myself have not been able to see any "nuclear" structure. Greef's account of his Hyalolampe (equivalent to my Pompholyxophrys) runs thus:—"This form is surrounded by a beautiful siliceous shell which appears to be composed as if of individual glass-globules laid upon one another. At first glance I imagined that I had before me an alveolar vacuolar (schaumiges) sarcode-net, but I soon satisfied myself by an examination of the contours that this was of the former structure." He then goes on to say that he satisfied himself of the siliceous nature of what he regards as a true *perforate case* (Gitterhaus; Gittergehäuse) by the application of acetic acid, caustic potash, and even sulphuric acid, and he would refer this form, along with Clathrulina, to the Ethmosphærida. Now, if I am at all right in believing our forms to be one and the same, I very deferentially think this account of the structure is inaccurate; as I have already described I do not think this form is surrounded by a "Gittergehäuse"; the globules are free and separate, included by a delicate sarcode layer, and sometimes (I think) appear even collapsed; they are not comparable to the solid, fenestrate, true "Gittergehäuse" of Clathrulina, and, in fact, this form in the possession of these structures stands alone, so far as I can see. The creature is not at all uncommon in our moor pools, and I only wonder it has hitherto been overlooked; it is, however, minute, and never to my eyes has shown any stage of development, and but seldom shows any incepted food, though sometimes, indeed, distorted by an unusually large morsel and the rejectamenta discharged by a sudden effort, which could hardly be the case if enclosed in a solid or connected "Gittergehäuse." But as Greef has more to convey on the forms brought forward by him perhaps he may hereafter clear up my difficulties and remove the apparent discrepancies.

Genus, *Heterophrys* (Arch.).

Generic characters.—Rhizopod composed of two distinct sarcode regions—the inner one or more dense, globular sarcode masses often bearing colouring granules—the outer forming a complete investment thereto, more or less coloured, not enclosing any spicula or differentiated structures, but giving off at the circumference marginal processes, and allowing the passage forth from the inner sarcode mass of numerous linear, elongate, granulferous, non-coalescing pseudopodia.

The Rhizopoda falling under this genus appear individually comparable to a Raphidiophrys or an Acanthocystis, the outer region destitute of spicula—to a Pompholyxophrys, the outer region destitute of the peculiar globular structures characteristic of that new genus—to an Astrodisculus, the outer region not condensed, but of a mobile sarcode, and the inner without the "central capsule;" so far differing thus from all these in a negative manner, and, further, differing from them in a positive manner, by the margin of the outer region giving off the characteristic processes. In the fringe-like border formed by the numerous fine linear processes of *Heterophrys myriopoda*, however, there is much resemblance to the same portion of Greef's form of Acanthocystis, unnamed, which, as I have mentioned, appears to me well distinguished as a specific form, figured on his Pl. XXVII, fig. 18. This genus differs from *Actinophrys sol* in the amount of differentiation in the body-structure expressed in the two sharply-marked strata of sarcode of which it is composed.

Heterophrys Fockii (Arch.).

Pl. XVI, fig. 3.

Specific characters.— *Outer region of a palish buff colour or nearly colourless, mobile, not homogeneous, but showing various lines, dots, granules, and inequalities, frequently changing in aspect and its margin fading off indefinitely, and giving off indefinite, variously figured marginal processes—inner region one or several orbicular sarcode masses of a light bluish-coloured tint, enclosing various opaque granules, colourless and of a brownish colour, and sometimes chlorophyll-granules, its margin sometimes exhibiting one or more pulsating vacuoles, and giving off numerous linear, colourless, granuliferous, non-coalescing pseudopodia; the compound groups sometimes cohere for a length of time, finally conjoined only by the persistent mutual fusion of the pseudopodia extending from one to another.*

Measurements.—Somewhat variable in size, diameter of inner globe averaging about $\frac{1}{1000}''$.

Localities.—Found in various situations in Co. Wicklow, Coik, Kerry, Westmeath, in moor pools, but scanty.

Affinities and Differences.—Distinguished at first glance from the form I associate with it under the name of *Heterophrys myriopoda* by its much smaller size, less green colour of the inner body, more highly coloured marginal region, which gives off irregular and fitful sub-triangular, indefinitely bounded projections, not subdivided into a very great number of hair-like linear processess. The marginal pulsaing vacuoles of the inner body, sometimes seen, render it like *Actinophrys sol*, but the conspicuous outer region distinguishes this form readily. As previously alluded to, I conceive mine to be most likely the same as Focke's "No. 1;"[1] but the fact that Greef does not allude thereto in his paper, though he figures a form which appears to me to be so very like it,[2] causes me now a little to waver. The form, however, which Greef figures he suggestively thinks may be either a young state of *Acanthocystis turfacea* (*viridis*, ejus), or perhaps a distinct species.[2] But I would very deferentially think, if he comes to the latter conclusion, it would be incorrect to place a rhizopod quite destitute of "skeleton" or spicula in the same genus with other forms characterised by the possession of these in a very marked degree. If indeed that figured by Greef be truly identical with my *H. Fockii*, I certainly would

[1] Loc. cit., t. xxv, [2] Loc. cit., t. xxvii, fig. 35.

not be at all disposed to regard it as a young state of *A. turfacea;* the very smallest examples I have ever seen of the latter have shown the radiating spine-like spicules, and I have seen examples smaller than the average of those of *H. Fockii.* In this form I have not been able to see anything to represent a "nucleus" or a "central capsule." I have in a previous part of this paper expressed my own dissent from the view that the outer boundary of the inner globe in such a form as this can be rightly looked upon as the representative of the "central capsule." It has sometimes suggested itself to me that this form might be identical with that figured by Carter,[1] which he provisionally considered might be a young condition of *Actinosphærium Eichhornii,* but whether I may be right or wrong in that conjecture, I conceive the structure of *H. Fockii* to be quite unlike even a young state of *Actinosphærium Eichhornii,* nor have I been able to see any nucleus as shown by Carter.

I perceive that Leuckart,[2] in giving the favour of a notice to my communication to the Dublin Microscopical Club, in which I first chronicled this rhizopod, suggestively puts it that it may appertain to Greef's genus Amphizonella, established by him in a previous paper,[3] but this is most clearly not so. My form is not at all referable to Amphizonella; in that genus the *Amœba-* or *Difflugia-like, nucleus-bearing* sarcode body is surrounded by a resistent so-called "capsule," still yielding enough to permit the exit of a few *finger-like* pseudopodia. Thus, that genus would differ from Heterophrys sufficiently widely to fall under a completely distinct group of Rhizopoda, no matter which of the hitherto proposed schemes of classification one might lean to adopt. There is doubtless somewhat more resemblance in Greef's new genus Astrodisculus to his previously established Amphizonella, to which, indeed, he himself refers; but still even they appear abundantly distinct, and I quite concur, so far as I may venture to express an opinion, that the forms respectively referable to Amphizonella and Astrodisculus demand being placed wide apart. I shall endeavour below briefly to recapitulate the characteristics as given by Greef in his recent paper of this latter new genus.

[1] 'Annals of Natural History,' vol. xiii (1864), pl. ii, fig. 23.

[2] 'Bericht über die wissenschaftliche Leistungen in der Naturgeschicte der niederen Thiere,' 1866, 1867, p. 270.

[3] Greef, "Ueber einige in der Erde lebende Amœben und andere Rhizopoden," in Schultze's 'Archiv fur mikroskopische Anatomie,' 1866, p. 323, *et seqq.*, t. xviii.

Heterophrys myriopoda (Arch.).

Pl. XVII, fig. 4.

Specific characters.— Outer region colourless, granular, comparatively rigid and unchangeable, passing off at the margin into a fringe-like border of innumerable linear hyaline processes—inner region forming a large, globular, hyaline, definitely bounded sarcode mass, enclosing numerous large chlorophyll-granules, with a few colourless ones, and giving off not very numerous, comparatively thick, slightly tapering, long, granuliferous, non-coalescing pseudopodia.

I believe I have seen a central stellate arrangement of lines similar to that described in *Acanthocystis turfacea* by Grenacher and Greef, but its examination deserves to be repeated before entering upon a description of this part of the structure.

Measurements.—Diameter of the total rhizopod (leaving the pseudopodia out of view, that is, from extremities of marginal processes) about $\frac{1}{200}''$—of the inner sharply defined globe $\frac{1}{380}''$ to $\frac{1}{300}''$.

Localities.—Found as yet in but one or two spots in Co. Wicklow, near Carrig Mountain, and in a boggy spot at lower end of Lough Dan (in the grounds of "Lake Park"), both very restrictedly; hence this form must as yet be considered very rare.

Affinities and Differences.—Distinguished at once from the preceding by its larger size, constantly possessing a copious quantity of chlorophyll-granules, its longer, stouter pseudopodia, and the innumerable linear acute processes bordering the outer sarcode region. The colourless outer region without spicula would at first glance distinguish this form from an example of *Raphidiophrys viridis,* which might possess but a single central globular sarcode body. The absence of the vertical spines at once readily distinguishes it from *Acanthocystis turfacea.* The curious resemblance of the remarkable marginal, slender, acute processes given off from the outer region to those represented by Greef in a new (?) Acanthocystis form [1] has been already drawn attention to.

Genus, *Cystophrys* (Arch.).

Pl. XVII, figs. 1, 2, 3.

Generic characters.—Rhizopod changeable in figure, sarcode mass of but one character, and containing immersed

[1] Loc. cit., t. xxvii, fig. 18.

therein more or less numerous cell-like structures, and giving forth slender marginal pseudopodia.

The generic type sought to be established here would be unnecessary if the two forms it is destined to contain possessed "central capsules," for in that case they would seemingly fitly enough fall under the genus Thallasolampe (Haeckel), which is destitute of "skeleton," admitting, indeed, that the contained cells might be assumed as homologous with the "yellow cells" of that genus. But, as before alluded to, the absence of the "central capsule" would altogether exclude our forms from the true "Radiolaria." There appears a certain resemblance to Strethill Wright's genus Boderia [1] (*B. Turneri,* Strethill Wright), and this supposition favoured rather by his figure than by his description. That form is described as consisting of "a simple mass of brown or orange sarcode, enclosed in a very delicate and colourless membranous envelope, from openings in which protrude long pseudopodial branches, generally three or four in number, but sometimes more numerous, especially in larger specimens." It is, I think, quite certain that in my forms there is no "membranous envelope" (however delicate), though, indeed, this character is not seemingly expressed in his figures, whilst his fig. 2 is supposed to show two examples "conjugated," and this indeed not prevented by the assumed outer "membrane." Further, the rounded bodies, superficially bearing some resemblance to those of my forms, are represented as "nuclei," or rather "ova," not as themselves nucleus-containing, outwardly bounded homologues (?) of "yellow cells." Judging from Strethill Wright's *description,* it is, I think, only possible, but not very probable, that either of my forms can be congeneric with his. Perhaps, should these lines ever meet his eye, he may, however, be able to throw a further light on the question.

If, as is probable, such truly cellular structures as those characteristic, for instance, of *Cystophrys Haeckeliana,* are to be considered as the representatives of the "yellow cells" of the typical "Radiolaria," then I think that Wallich's view that the homologues of those structures pervade all the Rhizopoda cannot be maintained.[2] I am unable to follow him in recognising the representatives of "yellow cells" in freshwater rhizopoda in general, for the granular and quasi-cellular

[1] "Observations on British Zoophytes and Protozoa," by T. Strethill Wright, M.D., in 'Journal of Anatomy and Physiology,' vol. i, page 335, pl. xv.

[2] Loc. cit, p. 70.

structures noticeable seem to lack the "cell"-characteristics present here and described for the "yellow cells" by Haeckel in his beautiful and elaborate Monograph. These must possess a rigid and firm membrane or "wall," granular (yellow) contents, and a clearly defined nucleus, and they must subdivide by internal self-fission, a new nucleus appearing in each half of the divided cell; in these points, then, the cells of *C. Haeckeliana* agree, with the exception of the wall only (not the contents) being of a yellowish colour. I do not yet see, indeed, that the ordinary structures met with pervading the general body-mass of certain rhizopoda can be truly said to come under the same category as the cells of my forms, and hence I think Wallich's views alluded to are not yet justified.

Further, I venture to think that Greef's suggested comparison of the green (chlorophyll) granules of *A. turfacea* with the "yellow cells" is not tenable. I take these green granules occurring in that form, as well as in *Raphidiophrys viridis, Heterophrys myriopoda, Pleurophrys amphitremoides,* occasionally in certain Difflugiæ, &c., as all one and the same thing—chlorophyll—either persistently or temporarily characteristic here, as in certain other lowly animal forms, and can no more be the homologues of "yellow cells" in the one than the other.

Cystophrys Haeckeliana (Arch.).
Pl. XVII, figs. 1, 2.

Specific characters.—*Sarcode body-mass very polymorphous, colourless, granular; inner cells rather large, orbicular, their walls yellowish, thin; contents bluish, granular; nucleus white, excentric; nucleolus dark, very minute; pseudopodia slender, irregular, variable in length, granular, more or less arborescent, and the branches occasionally inosculating.*

Measurements.—Size of examples variable. Diameter of inner cells about $\frac{1}{1500}''$ to as small as $\frac{1}{2000}''$.

Localities.—One or two pools in Callery and Carrig neighbourhoods, very sparingly rare.

Affinities and Differences.—Distinguished at once from the following by its large orbicular central cells, with pale (not reddish) nucleus, and its granular arborescent (not linear and unbranched) pseudopodia.

Cystophrys oculea (Arch.).
Pl. XVII, fig. 3.

Specific characters.—*Sarcode body-mass maintaining more*

or less of a rounded figure, though locomotive power rather active, homogeneous, of a slightly bluish tint; inner cells orbicular or sub-elliptic, minute, walls not definitely perceptible; contents hyaline; nucleus (?) varying from yellow to red, sharply bounded; nucleolus (?) dark, very minute; pseudopodia very slender, varying in length, but averaging from about that of the diameter of the body to about one half, straight, linear, hyaline, unbranched, radiating in many directions, and then sometimes crossing each other, but not inosculating.

Measurements.—Size of examples variable. Diameter of inner bodies about $\frac{1}{5000}''$.

Localities.—Pools in Callery and Carrig neighbourhoods (Co. Wicklow), very sparingly; rare.

Affinities and Differences.—Distinguished from preceding at once by the salient characters alluded to under this heading following the description of *C. Haeckeliana.* As mentioned under the similar paragraph following the description of *Acanthocystis Pertyana,* it will be seen that I hold like doubt but that form is identical with Greef's fig. 29 (l. c.), to which I have already adverted. This form is not named by Greef; hence it has no synonym.

Having made several allusions in the foregoing to Greef's new genus Astrodisculus, and as he likewise in his recent valuable paper makes some observations upon *Clathrulina* (Cienkowski), a short *résumé* of those portions of his communication may be of some advantage.

In bringing forward the genus Astrodisculus, however, he does not give, as yet at least, diagnostic characters; but as I glean them from his general account of the forms he refers thereto, I believe they are something like the following:

Genus, *Astrodisculus* (Greef).[1]

Body of two distinct well-marked regions, the outer a hyaline, "porous," sharply bounded investment, without any external process (which withstands the action of sulphuric acid), and gives passage to a number of fine linear pseudopodia emanating from the contained inner sarcode mass, which contains a globular, smoothly bounded "central capsule" (or sometimes several, *A. minutus, A. radians*), with variously coloured contents.

Greef describes the outer coat or marginal region of the form appertaining here as porous and siliceous. I myself have

[1] Loc. cit., p. 496, *et seqq.*

met at least one form which I am greatly disposed to suspect would belong here, but I know it too slightly as yet to venture to record it. But it strikes me that in the form to which I allude, the outer marginal region is distinctly flexible, giving way to certain circumscribed changes of figure of the rhizopod. If so, I can hardly suppose this outer region can be a siliceous skeleton. Nor does it appear to me porous, though showing an evenly and regular dotted appearance through its substance, and giving a passage to exceedingly fine pseudopodia. In this rhizopod I certainly did but consider that I had before me a form coming close to Heterophrys, but differing in the sharply bounded outer surface of the comparatively rigid marginal region not being mobile nor divided into processes.

If, however, Greef's interpretation be truly correct, then the nearest relationship of the genus would seemingly be to Clathrulina, which would differ by the very much larger "pores" to the globular "skeleton," and this being supported on a stipes, and in the absence of the "central capsule" presented by Astrodisculus.

But opposed to Greef's opinion as to the nature of this outer region is the fact he mentions, that not only do the pseudopodia project through it into the water (which they might certainly do through extremely fine pores), but also that the large characteristic red granules (in *A. ruber*) can pass in and out along the pseudopodia with ease. Now, this being so, it is not readily to be seen how they could do so unless considerably large "pores" existed, which, if present, must readily be seen, but no one of his forms reveals such. May I then, as yet, venture to suppose that the outer region (as I certainly thought in the Irish form I have in my mind's eye) is but a more dense and more hyaline sharply defined differentiated region of the sarcode body—more dense, I say, than the inner region, but not too dense to allow of the passage of not only the pseudopodia, but of the large granules at any given spot?

Of the newly established genus Astrodisculus, Greef records the following species; no strict diagnosis of their characters is given, but I venture to abbreviate his general account of them as follows:

A. minutus (Greef).[1]

Sarcode of inner body of a greyish-brown colour, en-

[1] Loc. cit., t. xxvii, fig. 30.

closing several minute round bodies (central capsules) of same colour, but of a deeper tint.

A. ruber (Greef).[1]

Sarcode of inner body coloured red by a number of granular red pigment corpuscles, and enclosing a sharply bounded central capsule, filled with a bright red, finely granular substance.

A. flavescens (Greef).[2]

Sarcode of inner body yellow, containing several brownish-red pigment granules, and enclosing a central capsule filled with a yellow, finely granular substance.

A. flavocapsulatus (Greef).[3]

Sarcode of inner body colourless, showing a number of pale, circular, dot-like granules, and enclosing a central capsule filled with yellow, finely granular substance.

As regards this species Greef suggests that the round dots seemingly in the extra-capsular region may actually be the superficially posed discoid bases of exceedingly fine and delicate spicula, comparable to those of Acanthocystis, but in the present instance he leaves this quite as matter of doubt. I would venture to think that the placing side by side in one genus of a spicule-bearing form with others destitute of spicules would be incorrect. But he distinctly attributes spicules to the last form he records under this genus, namely:

A. radians (Greef).[4]

Sarcode inner body colourless, and enclosing several (two to three) "central capsules," filled also with finely granular colourless substance; outer region of a slightly brownish hue, and containing several acicular, very slender and delicate radial spicula, reaching from the periphery of the inner body to the outer surface.

The presence of these spicula would, I venture to think, place this form in a distinct genus, and, in fact, very close to Acanthocystis. Indeed, I would almost query if the form

[1] Loc. cit., t. xxvii, fig. 31.
[2] Loc. cit., t. xxvii, figs. 32 and 32*a*.
[3] Loc. cit., t. xxvii, figs. 33 and 33*a*.
[4] Loc. cit., t. xxvii, figs. 36 and 36*a*.

here recorded might not be truly a state of Greef's *A. spinifera,* with the outer (sarcode) region more conspicuous than usual, and more than one "central capsule" present, and no yellow bodies developed. The tint of colour shown in Greef's drawing as belonging to the outer marginal region is seemingly the same, or nearly so, of that prevalent in the outer sarcode stratum of my own *Heterophrys Fockii,* or in *Raphidiophrys viridis,* where it is undoubtedly mobile and changeable, and no siliceous "skeleton" or "shell." If Greef's views were correct, then these spicula, contrary to analogy, would not be surrounded by sarcode by which deposited, but one siliceous structure penetrating and embedded in another.

It will not be thought out of place to endeavour here to present as brief an epitome as I can of Cienkowski's previous and Greef's later observations on the but recently described rhizopod, *Clathrulina elegans* (Cienk.), as, doubtless, any fresh views or new points as regards an organism seemingly so comparatively rare, and at the same time offering several interesting considerations with respect to its position and affinities, must be accounted of interest.

This pretty freshwater form was made the type of a new genus by Cienkowski,[1] and as yet has been recorded so far as I know, but by that author himself, from near St. Petersburg, and from two localities in Germany; by Haeckel, from near Jena; by Greef, from near Bonn; and by myself (since by others) from two or three localities in Ireland, and one in Wales. But I am greatly disposed to think its distribution is pretty wide, though always scanty and restricted to isolated spots. I had myself ventured to draw attention to it and to describe it at a meeting of the Natural History Society of Dublin, and that before Cienkowski's paper was published; but as he brought it forward far more elaborately than I could have hoped to have done, and as my paper could not have been put into type before the Part of the 'Archiv' containing his paper appeared, I gladly waived my nomenclature, and withdrew my paper from publication. Greef is, indeed, quite right that I had the priority of publication, and that the form brought forward by me before our Microscopical Club was indeed truly one and the same thing as Cienkowski's. On the first occasion, indeed, I overlooked the stipes, or rather did not, indeed, fail to see, but misunderstood

[1] Cienkowski, "Ueber die Clathrulina, eine neue Actinophryen-Gattung," in 'Archiv für Mikroskopische Anatomie,' Bd. iii, 1867, p. 311, t. xviii.

it, conceiving it to be some foreign filament; but a further examination of the specimens at command soon revealed the novel fact that we had indeed to do with a truly *stipitate* (*all but* Radiolarian) Rhizopod, and this error or oversight I was able to correct at the subsequent meeting.

The genus in question may be, I think, characterised as follows:

Genus, *Clathrulina* (Cienk.).

An "Actinophryan" Rhizopod, without a "central capsule," and enclosed within a hollow, globular, fenestrate siliceous "shell" (or "skeleton"), the pseudopodia radiating all around through its apertures, and which is borne aloft at the summit of a slender stipes, the latter attached by a somewhat expanded base to foreign objects, or one to another.

Clathrulina elegans (Cienk.).

Specific characters.—Body colourless, granular, vacuolar, very mobile, in a young state showing a pale central "nuclear" structure; the perforate "shell" when young pale and colourless, when older more or less brownish; the apertures roundish or subpolygonal, bounded by a kind of raised rim, thus producing a groove or furrow, varying in width, between them; the stipes in length two to six times the diameter of the "shell," colourless; the pseudopodia numerous, fine, often long, colourless, granuliferous, slightly branching.

Reproduction of two kinds—(1) by self-fission into two, and eventual passage forth through the apertures of the shell of the individualised sarcode bodies, which presently assume the inherent Actinophryan characteristics, reproducing the Clathrulina by development of stipes and shell; (2) by formation of motile (ciliated?) embryos, originating from a separately encysted condition within the "shell" of the sarcode body, mostly previously subdivided into several portions, each enclosed by a firm coat. These, after a period of rest (often long), permit the escape each of a motile monad-like embryo, showing a nucleus and nucleolus, which after a brief period passes out through an aperture of the shell and settles near at hand (not unfrequently upon the just quitted primary Clathrulina), and (like the individualised portion of a Clathrulina subdivided without passing into an encysted and embryo state) at once puts forth pseudopodia, develops a stipes and shell, and thus produces a new Clathrulina.

Cienkowski refers to what he calls a variety of *C. elegans* (designated as *minor*), which he considers marked by its paler colour and more hyaline appearance, its more delicate structure, the apertures indistinct, and perhaps most notably by an evident pulsation of certain of its vacuoles. But such examples seem not unusually to present themselves where the ordinary highly coloured and most marked examples occur, and I would venture to suppose they are but younger specimens, or examples more than ordinarily retarded.

Had I known the further figures of this species were forthcoming from Greef's hand, I would not have ventured to put forward the too stiff figure I have given on Pl. XVII, fig. 5.

Such will, I believe, convey a true conception of what this pretty rhizopod is. Cienkowski designated it as hardly distinguishable from *Actinophrys sol* within the stipitate shell; but Greef justly points out that it does not *exactly* bear out that comparison. An Actinophrys presents a truly globular figure, its circular outline not interrupted by the passage off of the evenly set and regularly distributed pseudopodia, and it is marked by the striking marginal pulsating vacuoles. Opposed to this the body of Clathrulina is more mobile, though of a general rather rounded figure; the pseudopodia varying in thickness, and some of them, expanding at the base, lend a more lobed appearance to the outline. Neither are the pulsating vacuoles by any means so frequent or so striking, if, indeed, they can be strictly attributed to this form. These in themselves may appear to many to be very small and trivial distinctions, but such, at first sight, not very striking idiosyncrasies soon attract the notice of observers bestowing a closer attention on these beings, and, as I think, are ere long seen in certain forms to put forward a not unfounded claim to be regarded as special and inherent.

Greef states that he has perceived in young and paler examples that (as in *Actinosphærium Eichhornii,* for instance) a differentiated "axile" and "cortical" region in the granuliferous pseudopodia can be made out, but he has not been able to satisfy himself that an uninterrupted connection exists between the axes and a vesicular "nucleus-like" central body, said by him to be *constantly* present, which I think, however, must be queried just as yet, though such an apparent structure no doubt sometimes shows itself, and I imagine it may be what I ventured to suppose in my, from time to time, casual remarks before our Microscopical Club might possibly be the representative of a "central capsule." It is to be remarked that none of Greef's figures (figs. 1, 6, 7) actually depict either this presumed central "nuclear" body or the axile substance in the pseudopodia. He, however, enters into an extended argument to show that this may be truly the case, too long to epitomise here, but very interesting and instructive.

A further point shown by Greef is, that the walls of the "cysts" show a decided resistance to reagents, and are seen to be superficially covered by very minute spinelets, giving a somewhat pilose appearance to the surface. I do not think the cysts seen in Irish examples have shown this characteristic. Greef argues that both perforate globe and stipes are doubtless (?) siliceous, for they likewise withstand the action of concentrated sulphuric acid.

Touching the systematic position of this interesting rhizopod, Greef, as did Cienkowski and myself, of course suggests its relationship to the Ethmosphærida, and that close to the marine Heliosphæra, justly remarking that if such a perforate "skeleton" were met with, as that possessed by this form, in the sea or fossil, no one would hesitate to place it amongst Polycystina proper. But then we know the living animal, and it has no "central capsule," unless, indeed, the questionable "nuclear" body be its representative, nor (less significant or important indeed) has it "yellow cells." Greef, indeed, suggests besides that this central body (I think not at all always present, or at least perceptible) may be, perhaps, rather the representative of the not always present, so denominated "inner vesicle" ("Binnenblase") of the typical marine Radiolaria. The stipes, too, is seemingly unique, and I had imagined the genus would have been better named in allusion to that character than to the fenestrate shell, a character pervading so very many of its marine relatives. The absence of the "central capsule" has indeed a possible parallel in one marine form, *Coscinosphæra ciliosa* (A. Stuart),[1] possible only, indeed, because even the form referred to may yet, according to Stuart himself, be seen actually to have a central capsule.

Greef finally makes some suggestions as regards the seeming inclination to formation of colonies presented by Clathrulina, if indeed the fact of the individuals sometimes mutually standing off from one another and attached to each other by the bases of the stipes, as they might be to foreign objects, deserves to be so called, and he builds a hypothesis on that circumstance. But I imagine there is in this fact no analogy to such an organism as Carchesium. I venture to think the younger examples orginating from the germs

[1] 'Zeitschrift für wissensch. Zoologie,' Bd. xxvi, p. 328, t. xviii.

evolved from the lower and first produced individual are merely to a great extent accidentally located attached to the latter, the locomotion of the germs ceasing and development commencing before they succeed in reaching a more distant or a foreign object on which to establish themselves. Something like what may perhaps be cited as a kind of growth in minute Algæ, forming a parallel to that of Clathrulina referred to, is that seen in *Sciadium arbuscula*, as a constant characteristic of the species, and in *Ophiocytium majus* as an exceptional circumstance. In the latter, the young plantlets produced by the development of the germs are sometimes stationed, at least temporarily, at the summit of the parent-cell-walls, or along its length, or one upon another, and attached by the lower pedicle-like extremity, just like the younger individuals in Clathrulina, and in one, as in the other case, it would look as if the germs establish themselves merely on the first solid support that became presented. Still, Greef would compare this occasional habit in Clathrulina almost to a kind of parasitism, and, pushing the matter further, suggests the idea that here the propinquity of the generations to one another is no accidental circumstance, but an adaptation for the purpose of securing such a proximity of individuals as would be essential for a possible or assumed sexual reproduction, and which would be otherwise denied to them, owing to this being a permanently fixed form on a rigid stipes. But if a sexual reproduction truly takes place (by "conjugation" or otherwise), the emerged Actinophryan bodies are just as free as any other rhizopods, and would seem to stand in quite an analogous position, and under similar circumstances, as regards any possible sexual mode of reproduction. On the whole, then, I venture to think the suggestions put forward by Greef in this regard, as, at least as yet, very hypothetical.

To appreciate as completely as the interest of the subject deserves all the valuable considerations and remarks put forward by Greef would indeed necessitate a careful perusal of his paper in full; and I am sure I ought to apologise if in so brief an epitome I have failed to indicate as accurately as ought to be the points put forward by him. Those who wish to pursue the subject further will have recourse to the original; whilst, perhaps, though I have nought myself to add to what has before been done so much better than I could have hoped, my allusion to it here, and incorporation of Greef's remarks, may not, I trust, be thought redundant.

I attempt to add below brief diagnoses of those forms on Pl. XX, which to my eyes assume a position to a great extent independent from the more common fresh-water rhizopodus genera represented by Difflugia, but which, like those more nearly allied to the Radiolaria, seem to me not yet sufficiently known or defined to be placed in special recognised groups or orders. Hence they must just follow without any attempt thereat; although, indeed, Claparède and Lachmann place their Pleurophrys under Actinophryna, my figures 1 to 6 seem to be as little comparable to that group as to Amœbina. I begin with—

Genus *Pleurophrys* (Clap. et Lachm.).

Pleurophrys spherica ? (Clap. et Lachm.), Arch.
Pl. XX, fig. 1.

Specific characters.—Large, orbicular, elliptic, or somewhat irregularly shaped. Body not filling the cavity of the test, containing a large granular nucleus; pseudopodia slender, slightly tapering, hyaline, pellucid, non-granular, rather straight, comparatively rigid, somewhat branched, very inert. Test brownish, composed of somewhat elongate and granular particles, agglutinated together by a common organic (?) substance.

Measurements.—So varied in figure as not to present any very definite distinction of length and breadth; the largest specimens probably as much as $\frac{1}{150}''$ in diameter.

Localities.—Sparingly in Co. Wicklow, not yet seen in other localities, but probably to be found in similar situations hereafter, when better sought for.

Pleurophrys ? amphitremoides (Arch.).
Pl. XX, fig. 2.

Specific characters.—Rather small, elliptic, or suborbicular, body seemingly filling the cavity of the test, and sometimes coloured by chlorophyll-granules; pseudopodia like the last, but somewhat more branched; test more or less densely covered by diatomaceous frustules or arenaceous particles.

Measurements.—The largest met with in length about $\frac{1}{410}''$, breadth $\frac{1}{500}''$, but variable in dimensions.

Localities.—Very sparingly in pools in Co. Wicklow and other places.

Pleurophrys ? fulva (Arch.).
Pl. XX, fig. 3.

Specific characters.—Minute, egg-shaped, body filling the cavity of the test; pseudopodia short, not straight, more or less branched, hyaline; test of a buff or tawny colour, and covered by hyaline quartzose granules.

Measurements.—Length and breadth respectively averaging about one half less than the preceding.

Localities.—Very sparingly in pools in same situations as foregoing.

Genus *Amphitrema* (Arch.).

Generic characters.—Rhizopod giving off two tufts of elongate linear branched pellucid pseudopodia, each tuft through an aperture at opposite ends of a test more or less covered by foreign arenaceous particles, the apertures provided with a rim-like neck.

Amphitrema Wrightianum (Arch.).
Pl. XX, figs. 4, 5.

Specific characters.—Body generally not filling the cavity of the test, (always?) including chlorophyll-granules; opposite tufts of pseudopodia unequal; the compressed elliptic test, with the foreign particles, more crowded at the margins, and often obscuring the extremely short necks.

No nucleus nor contractile vesicle detected, either in this or the two latter forms referred to Pleurophrys.

I have much pleasure in naming this form after my friend Prof. E. Perceval Wright, in whose company the first examples I saw were met with, and along with whom the most recently seen near Kenmare were taken.

Measurements—.Length of test about $\frac{1}{400}''$, breadth about $\frac{1}{500}''$, but slightly variable in dimensions.

Localities.—"Feather-bed Bog," Co. Dublin; "Glen-malur," Co. Wicklow; a boggy place on the road between Killarney lakes and Kenmare; as yet rare and local, but doubtless more widely distributed.

Affinities and Differences.—I am not aware of any monothalamian Rhizopod with two apertures for emission of pseudopodia, but there is no doubt a great affinity to such a form as *Pleurophrys? amphitremoides* (mihi), but I never could detect any neck to the test, however short in that form, yet in the present it is often obscured by foreign particles, and even one of the apertures may be shut up by their presence in abundance. The contents, however, are seemingly always green, and the larger foreign particles distributed to the margin of the test; the pseudopodia, too, are finer and longer than in *P. amphitremoides.* From the type or genus which seemingly must be admitted to be represented by Diplophrys (Barker), the present is distinguished by its test being covered by foreign particles, quite as decidedly, seemingly, as Difflugia from Arcella, as Pleurophrys from Plagiophrys, as the group Lituolida from the group Gromida. The whole aspect of the forms I have put forward under this genus and Pleurophrys seems to me to be quite distinct, as I have mentioned, from Gromida or the Difflugiæ.

Genus, *Diaphoropodon* (Arch.).

Generic characters.—Rhizopod with a nucleus, giving off rhizopodial processes of two kinds, one from the anterior end long, pellucid, and retractile, the other given off from the body, short, pellucid, and persistent, enclosed in a test formed of foreign particles loosely agglomerated.

Diaphoropodon mobile (Arch.).

Specific characters.—Rhizopod large, egg-shaped, nucleus large, granular in appearance; anterior pseudopodia often very long, much branched, hyaline, very contractile; marginal ones short fringe-like, hyaline; anterior extremity sometimes showing amarginal pulsating vacuole; test brownish, but formed of very heterogeneous particles (including protoccaceous cells) and diatomaceous frustules.

Measurement.—In length averaging about $\frac{1}{150}''$

Locality.—A single pool (in a single spot of it) in "Glenma-lur Valley," Co. Wicklow; hence as yet very rare.

Affinities and Differences.—No other rhizopod, I believe, shows the curious fringe-like processes, otherwise this form resembles some of those I have (be it but provisionally as it may) referred to Pleurophrys; the pseudopodia are, however, far more changeable, and more arborescent, and the foreign bodies on the "test" far more loosely aggregated than in Pleurophrys or any Difflugia. The forms meant to be represented by fig. 1 and fig. 6, are, perhaps, the only two that could be passed over the one for the other, but I need hardly insist on their differences being sufficiently important. The whole of the forms which are portrayed in my figs. 1, 2, 3, 4 and 5, 6, seem to me all to belong to a type quite and equally distinct from Difflugia as from Gromida.

Genus, *Gromia* (Duj.).
Gromia socialis (Arch.).
Pl. XX, figs. 7—11.

Specific characters.—Very minute, often occurring socially; body bluish, granular, with a distinct, sharply marked, white nucleus, containing a minute, dark nucleolus; pseudopodia elongate, branched, slender, reticulosely incorporated with each other, and often mutually with those proceeding from other individuals, and showing irregularly shaped expansions, and carrying along in a slowish current minute opaque granules; test hyaline, colourless, orbicular, or broadly elliptic.

Measurements.—Diameter from about $\frac{1}{1700}''$ to $\frac{1}{1400}''$.

Locality.—A single pool only in Glen-ma-lur Valley, hence as yet very rare.

Affinities and Differences.—The minute size, hyaline test, dense body substance, minute passage for the pseudopodia, white nucleus (with its nucleolus), and social habit, clearly mark out this form from other Gromiæ; its distinction from *Cystophrys Haeckeliana* have been pointed out, as they seem to me, in a previous part of the present communication.

On Imbedding Substances *for* Microscopic Section.
By M. Foster, M.D., Fullerian Professor of Physiology.

In Max Schultze's 'Archiv' (vol. clxiv) is a short paper by Professor Klebs, in which that accomplished microscopist recommends the use of strong glycerine jelly for the purpose of imbedding objects previous to the preparation of microscopic sections. I have tried this repeatedly, but having failed to secure the advantages said to be gained by it, have fallen back upon the paraffin process, which, by the bye, we also owe to Professor Kleb's ingenuity. As this method does not seem very popular in England, and is, morever, very badly described in Stricker's 'Handbuch,' I venture to give a note of the details upon which success in it mainly depends.

The process is most useful with objects which have been hardened in chromic acid or alcohol, and which are intended to be mounted in balsam; but it may be applied advantageously to objects of all kinds.

The first thing to be done is to choose a material corresponding in firmness with the object about to be cut. By melting together common solid paraffin (the paraffin candles sold at the shops answer very well) with the so-called paraffin oil, a substance may be obtained with almost any melting point you like, and setting, on being cooled, into almost any degree of firmness you please. A mixture of about one quarter or less of paraffin oil to three fourths of paraffin candle has appeared to me most generally useful.

Supposing the object to have been hardened in chromic acid, and to be about to be mounted in balsam, the way to proceed is as follows.

The object is taken out of the chromic acid and immersed for a few hours in alcohol. A cake of the paraffin mixture is then cut of a size suitable for holding in the hand; a hole scooped in it, and a small quantity of the melted paraffin poured in. The object, previously taken out of the alcohol, and its wetness removed, either with blotting paper, or by evaporation, is then dropped in. If the paraffin be sufficiently liquid the object will sink a little way in it, and the under surface will soon become completely coated with the paraffin. So much paraffin only should be used as is enough to cover the lower half of the object, the upper portion remaining uncovered. The exact position of the object must now be noted, in order that there may be afterwards no doubt of the plane in which the section is to be made. An arrow marked on the cake, or, if necessary, a sketch in ink, will prevent all difficulties. As soon as by the cooling of the paraffin the object has become fixed, more paraffin is poured over it until it is thoroughly covered in. The cake with the imbedded object is now thrown into spirit, and in a few minutes is ready for section, though it is, on the whole, better to let it stay in the spirit till the next day.

The cake must afterwards be carefully pared away, a little at a time, in the plane of intended section, until the object is seen to shine through. The sections may then be made either with a microtome or with a hand razor. If the cut surface be kept well wetted with spirit, and a layer of spirit be carefully carried on the upper surface of the razor, there will be no difficulty in floating the section on to a glass slide; the movement may be aided by a pipette or syringe. It is often useful to bevel down the edges of the cake, from time to time, as sections continue to be made, in order that as little paraffin as possible may be carried away on the razor with the section.

If the specimen be not intended for mounting in balsam, the paraffin may easily be washed away, at first with spirit, and afterwards with water, and the section stained, mounted in glycerine, &c., &c. There really is very little difficulty in getting rid of the paraffin, except where it has run into internal cavities.

If balsam-mounting be adopted, there is no difficulty at all. Coming directly after the spirit, a drop of creosote (common creosote, not carbolic acid, the odd ingredients of the former rendering it far more useful than the latter) clears the section up at once, and a few washings with turpentine gets rid of all the paraffin, and leaves the tissue quite ready for the balsam.

It is by no means necessary, as Stricker recommends, to apply the creosote and turpentine before imbedding. With care the cavities and spaces between the paraffin and the object may be reduced to a minimum; and even when they are formed, it is quite possible, in spite of them, to get excellent sections.

Nor need the method be necessarily limited to objects hardened in chromic acid or preserved in spirit. It is most useful with them, but may be employed without the intervention of any alcohol at all. Care, however, must then be taken to remove as much moisture as possible from the surface of the object, and to select a paraffin mixture of suitable firmness.

There is an incidental advantage of this imbedding process. An object may be imbedded, several sections made, and remainder of the cake replaced in spirit, in which, with a sufficiently instructive label, it may be preserved for any length of time, ready for other sections to be cut whenever they may be wanted.

The GREY MATTER *of the* CEREBRAL CONVOLUTIONS.

By Professor CLELAND, of Galway.

(With Plate VII).

THE microscopy of the grey matter of the cerebral hemispheres is a subject which has occupied in recent years the attention of numerous observers; but the accounts of different writers are sadly conflicting, and the layers which they describe are so variously and even arbitrarily enumerated, that it is difficult to compare their descriptions in detail.

The researches of Arndt,[1] however, claim special notice on account of their elaboration, the care with which he has collated the opinions of many previous writers, and the remarkable conclusions at which he arrives. My own observations, although far more scanty than I could wish, having led me to conclusions in some respects differing very materially from those of Arndt, and in others equally from those of Lockhart Clarke,[2] which unfortunately appear to have escaped Arndt's notice, I venture to give a short account of what I have seen, trusting that thereby at least others who enjoy better opportunities of obtaining fresh specimens may be aided in carrying the investigation further.

My observations have been confined to the marginal convolution of the longitudinal fissure in the upper frontal and upper occipital regions, and other convolutions in those neighbourhoods. Extremely weak chromic acid or bichromate of potass I have, like Arndt, found to be the best preservative fluids; but, unlike him, I find that a great deal can be made out even after preservation for more than a fortnight, a little alcohol having been mixed with the fluid to prevent mould. It is for the examination of the fibres that the freshest specimens are necessary. Carmine-dyeing is most useful in the examination of the nuclei, but the fibres are not brought into view by it, the axis-cylinders in this locality having no special attraction for the colouring matter, and the ammonia being destructive of the medullary sheaths. Turpentine has failed with me as with others. Bile I have found extremely useful in bringing the fibres into view.

I have further to add that, whatever my success has been, it has in great part been owing to the use of the "hold-fast cutter" invented and patented by Mr. Stirling, of the Anatomical Museum of the University of Edinburgh, an invaluable instrument, by means of which large and even sections may be made of the softest tissues. Happily Mr. Stirling has been prevailed on to publish an account of this instrument, which may be expected to appear in the next number of the 'Journal of Anatomy and Physiology.'

The best basis for a nomenclature of the strata of the grey matter of the convolutions still continues to be that of Kölliker, because, although imperfect, it is founded principally on distinctions visible to the naked eye, which admit of no

[1] Rudolph Arndt in Schultze's 'Archiv für Mikroskopische Anatomie,' vol. ii, p. 441, vol. iv, p. 407, vol. v, p 317.

[2] Lockhart Clarke, in 'Royal Society's Proceedings,' vol. xii, p. 716; and in Maudsley's 'Physiology and Pathology of Mind,' Second Edition, p. 60.

confusion. One must begin by acknowledging a stratum of texture on the surface of the convolution different from the main bulk of grey matter; then the pale band or couple of bands which are seen with the naked eye dividing the grey substance into other strata attract attention; and when both bands are present an obvious division into at least six layers is obtained. But the question at once occurs—Do the pale bands separate utterly distinct strata from one another, or is it not rather the case that while the grey matter undergoes changes of composition as it recedes from the surface, it is traversed by other structures at a varying level? As regards the deeper pale band the second explanation is probably true; its existence as a special stratum appears to be due very much to the separation of a series of horizontal fibres from a much larger tract of them which is immediately internal to the grey matter. The other light coloured stratum, which we may term the *primary pale band,* and of which I can speak more confidently as being more than an accumulation of transverse fibres, maintains a sufficiently definite position as related to the surface, but the strata which it separates are neither so unconnected nor of so different a kind as to give importance to Kölliker distinction of the "pure grey" and "yellowish red."

Arndt divides the first or outermost layer of Kölliker into two, a fibrous layer and a granular, with sparsely strewn, small, irregular nuclei, and delicate fibres taking all directions. His third layer, consisting of substance rich in nuclei which require a high power to show that they belong to nerve corpuscles with their chief poles not all pointed in one direction, combines with his fourth, which is characterised by pyramidal nerve corpuscles pointing to the surface, but still not much larger than the nuclei, to correspond (according to him) with the pure grey layer of Kölliker, and is separated by horizontal fibres, which he takes for the primary pale band, from the deeper part of the grey matter. This remaining part he calls at first the fifth layer, but subsequently divides into fifth, sixth, and seventh layers, to accord with Meinert; it contains pyramidal corpuscles of large size, especially in its more superficial portions. It is very noticeable, however, that in Arndt's figure of a vertical section through the different layers the strongest indications of horizontal fibres are represented in such a position that they would separate the largest nerve-corpuscles into two layers. This I have no hesitation in considering the true position of the primary pale band; but, undoubtedly, horizontal fibres exist much more superficially, and they may possibly sometimes form a slight indication of a pale band superficial to the primary one.

Lockhart Clarke, in his paper in 1863, enumerates eight layers, but in his more recent account he combines in one his first or fibrous, and his second or nuclear layer. His first layer as altered corresponds with the first, second, and, at least, part of the third layer of Arndt; his third and fifth are the pale bands; the sixth is the deepest part of the grey matter; and he terms the white centre of the convolution a seventh layer.

The superficial layer of nerve-fibres immediately beneath the pia mater appears to have been best described by Remak, who recognised that it was not evenly distributed over the surface, but was less developed on the vertex than on the base. It consists in great part of medullated fibres of different sizes, running horizontally on the surface of the brain.

That beneath this superficial network of medullated fibres there may be sometimes present a sparsely nucleated layer of tissue, as is described by Arndt as constituting what he calls the second layer, I am not disposed to deny, although it is noteworthy that both Lockhart Clarke and Meinert throw the tissues which constitute Arndt's first two layers into one layer; but I am satisfied that both the network of medullated fibres and whatever sparsely nucleated tissue may be sometimes below it, are in some places absent, and that then there is in contact with the pia mater a very densely nucleated layer.

This densely nucleated layer may be termed appropriately enough the *external layer of nucleated protoplasm.* Its nuclei, rounded in form, are most densely crowded in its most superficial part; while, as it is traced inwards, it may be seen to pass gradually into the next layer. Those nuclei, except in the deeper part of the layer, are not situated within nerve-corpuscles, however small; and the corpuscles which can be detected in the deeper parts of the layer are so minute as merely to invest the nuclei, except where the slender poles are placed. This layer appears to constitute the superficial parts of Arndt's third layer; and I am not prepared to decide how far his second or sparsely nucleated layer may not be included in it. I am tempted to say so because the somewhat similar dense nucleation of a deeper stratum, which I shall term the nucleated protoplasm of the primary pale band, has escaped Arndt's observation, notwithstanding his attention having been called to the subject by Meinert's account of "Körnerschichten.[1]" This external layer of

[1] Arndt, in Schultze's 'Archiv,' vol. v., p. 416.

nucleated protoplasm is threaded thickly with extremely fine non-medullated nerve-fibres coming up from the interior and extending in horizontal directions, but which I have not had the opportunity of studying sufficiently closely. It is manifestly enough the outer of the two densely nucleated strata figured by both Besser[1] and Arndt in the infant brain.

Passing more deeply a broader stratum is gradually reached, in which the nuclei are much sparser and can be distinguished as of two kinds, those which are free, and those which are situated within minute nerve-corpuscles of a pyramidal form, with the apex pointing to the surface; and it may be admitted with Arndt that many of the free nuclei are somewhat smaller and less easily stained with carmine than those within the nerve-corpuscles; but the difference in these respects is not such that a deduction can be safely founded on it. This is the fourth layer of Arndt, the superficial part of what Lockhart Clarke latterly calls the second layer.

Passing still inwards the nerve corpuscles are found to enlarge, and a considerable depth of tissue is reached, characterised by pyramidal nerve-corpuscles of greatly increased size, forming, beyond question, the fifth layer of Arndt; but if it be examined with great care, particularly in a carmine-stained preparation, there will be found, after passing a certain depth of these corpuscles, a defined band of closely aggregated free nuclei of small size, constituting the outer of the "Körnerschichten" of Meinert,[2] already alluded to as the nucleated protoplasm of the primary pale band.

This stratum I have never failed to find; it is as distinct from the texture above and below it as the milky way is from the rest of the heavens; but as the slide is passed across the field of the microscope, it is easy to understand how it might escape observation not specially directed to it. If difficulty be experienced in finding its position, it may be detected by holding up the slide on which is the section against the light, noting the sharp line which separates the primary pale band from the subjacent texture, and placing a speck of gum on the cover, so as to touch the deep margin of the line. On examination under the microscope, the band of dense nuclei will be seen touching the speck with its deep

[1] Besser, in Virchow's 'Archiv,' vol. xxxvi; Taf. vii.

[2] I regret that I have not yet seen Meinert's paper, published in the 'Vierteljahrschrift für Psychiatrie,' vol. i; but it is copiously referred to by Arndt.

margin; and in a suitably prepared specimen the speck will be seen to rest over a well-marked horizontal band of medullated fibres. Thus, the primary pale band may be demonstrated to owe its whiteness, as appears to have been appreciated by Meinert, to the dense nucleated protoplasm within it, and to be limited internally by those horizontal fibres to which it has frequently been supposed to owe its special character.

Immediately beneath the primary pale band pyramidal corpuscles may again be found as large as those superficial to it; but I think not always so. There seem to be always present here nerve-corpuscles of more irregular form, and the size diminishes the further we penetrate towards the white substance I am disposed to think that Meinert is correct in describing another densely nucleated stratum in connection with the second pale band when that is present; but there is certainly, also, in connection with it, a stratum of horizontal medullated fibres, such as there is in connection with the primary pale band.

The nerve fibres in the convolutions are of two sets, those which are vertical to the surface, and those which are horizontal to it. The vertical fibres, coming up from the interior, as soon as they reach the grey matter split up, as is well understood, into bundles; and in the intervals between these the nerve corpuscles and free nuclei are found, so that they appear to be packed in vertical rows. In the deeper layers the fibres are broad, the axis cylinders being coated abundantly with medullary sheath; but as they proceed onwards each fibre becomes gradually thinner, and by this means the bundles get narrower. Tracing them onwards through the outer layer of small nerve corpuscles, they are seen to spread out, so as to be uniformly distributed and no longer parallel; but they continue to make for the surface, and only turn round to become horizontal in the external nucleated protoplasm. The horizontal fibres have been termed arciform fibres by Lockhart Clarke. There is a dense stratum of them medullated, situated immediately beneath the grey matter. As has been already stated, there may be one or two strata of fibres separated from this, corresponding to the inner edges of the pale bands seen with the naked eye; also scattered medullated fibres are found in the intervals between these strata; and, at least in sections cutting the convolutions longitudinally, others may be seen passing very obliquely from one stratum to another. But more superficially there still continue to be disposed numbers of horizontal fibres, not so easily seen on account of the slightness or absence of medullary sheath, yet capable of being well demonstrated when their torn ends project from the edge of a section.

Lockhart Clarke in part attributes the decreasing diameter of the vertical bundles of fibres to some of the fibres turning round in the inner layer so as to become horizontal; but this is really not the state of matters which exists. The examination of extremely thin sections, which have been preserved for only a few days in weak chromic acid, or in a filtered mixture of oxgall and spirit, shows the vertical and horizontal fibres crossing one another with the utmost clearness and without the slighest tendency of the one set to pass into the other. Unfortunately, such specimens cannot be preserved, but they are very distinct while they last. The horizontal and vertical fibres in some instances communicate, but it is through the medium of nerve-corpuscles.

The typical, and by far most frequent, form of nerve-corpuscle found in the convolutions of the brain is the pyramidal; it presents the same structure, both in the case of the large and that of the small corpuscles, and has been described by the different recent writers on the subject. The apex of the pyramid is always directed towards the surface, and is prolonged into a nerve fibre, which passes right on to join—probably always—the bundles of horizontal fibres at and near the surface. I say that, probably always it does so, for although I have only once or twice traced such fibres to the surface, and seen the turning round, yet one may argue from the constantly vertical course which they are seen to pursue when traced for considerable distances from their origins in the deep strata. Arndt describes instances in which he made sure that this main process, or apex fibre, of the nerve-corpuscle was continued into a dark-bordered nerve fibre; and I am disposed to go further, and to say that for a short distance, at least, it is always dark-bordered. In the case of the larger corpuscles, the granular contents of the corpuscle are continued for some distance within the apex fibre.

Lockhart Clarke, as well as others, describes the apex fibre as "giving off minute branches in its course;" Arndt describes it as unbranching, except in the cornu ammonis. I believe that Arndt is right in considering that in the ordinary arrangement this fibre does not branch. I have certainly seen, as he also has done, an appearance of one or two minute branches given off laterally from it, while the main fibre continues its uninterrupted course; but such appearances seemed to be due to the accidental adhesion of portions of slender fibres not really in structural continuity with the fibre from which they seemed to spring.

Arndt further describes the turning round of the apex fibre, so as to be possibly continuous with vertical fibres coming up from the white substance; but of this it would be well to have further evidence before it is received as a fact; for the apex fibres are liable to be disturbed from their proper position in making the section; and a good many of the archings figured by Arndt look very much as if they were due to that cause. From the base of each pyramidal corpuscle, especially when it has been isolated, may be seen to come off several extremely delicate processes, which take different directions and are often branched.

Arndt believes that these basal processes are never continued into any nerve fibres, while Lockhart Clarke describes them as running "partly toward the centre to be continuous with fibres radiating from the central stem, and partly parallel with the surface of the convolution to be continuous with arciform fibres."

This is a very difficult point to make certain with regard to the typically pyramidal corpuscles; but my own observations entirely favour Lockhart Clarke's view. If Arndt had studied the position of these basal processes when undisturbed and embedded in the surrounding textures, he would have seen that they have not the rounded curves which he represents in his scheme, but are stretched, for the most part horizontally, and sometimes also vertically, so that it seems extremely probable that they are continued into fibres running in these different directions.

Even Arndt, who strains to represent all the nerve corpuscles of the convolutions as founded on the pyramidal form, has to own that this shape is altered, as if by distortion, in many instances, and, in point of fact, represents fusiform and otherwise shaped corpuscles which have come under his own observation. These various shaped corpuscles are most numerous in the parts subjacent to the primary pale band; and indeed Lockhart Clarke, in describing those medullated horizontal fibres which I have pointed out to be the inner limit of that band, states that "they are thickly interspersed with large and small cells of different shapes." Apparently fusiform corpuscles are not infrequent at this level, for the most part placed vertically, but sometimes also horizontally. It is more questionable, however, that these fusiform corpuscles are really only bipolar, as it must be admitted that when they are imbedded in texture, small processes given off from the sides are liable to escape detection. In one instance, in a tear of a section, I have seen, as I believe, three medullated fibres spring from one of these irregular corpuscles, one passing inwards, one to the surface, and one horizontally (Pl. VII, fig. 4). In another instance I have seen a similarly shaped corpuscle with three principal poles, which differed in this respect, that both the deep and the horizontal process were bifurcated (fig. 5). True pyramidal corpuscles are sometimes to be met with among others of different shapes in the apparent white substance beneath the broad band of horizontal fibres which limits the grey matter.

Hitherto we have dealt only with the arrangement of the nerve corpuscles, nerve fibres, and nucleated matrix; a few words may be added on the composition of each of these elements. As regards the matrix, I agree with the view of Henle and Merkel[1] that it is not proper connective tissue, but allied to protoplasm. It is not swollen, but made firmer with boiling.

Arndt, while right with regard to its protoplasmic nature, is doubtless mistaken with regard to its fibrous character; the fibres which he and others have described being produced by the coagulating effect of reagents. In sections which have been exposed to only weak reagents the fibrous appearance is absent; and in the densely nucleated parts a ball of protoplasm seems to surround each nucleus. Scattered through this basis are granules of a substance probably allied to cholesterine or protagon; but those granules are mere chemical deposits, probably run into more distinct granules after death, and certainly without structural connections. Masses of matrix, with granules in them, are no doubt often seen adhering to torn extremities of processes of the nerve corpuscles, just as Arndt has figured them; but it is the merest hypothesis, to my mind not a probable one, to suppose that those processes are structurally continuous with the adherent substance or its granules.

The medullated fibres of the convolutions are limited by no membrane round the medullary sheath; and that sheath consists apparently of a uniform substance, probably enough pure protagon, which softens and becomes semifluid or fluid on addition of various substances such as oxgall. The nerve corpuscles contain a similar, but not identical, substance, which has already been alluded to, and to which they owe their granular appearance. It runs into larger globules under prolonged subjection to the action of dilute oxgall, but remains entangled in the stroma of the corpuscle. In a

[1] Henle's 'Bericht über die Fortschritte der Anat. u. Phys. im Jahre,' 1867, p. 63, and 1868, p. 58.

specimen immersed for some days in dilute liquor potassæ, and reduced to a pulp, this substance was run into masses, one from each nerve corpuscle, and having a firm waxy appearance contrasting with the pulpiness of the rest of the texture. There are thus apparently three more or less waxy-looking substances in the grey matter of the brain, to which the attention of chemists should be directed to distinguish them, viz. the matrix granules, the medullary sheaths of nerves, and the granules of the nerve corpuscles.

I have been able to verify the appearance of striation or fibrillation which Arndt has remarked in the structure of some of the nerve corpuscles. I do not see, however, that this appearance is explained by the process of development which Arndt imagines from what he has seen in the infant brain; and when he proceeds to depreciate the importance of the nerve corpuscles, on the ground that they have a fibrillated texture, he is certainly very wrong. Contractility may fairly be taken as one of the most markedly vital properties of texture; and this property is exhibited in the highest intensity by a texture which can easily be made to show far more fibrillation than these corpuscles. He should have remembered that however fibrillated the texture of the nerve corpuscles may be, it is composed of albuminoid or protoplasmic, not of gelatinous substance; and it is in vain, in the present state of our knowledge, to attribute to the differentiated protoplasm of the nerve corpuscles a lower amount of vital property than to the comparatively undifferentiated protoplasm round about.

The present teaching of science, with regard to the functions of protoplasm in connection with the special properties of nervous textures, may, I think, be fairly summarized thus: that vital properties—to wit, irritability, contractility, sensibility, selection of food, reproductive power, and capability of development into differentiated vital tissues, are resident in nucleated corpuscles consisting essentially of masses of protoplasm; and the various differentiated tissues referred to, such as the nervous, muscular, and glandular, are specially developed so as to exhibit in perfection certain of the properties of those bodies from which they all originally spring. Taking this view, I see no objection to supposing that the nucleated protoplasm of the brain may foster and facilitate the actions of innervation;[1] but it seems most probable that

[1] In connection with the attributing of vital properties to the nucleated protoplasm of the brain, it is interesting to note the curious observation by Walther of amœboid movements in sections of a frozen frog's brain, referred to in Henle's 'Bericht,' for 1868, p. 60.

the more highly developed structures, the nerve-corpuscles, are those which regulate and primarily carry on the action.

It may further be allowable to suggest that, although in the nervous actions connected with sensibility and movement throughout the body, the travelling of impressions plays a most important part, yet it seems more probable that in the cerebral hemispheres such travelling has only the subsidiary end to serve of bringing all the corpuscles into communication, and that it is a condition of these, in some degree comparable with the contracted condition of a muscle, which is the physical element necessary for mental action; that the total amount of mental action at one time is thus dependent on the total amount of the physical action; and that it is very questionable if there be any closer connection between the special qualities of the mind and the structure of the brain.

On the EMBRYONIC FORM *of* NEMATOBOTHRIUM FILARINA, Van Ben. By Dr. EDOUARD VAN BENEDEN. Plate VIII.

IN his memoir on the Intestinal Worms, M. P. J. Van Beneden made known an extremely remarkable animal which he described under the name of *Nematobothrium filarina*.[1] It lives parasitically in *Sciæna aquila*, a fish which is found sometimes on our coasts and which is met with from time to time in the Channel and on the coasts of the Atlantic. But as it is always a relatively rare fish, one has not at all times the opportunity of studying the singular parasites which this animal harbours.

Nematobothrium differs considerably from all known worms. It has the external form of a Nematod, attains an enormous length, often more than a metre, and is always rolled up on itself, forming a regular ball, the volume of which varies between that of a large nut and an ordinary orange. This ball is lodged beneath the skin, which covers the region of the shoulder-girdle of the fish, in a regular closed cyst, which presents exteriorly the aspect of a voluminous tumour.

The Nematobothrium is not free in its cyst; it is lodged in a membranous tube rolled up like the worm which it lodges, which acquires adhesions on all sides, and from this

[1] P. J. Van Beneden, 'Memoir sur les Vers Intestinaux,' Paris, 1858, p. 107, pl. xiii.

arises the extreme difficulty which there is in disentangling the animal and isolating it throughout its length. It requires the exercise of unusual patience to succeed in isolating even a small portion of the body. In such conditions, it is easy to understand how difficult it is to study the organisation of an animal, so difficult that one cannot even distinguish with certainty its anterior from its posterior extremity. What place should be assigned to this animal in helminthological classification? It is not without doubt that M. Van Beneden has placed it among the Trematods. In assigning to it this place, he depended chiefly on, 1st, the extreme mobility of the extremity considered as cephalic, which extends and contracts, shrinks and enlarges successively as is observed in certain Trematods and some Cestoids, such as Caryophyllæus; 2nd, on the presence in the axis of the body of a contractile vessel, opening very probably on the exterior, and which must be a part of the excretory apparatus; 3rd, the analogy which this animal presents with certain Trematods, such as *Distoma filicolle* (Rud.), *Distoma Okenii* (Köll.), which lives in the *Brama Raii* in cysts similar to those in which *Nematobothrium* is found.

Nematobothrium appears then to be an exceptional Trematod, of at least a metre in length, having the external characters of certain Nematods, such as Filaria and Gordius, which, like it at the period of sexual maturity, become reduced, in a great measure, to a mere bag of eggs.

The knowledge of the embryonic form is an element which ought to be of great weight in the solution of the problem relating to the affinities of this singular animal. V. Carus, whilst placing Nematobothrium provisionally among Trematods declares that the position of this strange form cannot be determined certainly until the time when the embryonic form is known.[1]

It is this embryonic form which I propose to make known by this notice.

I will say a word to begin with concerning the egg. The egg is of extraordinary minuteness, of oval form; its long axis measures barely ·027 millimetres, its small axis reaches about ·020 millimeters. The principle that the number of eggs which an animal produces is indirectly proportional to their dimensions is, completely verified here; it is neither by hundreds nor by thousands, but by millions of eggs that our animal reproduces. The calculation is easy to make; knowing the number of eggs contained in a given length of the

[1] 'Handbuch der Zoologie,' von J. V. Carus und Ad. Gerstaecker, ii Bd., p. 480.

ovigerous tube and the diameter of the tube, we can calculate the number of eggs contained in any portion of the tube. This organ is folded up in every direction in the interior of the body, so that in the breadth of the body it crosses and re-crosses six or eight times. With these data I have been able to estimate approximately the number of the eggs contained in a piece of the worm of a decimetre in length at six or seven millions. One can judge in this way of the total number of eggs which the animal can contain at one time in the interior of its body.

The egg presents a very thick shell of a chitinous character, the colour of which varies from yellow to dark brown. The dehiscence of the egg takes place by the formation, near one of its poles, of a circular crack always very regular, which divides the shell into two very unequal portions (Pl. VIII, figs. 2 and 6). It is the manner in which the eggs burst in the majority of Trematods.

The length of the body of the embryo measures nearly double the long axis of the egg in which it is folded up, bent at its middle in such a way that the posterior half of its body is found to be applied against the anterior half. Its average breadth is about equal to half the breadth of the small axis of the egg. Its body is of an elongated form, and draws itself slowly along from before backwards. In front it is furnished with a distended part, probably of a muscular nature, on which a crown of bristles or hooklets is implanted, the form and disposition of which are very remarkable.

The anterior enlargement of the body, which we will call the muscular disc, is naturally divided by two diameters, cutting each other at right angles, at the centre of the disc, into four sectors, which are nearly equal. *The form and disposition of the hooklets is the same in two opposite sectors, different in two adjacent sectors.* If we were dealing with a crystal, we should say that it presented a bilateral symmetry.

In each of the sectors A and B (figs. 4 and 8), we counted seven hooklets, of which four larger are separated from one another by three others, which are smaller. All are arranged in a radiating manner on the disc, and diverge from the centre to the periphery. All these hooklets are nearly straight. The large ones present, nevertheless, in one point a slight enlargement, which gives them a certain resemblance to the bristles characteristic of the six-hooked embryo of Cestoids. I was not able to recognise the presence of this median swelling in all the individuals which I have observed, but that may depend on the position of the hooklets. The little prickles, which alternate with those which I have just described, are placed in the peripheral part of the disc. They are straight, very slender, and do not present, I believe, any median enlargement. In the sectors M and N (figs. 4 and 8), very minute hooklets are implanted of great delicacy, which are all disposed in a radiate fashion on the disc, and are situated at equal distance from one another, without our being able to discover any regular order in their disposition.

All these hooklets, those of the sectors A and B, as well as those of the sectors M and N, are implanted in the substance of the disc for the greater part of their length; their points alone are free, and project a little from the edge of the disc when seen in projection.

If the surface of the disc is examined with care, concentric transverse striations can be observed, which speak to the existence of muscular fibres (?), the presence of which appears to be indispensable, in order that the embryo may avail itself of the crown of hooklets.

At the centre of the disc is seen a small circular space, much darker, appearing like a black spot. It is exceedingly difficult to determine the nature of this organ, on account of the extreme minuteness of the embryo. It, however, appears to me scarcely doubtful that the central spot of the disc of Nematobothrium has the same signification as the organ which G. Wagener has considered as a true orifice in the embryos of several Trematoids and of various Echinorhynchi.[1] Is it a sucker? is it really an orifice? This point I have not been able to settle in the animal which now occupies our attention.

The disc is the part of the body of the embryo which presents the greatest consistence. The body possesses great mobility, and its length augments and diminishes in inverse proportion to its breadth, during the movements which the embryo executes.

Structure.—The body of the young Nematobothrium is covered by a cuticular membrane of some thickness. Under this cuticle a cellular mass is found, composed of clear and transparent cells of extreme delicacy, and of very small dimensions. There are also distinguishable small refrangent globules, of which some are fatty in nature, others mineral, probably calcareous.

G. Wagener[2] and other helminthologists have recognised

[1] G. Wagener, "Helminthologis che Bemerkungen aus einem Sendschreiben an C. Th. v. Siebold," 'Zeitschrift für Wiss. Zool.,' Bd. ix, pl. v and vi.

[2] G. Wagener, loc. cit.

in the ciliated embryo of many Trematods the existence of vessels representing the urinary apparatus of the Trematoda and Cestoida. I could find no trace of these vessels in the embryo of the Nematobothrium; and I have not been able in consequence to verify the relation which Claparede[1] has pointed out between the calcareous corpuscles and the origin of the urinary apparatus of Trematods.

Affinities.—As we have said, M. P. J. Van Beneden, relying on certain anatomical characters, has assimilated Nematobothrium to the Trematods, remarking, however, that certain facts would tend to make one consider it as forming a veritable transition between Nematods and Cestoids.

The embryonic form of Nematobothrium has nothing in common with that of the Nematods; and the knowledge of the embryo enables us to affirm that Nematobothrium does not belong to this group of Worms.

The embryo of all the Cestoids is characterized by a common form; at the moment of exit from the egg it is provided with six hooklets: of which two are directed anteriorly, two to the right, and two to the left. There are only some Cestoids peculiar to marine fishes, which, in place of three pairs of hooks, present but two. The embryonic form of the animal which occupies us diverges considerably from that of the Cestoids; and in spite of a certain resemblance in the form of some hooklets, we do not hesitate to declare that Nematobothrium is not a Cestoid.

According to the manner of their development the Trematods are divided into two great groups: 1st, that of the *Trematoda monogenetica*, comprising all the ectoparasitic Trematods and some endoparasitic Trematods, such as the *Polystomum integerrimum* of the frog; 2nd, that of the *Trematoda digenetica*, which presents all the phenomena of 'alternations of generations.' The first have, when born, the form of the adult, the second come into the world in a form which gives no indication of that of the sexual animal. The embryo of the Digenetic Trematods is generally ciliated at the moment of birth. However, Von Siebold observed that the embryo of *Distoma tereticolle* is deprived of vibratile cilia; and some time after G. Wagener observed the same fact in the embryo of *Monostoma filum*.[2] The learned helminthologist pointed out at the same time the existence, around the cephalic extremity of the embryo, of a crown of

[1] Claparède, 'Ueber die Kalkkorperschen der Trematoden: Zeitschr. für wiss. Zoologie,' Bd, ix.

[2] 'Muller's Archiv,' 1854, page 16, pl. ii. Note of Wagner in a communication by Lieberkühn, "On the Psorospermia."

hooklets, which he compared to the crown of hooklets of the Cestoid worms.

In his memoir on the development of the Entozoa, G. Wagener figures the embryo of *Distoma tereticolle* carrying in front a complete crown, composed of two score of striæ, which are, he says, like hooklets.[1]

I have myself studied the embryo of this Trematod which had been incompletely studied by G. Wagener, and which he has figured without bringing out clearly the distinctive characters which it presents.

The embryo carries round the anterior enlargement of the body four little plates, of a triangular form, similar one to another, and disposed symmetrically. Each of them formed by a thickening of the cuticle carries a system of hooklets. These hooklets are very small, and present, at a short distance from their point, a little swelling, which gives them a peculiar aspect.

These plates are supported by the anterior extremity of the body, which is greatly enlarged; where they are disposed at the extremity of two diameters, cutting each other at right angles. There is here, then, a *quadrilateral* symmetry.

Near the posterior extremity of the body exists a circular zone of little ray-like fine bristles, which completely surround the body of the embryo.

I could not distinguish in the cellular mass of the body any trace of digestive tube nor any appearance of excretory canals.

It is impossible to overlook the analogy which the embryo of Nematobothrium presents to those of *Distoma tereticolle* and *D. variegatum* and of *Monostoma filum*; and it seems to me evident that the animal which forms the subject of this note ought to be ranged by the side of those Trematods in the group of *Trematoda digenetica*, with non-ciliated embryos. Their embryonic form is characterised—1st, by the absence of vibratile cilia; 2nd, by the swelling of the anterior part of the body, separated from the posterior part by a circular furrow; 3rd, by the presence around the cephalic extremity of a series of hooklets or prickles, the form and disposition of which varies slightly in the three genera.

The embryonic characters come then to the support of the conclusion which M. P. J. Van Beneden had drawn from the study of the organisation of Nematobothrium, and

[1] G. Wagener "Beitrage zur Entwickelungsgeschichte der Eingeweide-würmer," page 25, pl. xx, in 'Natuurkundige Verhandelingen van de Holl. maatsch. der Wetenschappen te Haarlem,' 2nd Verz., 13th sect.

render it necessary to consider this animal as a true Trematod; but it is a Trematod presenting the external form of a Nematod, which attains more than a metre of length, and which lives in quite exceptional conditions.

These digenetic Trematods whose embryo is deprived of cilia cannot be considered as forming the transition to the Cestoids; and we must not suppose that in these embryos with prickles we have the homologue of the six-hooked embryo of the Band-worms. From the beautiful observations of Schubart and of Knoch[1] we know that in the Bothriocephalids the six-hooked embryo is not the first embryonic form, but that this embryo succeeds a ciliated form, from which it arises by metagenesis.

I have pointed out in my memoir "On the Composition and the Significance of the Egg" that in the Tæniadæ (*Tænia bacillaris*) a membrane is formed corresponding to this ciliated investment of the Bothriocephali; but it is constantly free from vibratile cilia. The hexacanth embryo is then both in the one and the other a *secondary* embryonic form, born by metagenesis from an embryo, sometimes ciliated, sometimes free from vibratile cilia. The ciliated embryo of the digenetic Trematods corresponds to the ciliated form of the Bothriocephali; and it is the sporocyst which corresponds to the hexacanth embryo of the Cestoids.

The embryo of *Distoma tereticolle*, of *Monostoma filum*, of *Nematobothrium*, is evidently the homologue of the ciliated embryo of the other Distomata, and consequently, in spite of their crown of hooklets, they represent in no respect whatever the hexacanth embryo of the Cestoids.

The following table sets forth the remarkable parallelism which one can establish between Trematods and Cestoids as regards their development:

Cestoids	Monogenesis		= *Caryophyllæus.*
	Digenesis.—The first embryonic form is	Ciliate	= *Bothriocephalus.*
		Non-ciliate	= *Tæniadæ.*
Trematods	Monogenesis		= Ectoparasites.
	Digenesis.—The first embryonic form is	Ciliate	= *Distoma, Monostoma, Gasterostoma,* &c.
		Non-ciliate	= *Distoma tereticolle, Monostoma filum, Nematobothrium.*

[1] Knoch, 'Naturgeschichte des breiten Bandwurms.' St. Petersburg 1862, 4to.

It is observable, however, that the presence of vibratile cilia in the first embryonic form is a character of little importance, and cannot serve as the basis of a natural classification.

Remarks *on* Opalina *and its* Contractile Vesicles, *on* Pachydermon *and* Annelidan Spermatophors. By E. Ray Lankester, B.A. Oxon. Plate IX.

In examining the anatomy of oligochætous Annelids, I have necessarily met with certain species of Opalina, that curious mouthless genus of Infusoria the life history of which, like that of so many of the class, is as yet quite unknown. Some notes on the *Opalina Naidos* of Dujardin, which abundantly infests the *Nais serpentina*, may not be uninteresting; at the same time, I offer some evidence as to the nature of the bodies which Professor Claparède considered to be Opalinoid parasites, and termed *Pachydermon*, figuring them from two species of the oligochæt *Clitellio;* but which I think, from the characters of some observed by me in another worm (namely, *Limnodrilus*) must be considered as packets of spermatozoa or spermatophors.

Opalinæ.—The genus Opalina has sometimes been made to include those pyriform ciliated animalcules which swarm in the rectum of the common tadpole and in similar situations, called *Bursaria Ranæ* by Ehrenberg; but these forms should rightly be separated from those so frequently found in both marine and freshwater Annelids, from which they differ materially, as pointed out by Claparède in his work with Lachmann on the Infusoria.

The simple structureless body of these first-named parasites has really very little in common with Opalina, properly so called—an abundance of highly refrangent granules being the only differentiated portions of its substance (Pl. IX, fig. 9), no trace of the nucleus and contracted vesicles, nor of the furrowed cuticle of true Opalina being observable. It is not improbable that these swarming ciliated flakes of sarcode—for they are nothing more—may undergo subsequent metamorphosis of the most extreme character; but what is true of them does not apply to veritable *Opalina*.

Opalinæ of the type I am about to describe have been observed by Dujardin, Schultze, Schmidt, Stein, and Claparède in various worms. Thus we have *Opalina Naidos*,

Dujardin, in Nais; *O. polymorpha*, Schultze, in *Planaria torva; O. lineata*, Schultze, in *Nais* (?) *littoralis; O. rècurva*, Claparède, in *Planaria limacina; O. prolifera*, Claparède,[1] in a supposed marine Nais; *O. filum*, Claparède, in *Clitellio arenarius; O. ovata*, Clap., in *Phyllodoce; O. convexa*, Clap., in another *Phyllodoce; O. Pachydrili*, Clap., in *Pachydrilus verrucosus; O. uncinata*, Schultze, in *Planaria ulvæ; O. armata* and *O. falcifera*, Stein, in *Lumbricus terrestris* and *L. anatomicus*—these three species bearing hooks are placed in the genus *Hoplitophrya*, Stein; whilst another species, formerly known as *Leucophrys stiata*, is called *Anoplophrya lumbrici*. At the same time, *Hoplitophrya armata* is supposed by Stein to be a later development of *A. lumbrici*.

Though they are common enough there appears to have been doubt as to the character of Opalinæ; first as to the nature of the vesicles which they possess, and secondly, as to the nucleus. Schultze described a series of contractile vesicles in his *O. lineata;* and Claparède subsequently saw vesicles in another *Opalina*, but could not succeed in witnessing their contraction. In the species which he has most lately described, however ("Recherches sur les Annélides, Turbellariés, &c., observés dans les Hébrides"), he states that he has observed the contraction of the vesicles.

In certain species found in Planariæ a single long contractile vesicle has been said by Schultze to exist, taking the place of the series of smaller vesicles. A nucleus of various size and distinctness has been described in these true Opalinæ, but never a nucleolus. Stein describes the nucleus of *O. armata* as presenting oval embedded granules and rod-like bodies.

The two forms occurring in the Earthworm (*O. armata* and *O. fulcifera*), as well as *O. uncinata*, from a Planarian, possess a pair of recurved hooklets, which, situated at one end, are believed to furnish a means of attachment, as in some Gregarinæ (*G. Sieboldii*, Köll., from *Libellulu larva*). The tooth-like body of Dysteria (Huxley, 'Quart. Journ. Microsc. Sci.,' Vol. 5, 1857, p. 138) furnishes a parallel development in an adult ciliate Infusorian.

From the general statements with regard to Opalina, excluding from consideration the so-called *Bursaria Ranæ* and similar species, there seems to be no reason for supposing that they are anything but Infusoria; the supposition that they are a phase in the development of certain worms being

[1] This species is peculiarly interesting, since it is distinctly a segmented animal—as much as any *Tænia*—presenting a chain of incomplete zooids attached one behind the other.

curiously inconsistent with the existence of the nucleus, the contractile vesicles, and the ribbed ciliated cuticle of a typical Infusorian.

In Pl. IX, fig. 1, is drawn a fair average specimen of the Opalina infesting *Nais serpentina*. It is about the 1-200th of an inch long, flattened very considerably (so as to be band-like), and of an oblong form. Hundreds are often crowded together in the intestines of the *Nais;* and so tightly packed are they that frequently some specimens grow into an irregular shape, as seen in fig. 8. They do not vary much in size in the same worm; but in *Lumbriculus* I have seen what I believe to be the same species, as large as one fiftieth of an inch, and with thirty contractile vesicles.

Many of the specimens brought out on to the field of the microscope, by causing the worms to burst, are seen to be in the act of transverse fission (fig. 7); and their increase in size appears to be prevented by this removal of a portion of the individual. Those that have been recently separated by transverse fission have a pointed anterior extremity, which gradually assumes a flat and blunted character as growth proceeds. Usually when a Nais is placed on a glass slip, and squeezed beneath the thin cover, so as to cause the *Opalinæ* to be extruded, it is noticed that, though at first they move rapidly about the field of the microscope with a regular and active movement of their long cilia, yet slowly and surely these movements become weaker and *intermittent*—the cilia altogether ceasing and then *resuming* their action; until at last movement ceases altogether. When first extruded the observer will have noticed the great transparency of the Opalina, and will scarcely have been able to define the large nucleus (fig. 1, *n*) indicated by a somewhat darker tint; but he will have caught sight of the row of globular spaces usually confined to one side of the body, and will probably have watched in vain for their contraction.

Now, when the ciliary motion has ceased, it will be seen that some of the globular spaces have acquired a great size (fig. 2); whilst the nucleus has become very distinct, being defined by a broad space running round it, and having the same pink or purplish tint which is so characteristic of the contractile vesicles of all Infusoria. The death of the *Opalina* has been caused by its sudden introduction to fresh water—the distension of its contractile spaces, and the formation of the cavity between the nucleus and peripheral tissue being equally caused by the rapid endosmose of water. It seems very probable that this action of pure water upon an organism adapted to live in intestinal fluids accounts for the difficulty which many have experienced in witnessing the contractions of the contractile spaces in *Opalinæ*. It also accounts for the oblong vessel surrounding the nucleus in some of the species figured by Stein, as it most certainly *produces* this appearance in the Opalina which I have studied; and it also explains the nucleus-membrane as distinct from nucleus-content of which Stein speaks.

The long vessel of *Opalina Planariæ* described by Schultze may possibly be also due to such a separation of the inner and outer layers of the organism by the imbibition of water, since in my Opalinæ the cavity so produced had most closely the appearance of a long vessel, for which I mistook it at first. In the Opalina of Clitellio I observed and drew a long vessel, produced, I now believe, in this way. The imbibition of water, besides distending some of the vesicles, may cause them to run together, and form much larger lacunæ than ever exist in the living animal, as in fig. 2, *y*.

To avert the death and rapid post-mortem changes of the Opalinæ, it is only necessary to avoid using water and let the parasites, when extruded, remain in some of the fluids from the worm. If arranged in this way they will be seen as in fig. 1; and the rapid contractions of the series of globular cavities arranged on one side of the creatures may be watched when once the eye has got accustomed to their movements, and very few more beautiful sights can be presented to the observer.

As the Opalina rolls slowly over on to one side, the attention may be fixed on one of the globular cavities, which appears like a small pink bladder floating in the perfectly colourless sarcode of the Infusorian. Suddenly as you watch this bladder—so suddenly that you are almost startled—the bladder is gone; but almost immediately in its place a very minute spot is seen, which slowly increases in size, and ultimately proves to be the same cavity reappearing.

Dr. Moxon having recently, in the 'Journal of Anatomy' (May, 1869), written on the subject of the contractile vesicle of Infusoria, advancing arguments in favour of the view that it opens externally, I may here point out certain appearances which are visible in Opalina, and which seem clearly to accord with this view; though, at the same time, I should say that there can be no doubt about the opening of the vesicle, after the paper of Dr. Zenke, noticed in this Journal two years since. In fig. 2, in which the vesicles are distended by an excessive endosmose, they are seen to have a lemon shape, more or less, the point being nearest the cuticle. This is exactly the form presented by the vesicles (or rather *cavities*, for they have no proper membrane) just before disappearing in the living Opalina, if seen in profile, and seems to favour the idea of their opening externally. When the cavity is reappearing, after collapse, it has a globular form from the very first, increasing gradually at its periphery, but maintaining its globular form, as seen in fig. 6, *a*, *b*, *c*, *d*, *e*, until the full dimensions are attained. On some occasions during the collapse I have observed that this takes place in a totally different way, the opposite walls closing together, as in *f*, *g*, *h*. This mode of expansion and contraction clearly agrees with the supposition that these pulsating cavities are yielding points in the sarcode substance, which are slowly distended by the accumulation of fluid; the thin external wall ultimately bursting and allowing the fluid to escape; and then the adjacent parts unite again at once as sarcodic matter is known to do, and a fresh accumulation commences at the old place. The formation of elongated cavities surrounding the nucleus, and having all the appearance of the regular "vesicles," as well as the fusion of adjacent cavities, and the continued distension of others after death, by the excessive endosmose of liquid, all seem to favour the same theory of the formation and mode of action of the contractile cavities.

The cavities vary much in number in the Opalinæ I studied; very rarely they occurred on *both sides* of the body, but were generally in a series of from five to ten on one side only. Those in immediate proximity to one another appeared to contract in succession, but there were sometimes two or three points of departure, as it were. Frequently a gap occurs in a series, which evidently ought to be filled up by the reappearance of a collapsed cavity; but after long watching it does not reappear. The time of contraction and expansion of the same cavity varies, but the collapse occurs a little less frequently in active Opalinæ than twice a minute; thus one cavity or other is almost constantly contracting.

So much with regard to the "cavities." The substance in which these cavities are placed is almost homogeneous and colourless, but in the living Opalina a layer of fine highly refracting granules is seen to underlie this (*g*), and some of the fine granules are scattered in the cortical substance. The layer of granules marks out and bounds the great nucleus *n*, which in the living creature is not very different in appearance from the rest of the structure. A very slight addition of acidulated water, however, brings out the nucleus very distinctly, as in fig. 3. It is also rendered very obvious if the Opalina is allowed to die in water, a cavity forming then between it and the layer of granules belonging to the deeper part of the cortical substance; which, as remarked above, might be mistaken for a normal structure, and I think has been by Schultze and Stein in other species. With a good one eighth a very definite structure is revealed in the nucleus after treatment with dilute acetic acid. Minute round nuclei are embedded in its substance exactly as has been so often figured and described in the nucleus of many Infusoria. In the face of this structure it is difficult to understand how the Opalinæ can be refused a place among Infusoria proper, or supposed to be stages in the development of worms.

The cuticle is sharply ridged in this species of Opalina, as in other *true* Opalinæ, which is not the case in the so-called *Bursaria Ranæ*.

The Opalina infesting *Clitellio arenarius* is not unlike the one from *Nais serpentina*, but it is much longer. I have observed that from Clitellio in the Isle of Man. Claparède has found it also, and termed it *Opalina filum*. The anterior extremity presents a broad emarginate area, which is not seen in *O. Naidos*, nor in the very long specimens of *O. Naidos* which I once found in *Lumbriculus* at Hampstead.

Perhaps the most interesting form of *Opalina* is that named *O. prolifera* by Claparède, observed by him in a small oligochæt. This form was seen by him reproducing by transverse fission, but in such a way that *several partially separated* buds remained attached in a chain. In fact, we have here a chain of imperfect zooids, forming a segmented organism, exactly as a segmented worm is formed.

Pachydermon.—In fig. 10 is drawn a structure which I found in the spermatic reservoirs of a new species of Limnodrilus, a genus of Oligochæts allied to Tubifex, and established by Professor Claparède. I found three of these curious structures in each reservoir of a Limnodrilus which had recently undergone copulation, as was indicated by the condition of its copulatory organs; whilst in those which had not copulated I never found them.

If this is a distinct parasitic organism it evidently belongs to the genus Pachydermon of M. Claparède, two species of which he has described in his 'Recherches sur les Oligochétes,' from two specimens of the genus Clitellio (Oligochæta). It is also clearly identical with the appearance figured and described by the late M. Jules d'Udekem from the spermatic reservoirs of Tubifex; but I most assuredly cannot regard any one of these structures as indicating an Opalinoid, as does M. Claparède, for whose opinion I would, however, express the very greatest respect. There is no trace in those I obtained from the Limnodrilus of any distinct cuticle, of any contractile vesicle, nor of a nucleus-like structure; neither does the dense hair-like fringe exhibit any movement like that of cilia, though I watched the supposed organism when within the spermatic sac. M. Claparède observed languid ciliary movements when the Pachydermon was placed in salt water, which is not what we should expect in the case of an Opalina. On the contrary, I think these remarkable structures must be regarded as spermatophors—aggregations of spermatozoa, perhaps cemented and worked into this shape through the secretion and action of the spermatic sac. The dotted appearance of the central portion of the spermatophor is caused by the aggregated heads of the individual spermatozoa; whilst the dense fringe is due to their pendant and interwoven filaments. As indicated in the drawing, the filaments do not all stand out in one direction from the more central portion of the mass, but are crossed and *interwoven*, a circumstance which is not indicated in M. Claparède's figures of *Pachydermon.*

If one of these masses be broken by pressure, very conclusive evidence is obtained that we have not to do with an Infusorian, but that the mass is composed of aggregated filaments such as spermatozoa. No cuticle is ruptured by the pressure, and no differentiation of the supposed sarcodic material is disclosed, but simply a felted structure, composed of innumerable filaments. In fact, everything that I was able to ascertain with regard to these bodies tended to show that they were simply masses of spermatozoa woven together and agglutinated in that very remarkable manner in which we know spermatophors are produced in other cases. Spermatophors have been described in the Polychæta (by M. Claparède himself), and hence we may fairly expect them in the Oligochæta.

M. Jules d'Udekem regarded the bodies which he observed as connected with the formation of the egg-capsule, having mistaken the spermatic reservoirs for an egg-capsule secreting gland. Similarly he described long filaments in the spermatic reservoirs of Stylaria, as destined to strengthen the egg-capsule of that Naid.

In the spring (of 1869) I made some careful studies of the genital organs of *Nais serpentina;* and in the enormous spermatic reservoirs which develop in that Annelid at a late period of its sexual history, I found long coiling filaments, having a fibrous structure (figs. 11, 12). These, no doubt, are identical with the filaments seen by d'Udekem in Stylaria, and I am inclined to consider them also as spermatophors; identical thus in their nature with the Pachydermon of Claparède. In the spermatic reservoirs of *Nais serpentina* were also a few floating spermatozoa and very remarkable flat rhombic crystals of small size, which disappeared on the addition of weak acetic acid. If this interpretation of Pachydermon and of the coiling filaments of the spermatic reservoirs of Nais be correct, we may state, as a new and not unimportant fact, that spermatophors are developed in the following genera of oligochæta, viz. *Clitellio*, *Tubifex*, *Limnodrilus*, *Nais*, and *Stylaria*.

REFERENCES. *Opalinæ.*

Schultze.—Beiträge zur Naturgesch. der Turbellarien. Greifswald, 1851.

Stein.—Die Infusions thiere auf ihre Entwickelungsgeschichte untersucht. Leipzig, 1854. P. 181.

Stein.—Der Organismus der Infusions thiere. 1ste Abtheilung, 1 vol. folio. Leipzig, 1859. P. 91.

Claparède et Lachmann.—Recherches sur les Infusoires. Geneva.

Claparède.—Recherches sur les Annélides, Turbellariés, &c. Observés dans les Hebrides. Geneva, 1860.

Pachydermon.

Claparède.—Recherches sur les Annélides, Turbellariés, &c. P. 88, Pl. IV, fig. 1 and 1*a*.

Claparède.—Recherches Anatomiques sur les Oligochêtes. Geneva, 1862. Plate IV, fig. 12.

D'Udekem.—Hist. Nat. du Tubifex rivulorum. Mem. de l'Acad. de Belg. Tom. 26. 1855.

D'Udekem.—(As to Stylaria). Développement du Lombric terrestre. Mem. de l'Acad. de Belg. Tom. 27. 1856.

Claparède and Mecznikow.—(Spermatophor of Spio). Beiträge zur Kentniss der Entwick. der Chætopoden. Zeitschr. fur wiss. Zoologie. Bd. XIX.

On MICROSCOPIC ILLUMINATION. By JOHN F. HIGGINS, A.M., M.D. Read before the American Microscopical Society of the City of New York, October 12th, 1869.

THE subject of illumination in the use of the microscope is one upon which much has been written, especially in connection with the merits of various accessories, and yet one the understanding of which is far from general, and, moreover, in reference to which microscopists of acknowledged ability and attainments hold most opposite opinions. In endeavouring to ascertain what are really the facts in the case, and by what rule or rules we should be guided, it becomes necessary, laying aside all previous bias or hitherto conceived opinions, to commence *de novo* with first principles, and guided by known and absolute laws of light and vision, in conjunction with a careful, complete, and unwearied series of experiments, work out the much desired end.

Rays of light impinging upon an object, it is well known, are disposed of either, first, by regular reflection, as it is termed, in which the angles of incidence and reflection are the same, and which reflection, bear in mind, in no wise renders such object visible; secondly, by absorption, a manner of disposal not concerning the microscopist; or, thirdly, by irregular reflection, in which each superficial atom of the object becomes itself a focal point from which pencils of light diverge in every direction, and which reflection is totally independent of the form or angle of incidence of the bundle of rays falling upon such object. In truth, the particles or molecules of matter composing the surface of an object become original centres, from which light emanates and is reflected, or, to use here a term far more proper, is radiated in every direction. It is by means of this irregular reflection—a reflection, observe, governed by no law of incidence—a reflection, in fact, in which the object becomes a new sun or centre—it is by this reflection (unhappily termed irregular reflection), and by this alone, that objects are rendered visible to us. Such visibility (laying aside the varied capability of diverse objects for such reflection) depends then solely and entirely upon the reception by the object, from some original source or centre of light, of a sufficient quantity of rays, and except in the intensity of illumination, which will diminish in equal ratio as the size of the angle of obliquity of such surface to the direction of the rays is diminished, is totally independent of the angle of incidence of such rays, or of their parallellism, convergence, or divergence.

In the microscopic observation of transparent objects we have at our disposal and make use of two principal modes of illumination—the one central or direct, in which the illuminating pencil is a continuation of the optical axis of the instrument—the other oblique or eccentric, in which the pencil of rays forms an angle of greater or less degree with such optical axis. Various microscopists, writing upon the subject, advise, some, the use of parallel rays; others, the use of convergent rays, and, again, others the use of the one or the other, as on trial answers best. This variance of opinion among microscopists undoubtedly arises from the fact that, as we have noticed in speaking of the cause of the visibility of objects in general, outside of the microscopic field it really is immaterial, in so far as the mere visibility is concerned, and makes not the least difference whether your illuminating pencil is convergent, parallel, or divergent (an augmentation of the intensity in each case, however, being necessary); and persons having become accustomed to the one or the other, have taught and given out to the world that such was the true mode of illumination. Those advocating the use of either as on trial may be found expedient, testify by their indecision, I regret to say, to a want of that nicety of discernment, so eminently needed in microscopic research.

And, firstly, of central or direct illumination. That a slightly convergent pencil of light, as *e.g.* obtained from the concave mirror, is the one most suitable for central or direct illumination, is a truth that I think both theory and trial will fully support. That such is the case arises from the fact that, by a convergent pencil, we are enabled to condense a greater amount of light upon the object, and so give greater sharpness and distinctness to the impressions made upon the retina. And, again, in a pencil of rays thrown upon an object, a part only are irregularly reflected, thereby rendering the object visible, the other, and frequently a major part, passing by and around the object, and falling upon the surface of the objective, and then and there becoming obedient to all the laws of refraction the same as those emanating from the object itself. Now, the course of a convergent pencil of light in its refractions, while passing through an objective, differ widely from that of the divergent pencils emanating from the object—of course much more so than that of parallel rays—and consequently the dimness, obscuration, and want of definition resultant from interference, and from other rays entering the eye besides those emanating from the object, is more or less removed. It is with reference to this very point that we are instructed that our pencil of rays should, in as far as possible, be only of a size equal to the illumination of the object; and no greater proof can be had of the interference caused by rays thus entering the objective, other than those emanating from the object, than the far more excellent definition observed on the interposition of a diaphragm in close apposition to the slide, with an opening only sufficiently large for the illumination of the object. A series of carefully conducted experiments, and an impartial judgment based thereon, also prove the choice of a moderately convergent pencil for central illumination to be incontrovertibly correct.

The qualification of the convergence is necessary, since in the use of convergent pencils of large angles we are *pro rata* departing from direct and verging upon oblique illumination. Assuming, then, the correctness of what both theory and practice teaches, the microscopist will choose for central or direct illumination a convergent pencil of rays of, however, not too great an angle, and may on option stop out, and probably with good effect, the central or rays parallel with the optical axis of the instrument. The focal point of the illuminating cone, however, is not that which the object, as at first sight would seem natural, should occupy. Careful observation will show that when the object is situated somewhat within the focus of the illuminating pencil, the best result is obtained, and this separation or distance between the object and focus of the illuminating pencil increases in inverse ratio with the power of the lens.

Next as to the most suitable means for effecting such illumination. Universally almost, it may be said, the concave mirror is that which is furnished by the maker for such purpose. The use of this, however, is objectionable, by reason of the double reflection always attendant upon the use of any mirrored surface, of no matter what form, viz., the primary or greater reflection from the silvered surface of the back, and the secondary or lesser reflection from the surface of the glass itself, this double reflection causing two sets of rays to enter the objective, in addition to those emanating from the object, which militate greatly against the sharpness of definition observable on a more proper illumination. Of this one can easily satisfy oneself, more especially when working with high powers. The various achromatic condensers, kettle-drum illuminators, and other addenda, condensing light at a large angle upon objects, do not here claim consideration. In the use of such accessories we are not employing central or direct illumination; but, on the contrary, oblique or eccentric—a mode of illumination which subsequently will claim our attention. It may, perhaps, be as well, however, here to mention that such oblique illumination differs from oblique illumination as usually understood, in that it is omni-lateral, or, in other words, oblique from every side, whereas, in general, when speaking of an object as viewed by oblique light, unilateral, or light falling upon the object from one side, is meant.

The means or instrument best adapted to its accomplishment is, without doubt, the achromatic lenticular prism of Abrahams, or that of Messrs. Powell and Lealand. This consists of a rectangular prism of flint glass, with one of its equilateral sides rendered concave for the reception of a double convex lens of crown glass, and mounted in such a manner that to it every possible position can be given. In the prism, as constructed by Abrahams, of Liverpool, the focus of the lens is 4 inches. In that as constructed by Powell and Lealand, 2½ inches. I do not know that there is any choice between them; the selection should depend more on the convenience with which the one or the other can be used, in connection with the stage of the instrument. Although admitting, if desired, of being mounted upon the mirror arm of the microscope in lieu of the mirror, still you will find it preferable, and affording greater facility of manipulation when mounted upon a separate and independent stand, consisting simply of a perpendicular rod with foot, as in the case of the ordinary bull's-eye condenser. The English mode of mounting, however, upon such stand is not as perfect as it might be, and is far excelled by that of Mr. George Wales, a member of our society. In use parallel rays, and if by artificial light, rays rendered sufficiently parallel for the purpose by removal of tne light some twenty or more inches from the prism, in preference to the interposition of a lens or lenses, with a view to the same effect, are caused to fall upon the surface of the lens, and being converged then upon the diagonal surface of the prism, total reflection takes place, and emerging at right angles from the remaining side of the prism, the rays are brought to a focus at a point beyond or distant from the object, dependent upon the power of the objective. By the combination of flint and crown glass, achromatism is secured, and the total reflection which takes place from the diagonal side of the prism not only enables us to avoid the secondary reflection present upon the use of the mirror, but, as is well known, gives a purity and intensity of illumination unexcelled.

By such mode of illumination one is enabled, *e.g.* with a 1-5th in. objective, to bring out the markings upon the Podura scale, with their central light streak or interspace with a sharpness, clearness, and depth of definition unexampled, and leaving nothing more to be desired.

A word in reference to a want of care in central illumination before leaving this part of our subject will probably be excused, and may not be out of place. I refer to the too frequent use of a beam of light, not exactly centric, or, in other words, not truly coinciding with the optical axis of the instrument. This is easily determined by the equality in width of the "diffraction band" encircling the periphery of the object which demands attention, and should not be disregarded.

To those unprovided with the aforementioned prism, and indisposed to incur the expense (about $20), I would recommend, as the next best adaptation, the use of an achromatic converging lens, of a focus of from 1½ to 2 inches, interposed between the mirror and the stage, carefully centred with the optical axis of the instrument, and used with the plane mirror. A double combination of two plano-convex non-achromatic lenses of like focus, similarly placed, is advantageous, but will not equal the achromatic combination.

In oblique, or as usually termed by Continental microscopists eccentric illumination, we have to consider, not only light as radiated from the object, together with the refractions to which the pencils of rays are subjected in their passage through the combinations of the objective, but also are compelled to bear in mind the effects of shadow and relief, as influenced by the parallelism or convergence, as the case may be, of such illuminating pencil. A beam of light falling obliquely upon an object causes an obscuration, darkness, or shadow (technically termed umbra), to be formed upon the opposite side of the object, the depth and intensity of which depend *ceteris paribus* upon the intensity of illumination. Now, it is found that the line of demarcation between the shadow and light is not a marked, well-defined line, but that a lesser shadow, termed penumbra, shades or tones down the separation. This penumbra is greatest, and in just so far lessens the size of the shadow, when a convergent pencil of light is brought to bear upon the object; and, therefore, *à priori*, we would infer that illumination by parallel rays would be preferable for microscopic purposes.

In fact, Mr. J. B. Reade, President of the Royal Microscopical Society of London, has but only recently published a series of articles claiming the perfection of microscopical illumination to consist in the employment of parallel rays. His results, however, are undoubtedly due—1st, to the use of a reflecting prism in place of the mirror, whereby he loses or avoids the secondary reflection always present on the use of the mirror, and which is so antagonistic to true definition; 2ndly, in having been accustomed undoubtedly to the use of the accessory known as the achromatic condenser, until he himself invented and, if I am rightly informed, patented a far worse device, viz. the kettledrum illumination. Now he informs us that such accessories are wrong in both principle and use, and can hardly find words sufficiently expressive of the great superiority of his present mode of illumination

This mode, he informs us, consists in the use of a prism, rectangular or equilateral, in place of the mirror; and he ascribes the superiority of the results which are thus obtained to the use of parallel rays thereby thrown upon the object.

I have used Mr. Reade's prism, and can bear witness to its superiority over mirror, achromatic condenser, or kettle-drum illuminator, but do not agree at all with him as to the cause of the result; on the contrary, both reason and trial demonstrate most markedly to the unprejudiced observer that, by the addition of a lense to said prism, constituting at once the achromatic lenticular prism (as recommended above when treating of centric illumination) we gain most decidedly and then really seemed to have at last reached what may be now truly termed the very perfection of microscopical illumination.

In returning to the consideration of the shadows of objects, as cast by oblique light, we can at once see the reason for this. Although, as previously stated, the penumbra or lesser shadow, interfering in proportion to its extent with the umbra or true shadow, is less when parallel rays impinge upon an object, than when convergent rays are thrown thereupon; still, it will be found by experiment outside the microscope, that a moderately convergent pencil of light, owing, in a great measure, undoubtedly, to the greater intensity of illumination, will cast a blacker, sharper, and better-defined shadow, though somewhat smaller than is obtained when a parallel pencil is used to effect the same; and identically the same result is present when used in microscopical illumination. It may be said that I alone am able to see this; but the testimony of many able, experienced, and well-known microscopists bear me out in my assertion.

It is but recently that I had the pleasure of exhibiting this mode of illumination to such men as Messrs. Frey and Mason, members of our society; and still more recently at a reunion at my house, to Mr. Ward, Drs. Rich and Allen, and Professor Edwards, all members of the American Microscopical Society of the City of New York; and by the additional presence of Professor Hamilton Smith, of Ohio, the knowledge of whose skill, experience, and judgment is not limited to the microscopists of this country alone, but is equally known and accredited upon the other side of the water, I felt yet more strongly that the truth, either *pro* or *con*, as it might happen, still that the truth, and nothing but the truth, would be elicited. It is, I say, to such men that I have had the good fortune of exhibiting this or, as Mr. Reade would say, "my mode of illumination," and the testimony of each and all has been outspoken at once freely in its favour.

Professor Smith, on viewing the resolution of the *Surirella gemma*, with a $\frac{1}{15}$ immersion made for me by Mr. William Wales, of this city, had no hesitation in acknowledging the great beauty and superiority of the same; saying that he had never seen it better. When we take into consideration that, with Hartnach of Paris, with Lobb of London, and various other foreign celebrities, Professor Smith has critically viewed this same shell, as yet the most difficult of test objects; (for Professor Smith denies a true and indisputable lining to *Amphipleura pellucida,* although these very men attempted to show it to him); and then hear him tell us that the exhibition of this shell, as shown by the $\frac{1}{15}$ immersion was superior to anything he had yet seen—can we wonder for a moment at his additional remark that it must be due, more or less, to this mode of illumination?

A $\frac{1}{40}$ immersion of Mr. Wales's make, the property of Dr. Rich, was also exhibited; and although it was impossible, by reason of not having an object provided with a cover of the necessary thinness, to judge fully of its merits, still it was observed that the quantity of light furnished by this simple mode of illumination was amply sufficient.

A recapitulation is hardly requisite, as the inference is probably already drawn that for oblique or eccentric illumination a moderately convergent pencil of rays is that best adapted to the purpose, and that the means of obtaining such pencil is through the agency of the same achromatic lenticular prism, as recommended, and described when speaking of central or direct illumination. To those provided therewith and desirous of trying for themselves the effect of parallel rays, and not wishing to purchase a separate prism for such trial, I would state that, by placing the lamp or source of illumination at a distance from the lens, equal only to the focal distance of said lens, the emergent rays become parallel, and may be used, especially with low power, with great satisfaction and pleasure. The intensity or amount of illumination thus obtained is far greater than that received from the plane mirror being even sufficient for a $\frac{1}{15}$.

It may not be amiss, in conclusion, to consider briefly the source of illumination itself. The St. Germain student-lamp is, *par excellence,* the lamp of the microscopist. The chimney of this lamp should be encircled with an external metallic tube, having, at the point coinciding with the constriction of the glass chimney, two openings placed opposite to each other. These openings should be one and a quarter inch in diameter, and proceeding from each a tube two inches in length, the one at right angles to the larger or main tube, and the other forming an angle with the lower part of the same of sixty degrees. These projecting tubes are adapted to and should receive a paper or cardboard-tube of about 12-15 inches in length. The mechanical connection of the above with the lamp itself should be such that, on elevation or depression of the lamp, the whole arrangement should move at the same time and in unison. Such adaptation and arrangement virtually is that exhibited by me at a previous meeting oi the society, and is manufactured by Westmon, of Centre Street.

By this conversion of the source of illumination into, as it were, a dark chamber, with simply an outlet or conduit for the passage of the rays to the mirror or prism, you avoid not only the lighting up of the room with rays thrown from your lamp in every direction, and thus receiving upon the retina a mass of confused impressions which interfere greatly with observation (and in this connection it will be at once seen that the use of a hood attached to the eye-piece with a similar view, always disagreeable and inconvenient to the microscopist, is removed), but in addition are saved all trouble of so moving or manœuvring your lamp as to avoid the accident of incident rays upon the object, an occurrence totally destructive of and annihilating all attempts at definition or resolution.

If the edge of a flat flame as preferred by some is desired, by means of a diaphragm, with slit, affixed to the inner surface of the metallic chimney, this can be obtained. For regulating the intensity of the illumination, the use of the diaphragm placed beneath the stage for such purpose by the makers is not to be recommended. The alteration of the distance of the lamp from the mirror or prism, inasmuch as the intensity of light issuing from a luminous point diminishes in the same proportion as the square of the distance from such point increases, will be found a preferable way of effecting the object. Dark-ground illumination, a variety merely of oblique in which only rays emanating from the object are acted upon—the illumination of opaque objects, and illumination in connection with the polariscope, I must pass over or reserve the consideration of for possibly another paper, as I fear I have already trespassed too far upon your patience and time.

On the Relation *of* Assimilation *and* Secretion *to the* Functions *of* Organic Life. By J. Gedge, M.B. Cantab., M.R.C.S.

Physiology suffers much from the want of definite nomenclature. Organic processes specifically identical appear generically distinct, in consequence of the separate terms used in their description.

It is on this account that I would inquire into what is meant by secretion and assimilation, with a view to show that, if our modern notions of physiology are correct, these processes can no longer be described as though they had nothing in common with the ordinary functions of organic life.

Secretion as performed by the gland—the station where this function is somewhat isolated—will be first considered. Cells are the acknowledged agents in this process, as shown by the indiscriminate use of the terms "secreting-cell" and "gland cell," and as these histological units when they have arrived at maturity all share alike—all perform a similar office—we may limit our consideration of the function of the gland to that of a cell.

The old notion of cells was that they were vesicles of varying size, having various contents. But now it is allowed, even by those who will go no further, that the vesicle must, during at least part of its existence, be furnished with a nucleus which has definite chemical and physical characters. In this country, as is well known, we are no longer fettered by the vesicular notion of the histological unit: no longer imprisoned by the constant presence of a cell-wall. That the cell should have been universally considered to be vesicular in the early days of histology is not remarkable, when we consider the comparative ease with which such peculiar cells can be demonstrated: the imperfection of the optical instruments then in use not enabiing them to study carefully any but the large histological units met with in the vegetable kingdom. Furthermore, it cannot be regarded as surprising that the cell first described should have been considered the type, and that the attempt should have been made to reduce all other cells to the same formula. Besides, in animal cells where no cell-wall existed, an apparent one could often be made by the action of chemical reagents. Toy-cells have been constructed by Mr. Rainey and others, having the conventional vesicular character with the indefinite cell-contents simply by mingling certain chemical compounds.

The cell-wall then must be understood to be simply a complication of cell-structure, which being always absent in the healthy gland-cell and almost all other animal cells, cannot fairly be regarded as a part of the histological unit. But the discussion of this superadded structure, and its histological homology, must be reserved until we have entered in greater detail into cell-physiology.

The term *cell* is here used as a technicality for that elementary part or histological unit, which consists of *nucleus* and *plasm;* the nucleus being always distinguishable by those chemical or physical properties first shown by Dr. Beale to be possessed by it in common with the germinal matter of the ovum.

Formerly the term nucleus was used very vaguely, size being generally the only diagnostic character. Thus we find even Virchow continually speaking of cells, when correctly (as we now know by the nuclear test) he ought to say nuclei.

When we examine the tissues of the embryo, we find them crowded with nuclei, which have grown from the germinal matter of the ovum, and if this examination takes place when the parts differ from one another in form rather than texture, we find an aggregation of nuclei separated only by a small quantity of interposed plasm. The mass may be broken down into what have been called cells, but there is no true specialisation into these structures; the nuclei bring away the surrounding plasm only in the same way that the stones in a wall that has been thrown down bring away adhering mortar. Soon, however, in the tissues that are to become simply cellular the embryonic plasm is, for a given distance around the nucleus, so acted upon—perhaps by insterstitial addition—that a small portion is localised. This localised unit of nucleated plasm is the typical cell, as we find it throughout its active life in the gland. But in many epidermal structures, though best in the cæcal extremities of certain glands among the lower animals, we find this same process of cell-formation going on through life. Glancing at such a structure in the deepest layer, we again see the nuclei crowded together, at first hardly separated, but as we ascend, gradually getting more and more spaced by intervening plasm, which as gradually alters in texture until we have specialised histological units. And after the cell, has arrived at maturity, as surely as the relative proportion of the nucleus to the plasm diminishes, so surely does the working power of the cell decrease.

Now, having completed the anatomical life-history of the cell, it becomes necessary to consider its function, secretion. Formerly this process was regarded merely as separation.

The blood was supposed to contain all the complex substances poured out by different glands, and it was simply the duty of the gland-cell to choose from the blood, as it flowed by, the particular substance that it was required to separate. Now, however, the gland-cell is known to fulfil a more important office. Formation rather than separation, would appear to be the function of the gland-cell. It would seem to attract the nutrient material in its vicinity, and with unknown machinery to make use of chemical affinity to construct elaborate compounds. But we cannot gauge the action of a gland by merely analysing the secretion poured out, for there can be little doubt that all glands compound for the blood as well as for their own secretion. This has been shown to be the case in the liver, and we have no right to consider this gland as an exception. But besides thus refunding secretions of a complex nature, we must remember that an equivalent debasing action must take place, and as the cell itself must be restored, more fuel than is at first sight apparent is required. In some cases these debased materials are poured out in the gland's secretion; in others they are returned to foul the blood until it has passed through one of the refining glands.

We acknowledge that all this work emanates from the cell. Can we no further localise it? I think we can. In certain cases we are able, from the insolubility of the secretion, to observe the position it occupies when it is first formed. In such cases we see the secretion within the plasm. The exact site it there occupies is not perhaps of much importance, since we are aware of the existence of outward currents in the plasm. Still, certain observations that will be mentioned directly afford evidence of its being formed in actual contact with the nucleus. If this be so, it matters little whether in one case the secretion is seen to be intranuclear, in another adnuclear, or in a third circumnuclear, for it must be remembered that our nuclear matter is shut off from the plasm by no partition wall. In studying the physiological anatomy of the mammary gland during the secretion of milk, I think I have been clearly able to understand how some of the fallacious notions concerning the cell and its function have arisen.

In all cases when lactation first sets in, we find in the milk certain structures known as colostrum-corpuscles. These have been correctly regarded as altered gland-cells, and they can be shown to be vesicular; to consist of cell-walls and cell-contents, with a nucleus or a remnant of one. But if we examine the secretion when the gland is in full working order, we find no colostrum-corpuscles, no cell remains of any kind. On examining the gland itself this may be accounted for.

Here, as elsewhere, the more quickly grown the cell, the more fluid the plasm. The plasm of these cells is but little firmer than treacle, and offers but little resistance to the outward passage of the newly secreted globules; and it is probably in this outward passage that each globule gets its water-proof albuminous coating—the so-called haptogenic membrane. The nature of the colostrum-corpuscles may now be understood. They are the old superficial cells which had long been dormant, and become rigid with old age. They had set to work to perform their function to the best of their ability, but eagerness would not make up for incapacity, and though choked almost to bursting with their anxiety to succeed, they had to yield their places to their more active juniors. Virchow has jumped to the conclusion that milk is always secreted in colostrum-corpuscles, only that in the acute action of the gland the cell-wall ruptures, and the mass is more rapidly broken up. I should like to ask the learned Berlin professor whether it is usual in his histological experience to find rapid disintegration of young and active nuclei?

In the embryonic formation of adipose tissue we may with ease observe the adnuclear secretion of oil; the globule continuously enlarging until the plasm is compressed into cell-wall. This latter structure is here, as in the old mammary cell, the remnant of plasm, and wherever it occurs, however far separated from the nucleus, it may still fairly be considered as histologically homologous with plasm. But it is not always due to the outward pressure of cell-centents, for, as we see in the vegetable, the plasm may be in part absorbed (*i.e.* metamorphosed), and afterwards thickened by linings. The plant, in fact, has no other means of laying aside such excrementitious secretions as lignin, except by secondary deposit. The cell-wall is, moreover, sometimes altered by interstitial changes where the plasm has never been altered into cell-wall.

In the pathological condition known as fatty degeneration we may often study the secretion of oil with great ease. Lately I met with a case of this disease in the liver, where instead of each cell containing a multitude of small oil-globules shut off from one another by plasm, the oil was secreted in each cell in one drop.[1] So that in different cells the different stages in the formation of adipose tissue were exactly imitated. In some of these cells no plasms existed except in the form of cell-membrane, and as the nucleus was driven to the surface of the cell, the resemblance to the adipose vesicle was com-

[1] I have since met with this variation again, and I am inclined to think that it is caused in part by rapid, continuous secretion, and in part by a difference in the proportion of the mixed fats secreted.

plete. I have observed too in the early stages of fatty degeneration of muscle that the oil-globules are disposed around the nuclei, whence they spread so as soon to occupy an internuclear position in the axial canal, which is often as distinct in the fibres of the heart as it is in the muscles of insects.

These cells which thus retain their secretion ought no longer to be spoken of simply as cells. They are glands and reservoirs all in one, and I would propose to distinguish them from other histological elements by calling them *store-cells*. Structureless membrane or cell-wall is often formed out of the original plasm of a cell during the process of free-cell-formation, and such structures as long as they remain entire are known as *cell-capsules*. These bodies get more and more common as we descend in the animal and vegetable kingdom, but among the higher forms they will probably be found to be far more common than is at present believed, particularly in connection with the development of those glands which at present are so daringly described by a class of stratification-physiologists who evolve diagrams with a levity approaching to the imponderability of their consciousness.

Homogeneous or structureless membrane is also found investing the muscular fibre, and here, too, it must be regarded as a remnant of plasm. Any one who has carefully studied the formation of muscle in the nucleated plasm of the embryo cannot have failed to observe that the sarcous matter is stored much in the same way as oil in adipose tissue. In a very similar manner is elastic tissue formed. At a comparatively late period no trace of this tissue can be found. Then, within the plasm beside the nuclei we recognise its first formation. Soon the fibres increase in size, and the nuclei get further apart. At length the nuclei dwindle, and many disappear, though some are left among the sheathing to perform the very slight office required of them in so very permanent a tissue. These observations lead me to conclude that sarcous matter and elastic substance are as much secretions as the oil of adipose tissue or the calcareous matter of shell.

But let us return to the cell. At present we have only treated of secretions insoluble in the fluids of the body, and we have been obliged to give such prominence to this class in consequence of the obvious difficulty in getting evidence of the whereabouts of other cell-secretion. Still, we occasionally get an opportunity of seeing bile in the plasm of the liver-cell, and sometimes we observe other secretions exuding from the plasm of cells.

From what I have seen of the so-called "diaphanous corpuscles" or "colloid bodies," as well as from carefully-recorded observations, I think there can be but little doubt that these bodies are globules of albuminous secretion, and from having seen them around a naked nucleus, and at other times exuding from plasm, I think we may fairly consider that here we have evidence again of intra-nuclear secretion.

It is by a process similar to that of secretion that nutriment is first received into the system. It is through the spongiole of the rootlet and the villus of the intestine, that the highest plants and animals gain means of adding to their tissues. In each position we find the same process going on. Cells, arranged to filter and secrete, invest these structures, and are from time to time renewed as they become clogged and inefficient. In the spongiole the old and useless cells are left to coat the rootlet as it lengthens. But such an arrangement would obviously be ill-adapted to alimentation in animals, and consequently the inefficient cells are shed, and no elongation of the villus takes place. But this first process of elaboration is not alone sufficient in the higher animals; much more remains to be done. A great tubular gland, having its distal extremities stretching into these villi, receives this secretion, and retains it for further elaboration. This gland is furnished with free and naked nuclei, which elaborate the secretion in which they live, and at the same time reproduce their kind, growing and dividing, by likening to themselves the surrounding fluid. This is assimilation.

These nuclei, however, though so long as they remain in this gland they never invest themselves with plasm, still perform the function of secretion by forming new products in the surrounding liquid. Some of these nuclei pass out with the fluid they have elaborated into another tubular gland, and now no longer as lymph-corpuscles, but as white blood-corpuscles in a different medium, they commence a new life, and form another secretion. Whether this secretion invests them as it does in the lower vertebrates, or whether it is detached as soon as formed, depends, perhaps, more on rapidity of secretion than on the nature of the substance secreted. This second secretion is the formation of the red blood-corpuscle.

I have now attempted to show that nuclear growth is the true process of assimilation, and I have brought evidence to show that the cell-secretion is intra-nuclear, and that many tissues are mainly composed of stored secretions, histologically homologous with the secretion poured out by glands. In doing this I have foreshadowed what it is now necessary for the completion of my scheme to state distinctly, viz. that I consider the formation of plasm as distinctly a process of secretion as the storing up of oil. The formation of plasm is surely as much a secretion as the formation of fibrin. Just as the nucleus assimilates nutrient material and proliferates, so does the lymph-corpuscle; just as the nucleus secretes embryonic plasm, so does the lymph-corpuscle secrete fibrin; just as the nucleus secretes oil, so does the lymph-corpuscle secrete the red blood-corpuscle.

Certain minds will feel a repugnance to these notions, because substances of small chemical complexity are associated as histologically homologous to substances of great chemical complexity. But who can draw the line between silica, calcium carbonate, urea, oil sugar, albumen, and other protein compounds? Some, again, may object on the grounds of physical complexity. But surely we may see a certain gradation from slime to silk, to byssus, to elaetic fibre, to muscular fibre.

I have persistently marked out two groups of secretions—two broad classes; but as they go on in many cases synchronously, a difficulty in nomenclature arises, for so little is known of the directive agency manifested in secretion by means of which one secretion passes out at one side of a cell to go back into the blood, and another exudes from the other side to leave the gland, while at the same time plasm continues to be secreted, and the nucleus to grow. If my scheme of reconciliation be a natural one, it should be of the same value to the pathologist as to the physiologist. But the morbid variations of physiological functions require for their explanation a far more extensive knowledge of physiological chemistry than we at present possess. Still, in cases where we have only an exaggeration of healthy processes, we may, at least, expect to find fresh machinery. Prominent in this class are the abnormalities of assimilation manifested by nuclear growth. If any living tissue be torn, or otherwise injured, local congestion results. Assimilation, under these circumstances, goes on with unusual rapidity, in consequence of the increased supply of nutriment. The healthy nuclei surfeit themselves on the nearly stagnant blood, and suddenly grow and multiply. Soon we find a swarm of nuclei on the spot. These set to work like navvies after an accident to clear away obstacles, or fill up chasms, as the necessities of the case may require. These navvies, it is said, are brought by vessels, and disembarked on the spot, by orders emanating from a nerve-station; but with this view I am not disposed to agree. The nuclei are, however, there, and they may be seen working away at the removal of a piece of dead bone in necrosis, or excavating through elastic tissue in phthisis, or filling up a wound by granulation, reconstructing gradually a healthy tissue. When a large number are thus congregated together, the nutriment being handed on from the rear, we might expect, if the phalanx were thick, that those in front would come badly off, and such actually seems to be the case, for we find the front rank of ill-fed nuclei falling away as pus-corpuscles. During this rapid nuclear growth, we have, as a correlate of assimilation, abundant heat, whence the name inflammation. But under certain conditions this nuclear growth is arrested. The material that would have continued to form nuclei, and kept up the outward flow of pus, is now formed, by the nuclei there assembled, into a poisonous secretion, which passes into the blood. The presence of this blood-poson is known to the surgeon by the effect it produces as it flows through the nervous centres, causing rigors which remind us of malarious fever. The poison passes on, but not often without affecting the white blood-corpuscles—the fibrin-secreting nuclei. It is not generally, however, until it reaches the refining glands, that its effects become very serious. Then, if the dose be strong, the gland strikes work. The nuclei cease to pour out their normal secretions, and abnormal nuclear growth commences. At the same time a change takes place in the plasm, producing the appearance known as "cloudy swelling." This is soon seen to be due to the presence of fatty particles, and indicates an acute form of fatty degeneration, which may be caused by the nucleus playing upon the surrounding plasm, using a part for assimilation, and secreting the remainder as a less complex chemical compound. The degeneration, or metamorphosis proceeds, until at length the proliferating nucleus is free, nothing being left of the plasm except a little fatty detritus. This is the diffuse suppuration of pyæmia.

The physiological views that I have been advocating appear to advantage when we consider the life-history of cancer. These malignant growths I would arrange in three divisions. First, the gelatinous form, consisting simply of nucleated plasm, which may be an offshoot from any tissue, since it represents the primal condition of all tissues. Next, those groups composed of specialised elements, bearing no definite relationship to any normal structure. Here we have a complete remodelling of tissue on a new design. The remaining division includes those growths which bear a recognisable resemblance to the tissues whence they spring—a resemblance to the tissue in its rudimental state, but still at a stage after differentiation has commenced. From this last group, which is the least malignant, we pass by an easy gradation to simple hypertrophy.

The order in which I have arranged my divisions, like all arrangements in straight lines, is unnatural; but the groups seem to me to include all those growths known as cancers. But in neither of the three classes, speaking broadly, do we find any structure histologically in advance of gland-structure. And as these masses do not occupy themselves with storing oil, or sarcous matter, or elastic tissue, we are led to believe that as they grow they pour out a large quantity of fluid secretion into the blood, the effect of which we have some means of judging.

A growth of this nature makes its appearance in a healthy man; it may create no interference from its position, and its size may be inconsiderable; still we see disturbances in the nutrition—or rather assimilation and secretion—of every part of that man's body, and he wastes and dies. Are we to believe that death results from the drain on the system? I think not. The malignancy is not in proportion to the size of the tumour, and in chronic suppuration we have a far greater drain on the system, and yet life is not so quickly destroyed. Yet in this exaggerated nuclear growth there must be activity. Can one help suspecting that we are dealing with a poison-gland specially provided for elaborating and letting loose into the blood a secretion which produces mal-assimilation throughout the body?

Let us invert the picture, and imagine instead of a cancer an ovary or testicle. The blood flows through these glands, bearing with it secretions from every tissue of the body. Is is difficult to understand how the resulting product of the ovum and spermatozoon may inherit the peculiarities of both parents? The notion is, I think, preferable to Mr. Darwin's modern theory of Pangenesis.

It will be seen that I think it fair to suppose that every tissue has, in different degree, the function which we have assigned to the ductless glands. These vascular secreting-stations, I consider, ought to take a position physiologically (not histologically) by the side of adipose tissue. Fat, we know, is ordinarily of little use; but each store-cell of which it is composed has a permanent nucleus, and in times of dearth we find it unlading and furnishing fuel for further correlation. The ductless glands are principally remarkable for having the machinery for doing a vast amount of work; yet it has been shown we can get on without them. They are stations less necessary to the vascular system than ganglia are to the nervous system, but more necessary than lymphatic glands are to the absorbent system; they might to some extent be compared to London fire-engines standing ready for use in case of emergency, only requiring the order to get up steam. I should be curious to know how a patient without a spleen would fare—say in typhus fever.

I have more than once spoken of machinery, always using the phrase for nuclear matter; each nucleus would seem to be a battery, the construction of which we have at present no means of finding out. We see how forces can be correlated with its force, and we are thus able to study its affections; but we know no more of its construction than of that combination of matter which sustains chemical action, furnishing heat and magnetism in the earth's centre. Thermo-electricity and the hypothesis of Grothuss start us with ideas of what goes on during the incubation of an egg, but fall short of furnishing us with a parallel phenomenon. Physicists boast of our being able to construct such organic secretions as alcohol, oil, and urea, but this is quite beside the question. The day may come when the chemist in his laboratory may out of stones make bread, but I see no reason to think that he will even in that day do it with other than the comparatively clumsy apparatus with which he has constructed his alcohol and urea.

High-Power Definition *and its* Difficulties *and the* Visibility *of* Diatomaceous Beading. By G. W. Royston-Pigott, M.A., M.D. Cantab., M.R.C P., F.R.A.S., F.C.P.S., formerly Fellow of St. Peter's College, Cambridge. (With Pl. X.)

It is almost superfluous, in the first instance, to advert to the errors so well known to be developed by excessive amplification.

Indistinctness, haziness, confusion, and blurred outline and commixture of adjacent images, obliterating all that beauty or brilliance which is so charming a reward for carefully conducted observation.

In the following papers I hope to have the honour of describing some results of experiments carried forward for many years, which have for their chief object the increase of power and the advancing correction of residuary aberration, combining, also, a deeper penetration imparted to microscopic objectives, and greater distance between the delicate object lens and the object observed. In the following objectives of focal lengths—

1 inch	$\frac{1}{2}$ inch	$\frac{1}{4}$ inch	$\frac{1}{8}$ inch
$\frac{3}{10}$ "	$\frac{1}{10}$ "	$\frac{1}{80}$ "	$\frac{1}{120}$ "

were measured intervals between the front lens, or rather its setting, and the covering glass; and in proportion as the tube is lengthened in general, this small space diminishes as the power so gained increases, and except the covering-glass is extremely thin with the higher powers, the objective dangerously comes down with pressure upon it in the act of attaining distinct vision.

"Penetration," or the actual depth to which distinct vision reaches through a transparent object, is one of the most essential conditions for correct observation, and the extreme shallowness of the highest powers is one of the most formidable difficulties encountered in the interpretation of microscopical phenomena. The gorgeous brilliance, depth, and contour of illuminated objects (as the heads of insects) under a 3-inch objective form a striking contrast with the effect of an eighth, which presents to the eye sections formed by the focal planes of vision.

As power is increased, of course the front glasses approach the object until with the 1—16th the separation diminishes to three or four thousandths of an inch. Difficulties of observation now increase. A great variety of interesting objects already collected cannot be seen at all. Opaque objects are out of the question. With such a power the front lens is generally in contact, or nearly so, with the protecting covering glass. It is almost impossible to insinuate an illuminating ray between it and the object.

But even with the best arrangements difficulties of definition in individuals resemble the personal errors of astronomers eminently depending upon personal idiosyncracy. The actual powers of vision in the abstract are still debated. Certain persons within my own acquaintance have distinguished objects of a size and at a distance quite incredible—as the positions of Jupiter's satellites and the time by a distant clock.

In the microscope, to an eye adjusted to ten inches distinct vision, at a power of 1600 diameters, the diameter of a black spot magnified 1600 diameters, whose image presents the visual angle of $2\frac{1}{4}'$ is found by the expression,

$$\frac{d}{10} \times 1600 = \sin 2\tfrac{1}{4}' = \cdot 0007272,$$
$$\text{or } d = \cdot 00000454;$$

which is about the 200,000th part of an inch; but it is extremely doubtful whether this power can *define* so small a quantity as this, therefore it is safe to assume that an object presenting so small an angle as two minutes in the field of a microscope cannot be defined clearly; as to whether its outline be round or square, is totally beyond the powers of human vision.[1]

But such minute particles may be capable of casting a shadow, crescentic or otherwise, and a continuous line of them, if sufficiently brilliant, would sensibly affect the retina, but without admitting ocular disintegration. Further, parallel lines of such particles crossed by parallel lines slightly inclined to the former, and nearly in the same place, give rise to diffraction of a singular character.

I viewed a set of parallel wires against the light, behind which other lines, also parallel, could be seen; rotating gradually, the first set, suddenly at an angle of 15 degrees, the latter appeared broken and no longer straight. (Pl. X, Fig. 17.)

Some difficulties of definition may be thus illustrated:—If very fine lines cross each other at a general parallelism amongst each other, the slightest deviation from equidistance causes beautiful waves of interference, taking an infinite number of forms. The more nearly the two sets become parallel, the wider apart and broader appear the waves of interference. Figs. 1, 2.

But if these sets of lines are composed of spheroid particles, in contact and overlying each other, the difficulty of a distinct separation is enormously increased, overlapping, then causing the chief shadows.

If the aperture of the object-glass be diminished to 50°, two beads, exactly coinciding, transmit so little light as to become dark points and a series of intersecting lines. The dark parts form a black wedge-like marking, as seen in the Podura. Figs. 15, 16.

The dark lines or markings of interference become longer as the two sets of lines cross at a smaller angle. Figs. 1, 2.

When the lines intersect at any large angle, as 25° and

[1] Nobert's famous lines ruled on glass 112,000 to the inch can only be defined with Powell and Lealand's $\frac{1}{16}$th immersion lens.

upwards, all waviness disappears; the waves are displaced by chequers or lattice work.

Again, if the shadows all fall on one side a series of crescentic shadows in miniature simulate straight lines, and intersecting lines of shadows exhibit squares or hexagons according to the special arrangement. Figs. 4, 5.

By no conceivable torture of the rays can rhomboidal or triangular shadows really be produced by rows of spherical beads, yet such are often seen with high powers, as the 1-12th or 1-16th, falsely corrected.

If beads be arranged angularly, like a pile of shot, hexagonal lines appear, supplemented by a double set, at right angles to each other (Fig. 3). There are seven beads in contact, a central bead surrounded by six. A double line of indistinct shadow forms a hexagonal fringe, filling up, as it were, the interstices, whilst the extreme tenuity of the focal region presents merely a sectional view; but, with low power, a repetition of these shadowy lines or spaces exhibits continuous lines. Figs. 4, 5.

Another interesting illustration of difficult definition is fonnd in parallel rows of minute beads, placed on opposite sides of a thin membrane (Fig. 14). Unless the aberration be finely corrected, the images of the upper and lower set obliterate each other; but if they cross at varying angles, blank spaces are exhibited, variegated with markings more or less distinct; better definition shows striæ and cross lines resembling the rounds of a ladder. Figs. 6, 7, 8.

The first approach to a fine and brilliant definition is heralded by a red flashing of the striæ, resembling somewhat threads of amber shown upon a sea-green ground. Gradually the real structure can be brought into view by judicious illumination; and under direct sunlight alternate rows of pale ruby-coloured spheroids flash, in beautiful contrast with the emerald rows behind and partly hidden by those in front.

With a third or C eye-piece, beading is not in general visible so as to be *defined* as beading unless the diameter exceeds for any object-glass the 5000th part of its focal length, or the observer possesses extraordinary acuteness of vision. The magnifying power (ordinarily in excess of fine definition) is found for a given objective by dividing 100 by the focal length. Thus the highest powers at all practically useful are, for most of them—

1 in.	½ in.	¼ in.	⅛ in.	$\frac{1}{12}$ in.	$\frac{1}{16}$ in.	$\frac{1}{25}$ in.	$\frac{1}{50}$ in.
1000	200	400	800	1200	1600	2500	4000.

In interpreting results, it is necessary to proceed from the known to the unknown; and as we can only see minute objects through a medium, more or less false to truth, the selection of known actions and reactions can alone be a standard of truthful representation.

Thus the finest objectives are corrected to show a standard appearance (that of the podura) as the severest test known for penetration and definition, which is utterly untrue to nature; and with a false standard little progress can be made. For this reason I have sought the use of known objects of sufficient delicacy and refinement suitable for estimating and recording the actions and reactions of aberration, as a check and test of theory. See Fig. 16.

The finest glass threads offer beautiful phenomena as tests of the state of the corrections.

These may be obtained of excessive tenuity, and their tendency to adhere often give interspaces of the 20,000th of an inch, forming, as it were, an exquisite artificial slit.

(*a*) When the aberration is very improperly corrected, these threads transmit a white light, the spectral rays, being dispersed in the direction of the cylindrical focus of the thread are, by the confused direction of the aberrating rays of the glass, commingled so thoroughly as to unite them into white light, and create a general haziness; but so soon as the aberration is corrected they become illuminated with prismatic colours according to the particular plane of the focal vision.

Generally, according to the depth to which distinct vision can penetrate, a bright band of light traverses each cylinder, varying in breadth and variety of colour; the richness and distinctness testing the perfection of the corrections; whilst, if the plane mirror be used, fine longitudinal lines represent the cylindrical images of distinct objects, as the bars of a window. In this way the effects of eye-pieces three or four times the usual depth may be appreciated.

In the finest examples of microscopic definition hitherto known, as with some of the most exquisitely formed diatoms, as seen with a 1-16th objective and an immersion arrangement, their structure cannot be seen at all, unless the glasses are sufficiently corrected to show prismatic colours, especially the crimson rays which possess the most penetrating power, as seen in the colour of the solar disc in a London atmosphere; but in the study of the behaviour of minute spherules of glass or cylinders, which are more easily found, it is apparent that accurate aplanatism, obtained by skilful combination of achromatic glasses, should be true to nature, and therefore, as it is known that prismatic colours exist even in these minute refractions, a corrected glass should reveal their existence; different kinds of glass form an interesting study.

The beading of a great many microscopic objects, as diatoms and scales, is prismatic and more or less possesses dispersive powers; and they are the most beautifully and distinctly defined only when they begin to flash with the rays of the emerald and ruby. When these colours cannot be seen, in general their definition is defective.

But a still higher order of test is to be found in the observation of *artificial double stars* or triplets. Quicksilver carefully cleaned by pressure through several folds of soft leather, scattered on a black surface, may readily be divided into minute particles, and those about the 1-1000th of an inch form beautiful images. To produce double microscopic stars, two round apertures should shed brilliant pencils upon the globule; then, according to their distance apart, and from the globule, their images will appear to approach or recede from each other. At a certain point and angular distance from the minute, mirror these *double stars coalesce.* This effect measures the aberration. These apertures, arranged at equilateral distance, form beautiful triplets.

According to the investigation of my friend Professor Bashforth, who has kindly furnished me with a tracing of the curve taken by minute globules of mercury, the reflecting surface becomes more nearly spherical as the particle diminishes, and the upper surface forms the most accurate images. The aberration is better determined, therefore, on the upper surface of the mercury. As a rough approximation, two small lamps placed one foot distance from the globule, and two inches apart, may be first experimented upon with reflecting particles arranged from the 1-100th to the 1-1000th of an inch. The smallest globules form the more severe test, and the double stars will be easily divided as larger particles are selected.

Phenomena observed.

Generally diffraction rings coalesce and become oval as the drop is smaller. The shape of the flame is obscured; radiating rays ensparkle the images and conceal the definition.

Reduce the aperture of the objective. The images become fainter, but more distinct. Increase the distance between the front glasses of the objective, and so adjust for an uncovered object, and it is possible a fine definition of the flames may be obtained.

But the images of circular discs of light formed by two apertures, placed before a brilliant source of light, more accurately represent double stars, and triple apertures give beautiful triples, which can be made as close as possible by increasing the distance of the bright discs of light from the globule. This test is so severe that I have found the aberration corrected by an adjustment of the glasses by the screw collar requisite of extreme delicacy; so that a quarter of a division, fifty divisions representing one thread of the adjusting screw, spoiled the definition[1] of Powell and Lealand's one-eighth objective.

Diffraction Rings diminish to two, under very fine corrections.—I have counted seven rings in false adjustments. In the highest degree of correction one remains. There are generally several observed with the finest refracting telescopes surrounding double stars. Mercurial double stars and triplets form exquisite objects for testing the dividing power of the microscopic objectives, and they can be readily formed of any degree of closeness, and the separation of the images can be readily calculated. The beauty and precision of the glasses may be estimated upon exactly similar principles as that of telescopes, which are valued according as they will divide double stars from 5″ to $\frac{1}{2}$″ or $\frac{1}{4}$″ apart, under powers of 50, 500, or 1000 diameters, 100 being allowed in general for every foot of focal length, or every inch of objective diameter; whilst for microscopes the highest working power is given by dividing 100 by the objective focal length.

Triplets.—Triple artificial stars show, upon a minute globule, diffraction rings coalescing either circular, surrounding the triplet, or broken. As the corrections are more and more perfectly adjusted the triplets become less nebulous, more sharply divided, and defined as round separate discs.

A practised eye at once recognises whether the glasses are over or under corrected. The achromatic imperfections are seen well portrayed, yellow fog being the worst imperfection of all; good definition, sharp and clear, resembling distant objects vividly cut against the sky of a bright summer atmosphere, when loaded with perfectly dissolved aqueous vapour, is impossible with this kind of double fault embracing both chromatic and spherical error. A little colour, especially of the crimson hue, is the finest of all corrections for obtaining

[1] Screw 70 threads to the inch, $\frac{1}{4}$ division representing $\frac{1}{4} \times \frac{1}{70} \times \frac{1}{50} = \frac{1}{14000}$ of an inch change in the separating interval of the front glasses.

a sharp and brilliant definition, as seen with the immersion lens and 1-16th at the Quekett soirée of this year.

Measurement of the aberration.

The diameter of the mercurial drops having been carefully measured, they must be selected in diameter according to the power of the objective to be examined.

The following appear convenient relations:

Power	200	400	800	1600 diameters.
Objective	1 in.	$\frac{1}{2}$ in.	$\frac{1}{4}$ in.	$\frac{1}{8}$ inch.
Diameter of globule .	$\frac{1}{200}$	$\frac{1}{300}$	$\frac{1}{500}$	$\frac{1}{1000}$ of an inch.

Distance of bright discs one foot, separated by two inches interval. Adjustable stops being used the aberration will be measured by the *interval* at which their reflected double stars cannot be defined. As each particular globule varies its curvature according to an unknown law of capillary attraction, it is absolutely necessary to select globules of determinate size, and to place the discs at one and the same angle of incidence upon the globule. Glasses supposed to perform accurately reveal often egregious errors by the application of this method of research.

On the Visibility of the Beading of Diatoms and Lepidopterous Scales.

Nothing can be more vague than the estimation of size by the unassisted sight; and to avoid this vagueness a standard is of essential service, notwithstanding the extreme variation in ocular definition by different observers. The limiting visual angle probably ranges from 1″ to 6″. A bullet-mark upon a white target at 1000 yards represents, perhaps, the smallest object visible to the best riflemen, which (if an inch spot) very nearly represents 6″. But for all observing powers of the eye where comparison is required a much larger visual range is absolutely necessary.

On Ocular Power.—The beading of various objects, such as the diatomaceæ and lepidopterous scales, affords excellent means of testing the powers of the human eye. Let $\frac{I}{n \times 1000}$ be the diameter of a bead; P the magnifying power of the combined glasses; θ the visual angle subtended by an image of the bead, formed virtually at 10 inches' distance from the eye of the observer. Then it is manifest, as θ is very small, that—

$$\theta \text{ varies as P, if } n \text{ is constant.}$$

$$\theta \quad ,, \quad ,, \quad \frac{I}{n}, \text{ if P} \quad ,,$$

And thence compounding—

$$\theta = \frac{1}{\phi} \cdot \frac{P}{n};$$

Where ϕ is a constant now to be determined—

$$\text{let P} = 1, \sin 1' = \sin 60'' = \frac{1}{3438} \text{ nearly.}$$

Let the beads be 36,000 to the inch.

$$n = 36$$

$$\text{Then, } \sin \theta'' = \frac{\frac{1}{36,000}}{10} \times 1 = \frac{1}{360,000} = 0''\cdot573.$$

$$\text{Therefore, } ''0'573 = \frac{1}{\phi} \cdot \frac{1}{36}; \therefore \phi = \frac{1}{20'628}.$$

$$\text{Hence } \theta = 20'628 \times \frac{P}{n}.$$

Under 12,000 diameters, a bead 36,000th of an inch in diameter subtends an angle (as P = 12,000).

$$\theta = 20'628 \times \frac{12,000}{36}$$

$$= \frac{20628}{3}$$

$$= 6876''$$

$$= 1^\circ \,.\, 54' \,.\, 36''.$$

Assuming 6″ as the smallest defining angle for a spherical bead or string of beads of 1-36000th in diameter, the power can be found which represents a bead at this angle.

$$\text{For } 6'' = 20\cdot628 \times \frac{P}{36};$$

$$\therefore P = 10\cdot6 \text{ nearly.}$$

From this it appears that, with a perfectly aplanatic magnifying power of 10·6, such beads will appear under an angle of 6″. But no microscope has hitherto been able to distinguish them with this low power at such an angle.

These results can easily be tested by the well-known value of

$$\text{Sin } 1'' = \text{arc } 1'' = \cdot000004848.$$

The beading of the *Pleurosigma formosum*, one of the coarsest of the diomataceæ, and which varies from 32,000 to 36,000 beads to the inch, cannot well be distinguished with a less power than 250 linear; below this, striæ make their appearance, of a cylindrical character.

$$\text{Here } \theta = 20{\cdot}628 \times \frac{P}{36} = 20{\cdot}628 . \frac{250}{36}$$
$$= 143''$$
$$= 2',,23''$$

Perhaps very acute vision may detect them with a power of 174·5, in which case $\theta = 100''$ very nearly—

$$\text{For } \theta = 20{\cdot}628 \times \frac{174{\cdot}5}{36} = 100'' \text{ nearly.}$$

But at ten inches' distance a bead of this size really presents a visual angle of about ½ a second (0″·572).

Assuming, therefore, that 1′ is the smallest angle at which beading can be conveniently defined as such, unless they are represented with central black shadows, and that one tenth of this, or 6″, is even a limit of sphericular visibility, a table may be conveniently employed as a means of testing the aberration of objectives as follows:

Diameters of Beading.

Per inch.	Power.	Measure of visibility.
36,000	174·5	100″
72,000	174·5	50″
120,000	174·5	25″

But practically it is found that a power of 1200 most conveniently shows this beading, and even this fails to distinguish the *structure appearing in the spaces separating the beading.* The results would then take the following form:

Beading per inch.	Power.	Visibility.
36,000	1200	687″
72,000	1200	343″
120,000	1200	171″
240,000	1200	85″

We conclude from these tables that it is rather superb definition than great amplification which should render minute structure visible. A power of 12,000 applied to the Formosum displays the beading, indeed, of large diameter, but with an ordinary $\frac{1}{12}$ objective the aberration receives a proportionate exaggeration.

It should be observed that the third row is finer than

Nobert's 19th band of lines, 112,688 to the inch, which by the formula subtend an angle of 221 seconds, or above three minutes and a half, and are then counted with great difficulty. The second row represents nearly the *P. angulatum*, which no one, as I verily believe, can resolve with 174, at half a minute of angular subtense for the beading.

Visibility depends upon the aperture of the objective, the reduction of aberration, and the direction of the shadows :—

(1) *Aperture.*—The aperture of microscopic objectives varies from 170° to 10°, as usually constructed.

I may here be permitted to state that the first occasion upon which I discovered that the celebrated spikes of the Podura test scale might be resolved into beading, I observed them as black points, whether by solar or artificial light. The cause of this remained unexplained for some years, and I beg here to offer for consideration some experiments which appear satisfactorily to elucidate this singular difficulty.

Having obtained an apparatus constructed to reduce the aperture of objectives at pleasure, and applied to the "nose" of the instrument before the objective is screwed into its place, it was seen at once that beading of glass, formed of glass threads 3000th of an inch in diameter, presented different appearances according to the indications of the instrument.

Thus if the aperture of an excellent objective be reduced one half, the beads assume jet-black borders, so as to present a central light, surrounded by a black ring, and the breadth of the black ring diminishes as the aperture is increased. This affords a clue to the extraordinary dark-brown appearance of butterfly scales seen under objectives of small aperture. An assemblage of spherical beading is now presented with black rings, enclosing a light spot, which can only be detected by a surpassing beauty of definition. The objective which presented the black Podura beading was 1-6th focal length and aperture 60°, and in other respects a very imperfect combination of glasses.

(2) If the index of refraction $= \frac{3}{2}$, the maximum aperture required to define the image formed by a glass spherule is 83° 38′ for parallel rays. The focal length $= \frac{1}{2}$ radius $= \frac{1}{2}r$, and the focus is situated at a point distant from the centre $= \frac{3}{2}r$.

If a tangent be drawn from it to the sphere, and θ be the semi-aperture, it is manifest—

$$\text{Sin } \theta = \frac{r}{\frac{3}{2} r} = \frac{2}{3};$$

Full aperture or $2\ \theta = 83°\cdot38'$ very nearly.

(3) The reduction of aperture increases the breadth of the black annulus, until, if there is sufficient penetration, its apparent breadth, for direct illuminating rays, may be made to appear to be one third of the apparent diameter of the bead, and even much greater.

INDEX *to* DIATOMS *figured in the* 'QUARTERLY JOURNAL OF MICROSCOPICAL SCIENCE,' *from* 1853 *to* 1867. By C. J. MULLER, Esq.

Navicula (*continued*)—
angulosa, 1856; pl. 5, Trans.
macula, 1856; pl. 5, Trans.
pandura, 1856; pl. 5, Trans.
nitida, 1856; pl. 5, Trans.
incurvata, 1856; pl. 5, Trans.
splendida, 1856; pl. 5, Trans.
didyma, 1856; pl. 5, Trans.
clavata, 1856; pl. 5, Trans.
Gregoriana, 1857; pl. 3, Journ.
compacta, 1857; pl. 3, Journ.
liber, 1858; pl. 3, Journ.
lineata, 1858; pl. 3, Trans.
æstiva, 1858; pl. 3, Trans.
granulata, 1858; pl. 3, Trans.
forcipata, 1859; pl. 6, Journ.
Bullata, 1860; pl. 2, Trans.
Trevelyana, 1861; pl. 1, Journ.
Clepsydra, 1861; pl. 1, Journ.
spectatissima, 1866; pl. 9, Trans.
excavata, 1866; pl. 12, Trans.
Egyptiaca, 1866; pl. 12, Trans.
permagna, 1866; pl. 12, Trans.
Zanzibarica, 1866; pl. 12, Trans.
Jamaicensis, 1866; pl. 12, Trans.
strangulata, 1866; pl. 12, Trans.
rimosa, 1866; pl. 12, Trans.
Hibernia, 1867; pl. 5, Journ.
denticulata, 1867; pl. 5, Journ.
pellucida, 1867; pl. 5, Journ.
Wrightii, 1867; pl. 5, Journ.
amphoroides, 1867; pl. 5, Journ.
truncata, 1861; pl. 1, Journ.
Northumbrica, 1861; pl. 1, Journ.
hyalina, 1861; pl. 1, Journ.
cruciformis, 1861; pl. 1, Journ.
arenaria, 1861; pl. 1, Journ.
gregaria, 1861; pl. 1, Journ.
retusa, 1861; pl. 1, Journ.
indica, 1862; pl. 9, Trans.
Lewisiana, 1863; pl. 1, Trans.
Johnsoniana, 1863; pl. 1, Trans.
notabilis, 1863; pl. 1, Trans.
luxuriosa, 1863; pl. 1, Trans.
cistella, 1863; pl. 1, Trans.

Nitzchia—
Sigmatella, 1855; pl. 4, Journ.
socialis, 1857; pl. 1, Trans.
insignis, 1857; pl. 1, Trans.
virgataab, 1858; pl. 3, Journ.
arcuata, 1859; pl. 6, Journ.
macilenta, 1859; pl. 6, Journ.
lanceolata, 1360; pl. 2, Trans.
vitrea, 1861; pl. 2, Trans.

Odontidium—
tabellaria, 1854; pl. 4, Journ.

Odontidium (*continued*)—
speciosum, 1859; pl. 9, Journ.
punctatum, 1859; pl. 9, Journ.
Baldjickii, 1859; pl. 9, Journ.

Orthosira—
oceanica, 1860; pl. 6, Journ.

Omphalopelta, fig. 18, 1863; pl. 5, Trans.
moronensis, 1866; pl 11, Trans.

Peponia—
Barbadensis, 1863; pl. 5, Trans.

Pinnularia—
tennis, 1854; pl. 4, Journ.
undulata, 1854; pl. 4, Journ.
parva, 1854; pl. 4, Journ.
uncertain sp., 1854; pl. 4, Journ.
latistriata, 1854; pl. 4, Journ.
exigua, 1854; pl. 4, Journ.
divergens, 1854; pl. 4, Journ.
Stauroneiformis, 1854; pl. 4, Journ.
gastrum, 1855; pl. 4, Journ.
apiculata, 1855; pl. 4, Journ.
megaloptera, 1856; pl. 1, Journ.
dactylus, 1856; pl. 1, Journ.
pygmæa, 1856; pl. 1, Journ.
biceps, 1856; pl. 1, Journ.
linearis, 1856; pl. 1, Journ.
subcapitata, 1856; pl. 1, Journ.
gracillima, 1856; pl. 1, Journ.
digitoradiata, 1856; pl. 1, Journ.
Elginensis, 1856; pl. 1, Journ.
globiceps, 1856; pl. 1, Journ.
longa, 1856; pl. 5, Trans.
fortis, 1856; pl. 5, Trans.
inflexa, 1856; pl. 5, Trans.
acutiuscula, 1856; pl. 5, Trans.
ergadensis, 1856; pl. 5, Trans.
semiplena, 1859; pl. 6, Journ.
Hartleyana, 1865; pl. 6, Trans.
Arraniensis, 1867; pl. 5, Journ.
divaricata, 1867; pl. 5, Journ.
constricta, 1867; pl. 5, Journ.
forvicula, 1867; pl. 5, Journ.

Pleurosigma—
vallidum, 1854; pl. 1, Trans.
inflatum, 1854; pl. 1, Trans.
angulatum, 1856; pl. 13, Journ.
Hippocampus, 1856; pl. 13, Journ.
formosum, 1856; pl. 13, Journ.
compactum, 1857; pl. 3, Journ.
doubtful sp., 1857; pl. 1, Trans.
transversale, 1858; pl. 3, Journ.
marinum, 1858; pl. 3, Trans.
lanceolatum, 1858; pl. 3, Trans.
carcinatum, 1858; pl. 3, Trans.

The Cell Doctrine; its History and Present State, by JAMES TYSON, M.D. Philadelphia: Lindsay and Blakiston.

Whilst America is perhaps not doing her share of the work of original observation in the various branches of natural science, we have from time to time abundant proof that her scientific teachers are keeping abreast of the knowledge of the day. Dr. Tyson, who lectures on the microscope and physiology in Philadelphia, has produced in this book an admirable *résumè* of the history of the various theories and observations on the ultimate structure of organic beings. Whilst giving credit to the earlier observers with the microscope, for their views and observations, he rightly speaks of the researches of Schleiden and Schwann as having introduced a new era into the history of our knowledge of animal and vegetable structure. He passes successively in review the observations of Henle, von Mohl, von Baer, Beale, Bennett, and the views of Huxley. Without going into elaborate detail, he presents the subject in such a way that the student cannot fail to understand the exact bearing of the views of each particular writer, on the subject of what is called "cell-doctrine," or the "cell-theory." The author sums up his own views in the following passage.

"In conclusion, then, it may be stated, 1st, that the 'cell,' or 'elementary part,' originating only in a pre-existing cell, is the ultimate morphological element of the tissue of animals and plants.

"2nd. That the cell, contrary to the belief of the earlier histologists, and, indeed, many later observers, is *rarely vesicular* in its structure, but generally more or less solid throughout.

"3rd. That the cell is composed of 'germinal' or living matter which is central, and includes 'nucleus,' 'endoplast,' 'protoplasm,' and 'sarcode;' and of 'non-germinal,' or 'formed' matter, which is peripheral, and corresponds with 'cell-wall' and 'intercellular substance.'

"4th. That this germinal matter of the cell in a part or all of its substance, may assume a special morphological state, usually round or oval, commonly known as the 'nucleus' of the cell, which, when present, is always a young centre of germinal matter; but that in other instances both animal and vegetable cells may be complete without this special form of germinal matter or 'nucleus,' as in the non-nucleated amœbæ and protogenes primordialis of Hæckel, the non-nucleated monads of Cienkowsky, and in the leaf of Sphagnum, in such Algæ as Hydrodictyon, Vaucheria, and Caulerpa, and in young germinating fern.

"5th. That in consequence of these facts, it cannot be said that in the nucleus alone resides the power to reproduce the cell, since we find the nucleus not essential, but that in the germinal matter, of which, after all, the nucleus, when present, is but a part, resides this function.

"6th. That when the smaller body within the nucleus, usually known as the 'nucleolus,' is present, as it often is in complete cells, it is simply a younger centre of germinal matter than is the nucleus itself, and is the last formed portion of germinal matter, instead of being the oldest part of the cell, as originally taught by Schleiden and Schwann. And thus, according to the latest views, the whole process is reversed, the old order of succession being—1st, the 'nucleolus;' 2nd, about this the 'nucleus;' and finally about this the 'cell-wall,' which embraces the cell contents. Now, however, what constitutes the 'cell-wall' when present, is the oldest part of the cell; next in age are the so-called 'cell contents,' whether germinal matter or not; next the 'nucleus;' and, last and youngest, the 'nucleolus.'

"7th. That the formed material constituting the cell-wall and intercellular substance may be something chemically different from the germinal matter, or protoplasm whence it was converted, as the secretions of gland-cells, or may be a simple condensation of the exterior of the cell, as in the red blood-disc.

"8th. That the so-called 'free nuclei,' so often referred to by pathologists in their descriptions of minute structures, are simply masses of germinal matter, smaller than those to which the name cell is usually given, which, if time be permitted, will pass into perfect cells by the usual production of formed matter on their periphery; that they do not originate spontaneously, but from previously existing germinal matter. So, too, 'granules,' if they be composed of germinal matter, present the same attributes and endowments, arising from previously existing germinal matter, capable of growing, multiplying, and assuming all the characters of fully formed cells, but never originating spontaneously. Granules otherwise composed are *histolytic* (ιστος, a tissue; λυσις, a breaking), and *not histogenetic* (ιστος, a tissue; γενεσις, creation); that is, they result from the breaking down of tissue rather than go to building it up."

New Section Machine.—I beg to enclose sketch of a new Section Machine.

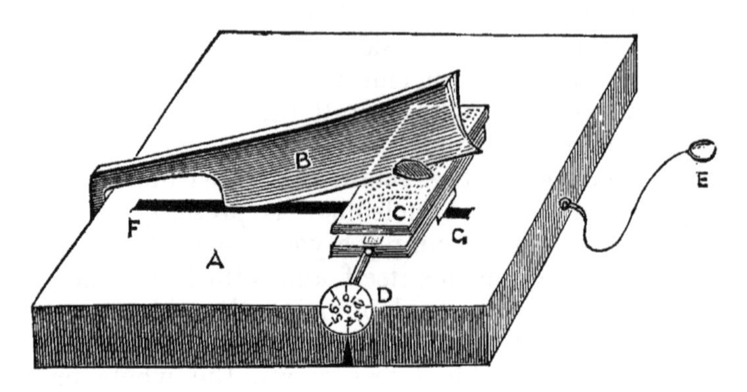

A. Mahogany base, about half size.

B. Knife blade inclined at an angle to the base, or two blades, as in Valentin's knife.

C. Cork stage upon which tissue rests, or is attached. This stage has a rectangular movement by means of the screw D, and may be moved to $\frac{1}{100}$ths of an inch.

Purposes of movement:

1st. To accurately adjust small objects beneath the edge of the knife when using single blade, and making sections of known thickness.

2nd. When using two blades, to place the object in position again without disturbing it, it has about half an inch rectangular motion.

E is the cord and ring by which the stage C is pulled from F to G. Upon attaching the little ring to a hook at the side, the stage remains as represented, but upon releasing it a spring pulls it rapidly as far as F, cutting in twain any object coming in contact with the edge of the knife.

I think the stage movement will be very valuable for specimens of some degree of solidity, such as those which have been subjected to chromic acid, &c. The screwhead (D) will be divided, so that the value of a part of a revolution may be known. Again, very small objects (such as the cephalic ganglia of beetles) can be accurately adjusted to the knife and cut with precision.—T. HAWKSLEY.

Corpuscular Blood-elements in the Urine in Bright's Disease.—Most general practitioners of medicine, when called upon to investigate an obscure case of Bright's Disease, have probably felt the want of some more definite guides to the exact condition of the kidneys than those furnished by the proportion of albumen in the renal secretion, or the character of the tube-casts which the urine contains; and although it seems at first sight an arduous task to attempt supplying the student with any additional aid beyond that afforded by the admirable manuals published by Doctor Dickinson, and, more recently, by Doctor Grainger Stewart, yet my connexion with the hospital has enabled me to carry on some researches which, enlightened by the new discoveries in regard to pus and mucus, may serve to render certain doubtful cases of this affection somewhat clearer than they would be without such assistance. It seems remarkable, indeed, how meagre are the references to the diagnostic importance of blood in the urine made in the monograms above mentioned; the former of these gentlemen observing in respect to casts (p. 18):—"If pus [white blood] cells are included, the inflammatory or catarrhal state has taken such hold of the tubes that the epithelial cells are replaced by pus-globules. Blood-globules will show that there has existed enough congestion to rupture the Malpighian capillaries." While the latter dismisses the subject with little more than the statement (p. 15):—"And, lastly, some [casts] are found in which blood-corpuscles in varying quantity are present. Along with the tube-casts, and sometimes in large quantity, blood-corpuscles altered by the action of the fluid in which they lay [*sic*], are found."

As, however, the important advance in pathological science, to which I have alluded, is still spoken of in this country under the title of "Cohnheim's alleged Discovery," it may not be unnecessary to advert briefly to its merits and the testimony which supports it. Dr. Cohnheim, as the readers of this Journal are aware (see number of this Journal for Oct., 1869, pp. 549—552), first published his theory of inflammation, and detailed the original and ingenious experiments from which it was built up, in a leading article in 'Virchow's Archives' for September, 1867, which soon attracted everywhere the notice of histologists. According to Cohnheim the process of pyogenesis consists, first, in a partial interruption of the flow of blood by which the red corpuscles move more slowly through, or almost block up, the capillaries, while white globules adhering to the parietes of the vessels arrange themselves in a layer upon this inner surface of the walls; and second, in the "wandering out" of these white blood-cells through the stomata, demonstrated by Recklinghausen in the walls of the finer blood-vessels by virtue of that amœboid movement which is one of the most remarkable attributes of the white blood-corpuscle, and so aptly illustrated by an English commentator on Professor Huxley's lecture upon protoplasm, when he explains the process of an amœba taking a minute diatom into its substance for food, by comparing it to a lump of dough growing of itself gradually around an apple to make an apple-dumpling: the white blood-corpuscles which have thus wandered out then constitute with exuded serum that yellow fluid, so long known under the name of pus, and hitherto generally supposed to be a product of the breaking down of tissue. In support of this doctrine experiments upon frogs and rabbits paralysed by woorara are described in which the mesentery of the animal being exposed and spread out upon the field of the microscope, multitudes of white corpuscles were *seen* in all stages of transit from the interior to the exterior of the vascular walls, in which latter position they constituted ordinary pus-globules.

Of course such a novelty in medical science has met with numerous assailants, among whom the most prominent seems to be Prof. Holman Balogh, of Pesth, who, in an article in 'Virchow's Archives' (Erste Heft, Band xlv, S. 19, u. s. w.), asserts that in spite of the most prolonged and careful attention, not once could he see the transit of the white blood-cells through the stomata in the vascular walls, which he thinks, if they exist, are such minute pores that they can give passage only to fluids. His observations are, however, sharply commented upon by Dr. A. Schklarewski, of Moskow, in the following volume of the 'Archives' (Band xlvi, Hft. 1, S. 116), and Cohnheim's experiments appear ('Transactions of Pathological Society of London,' vol. xix, p. 467) to have been repeated before the London Pathological Society in April, 1868, by Dr. H. Charlton Bastian, of London, with entire success. In our own country, Lieut.-Col. J. J. Woodward, Surgeon U.S.A., stated during a lecture at the Philadelphia College of Physicians, May 31st, 1869, that the experiments of Cohnheim had been tested under his direction in the Surgeon-General's Office at Washington, and that he had found the description of phenomena singularly accurate; the observations on frogs being fully corroborated, as far as they had time to repeat them, in every particular, and Dr.

William Norris, of this city, but for some years past residing in Germany, in an article now in press, detailing observations made, chiefly on the corneæ of frogs, in conjunction with Prof. Stricker, of Vienna, while maintaining that some of the corpuscles of pus originate in the proper cells of the tissue, admits as indubitable that many are in reality white blood-globules which have made their way through the walls of the vessels, as Cohnheim describes.

It has been urged, however, by some assailants of this doctrine, that even admitting, for the sake of the argument, Cohnheim's views on inflammation to be correct as regards the inferior animals, upon which his experiments were tried, there is no proof that the same ignoble process of suppuration affects man, a creature of such far higher attributes; but on this point I trust that my own experiments, published in the 'Pennsylvania Hospital Reports' for 1869, will be found conclusive. By diluting a drop of my own blood upon a slide, with pure water introduced at the margin of the thin glass cover, and thus reducing the liquor sanguinis to the specific gravity of the saliva, I found it quite possible to watch every step of the change, in which by mere distension the white blood-cell is converted into the salivary corpuscle, with its one, two, or three nuclei, its actively revolving molecules confined by a cell-wall of exceeding tenuity, capable of presenting all the phenomena of deep-staining of the nuclei with the entire cessation of movement on the addition of aniline dye. In like manner, when the liquor muci and liquor puris are similarly diluted their corpuscles are also seen for the most part to be converted into salivary globules, and I infer, therefore, that we may regard the strong presumption afforded by Cohnheim's experiments upon the rabbit as established into a fact, and conclude that most (at any rate) of the corpuscles of *human* pus are simply white blood-cells which have wandered out through the vascular walls.—Joseph G. Richardson, M.D., Philadelphia.—*American Journal of Medical Sciences*, January, 1870.

QUARTERLY CHRONICLE OF MICROSCOPICAL SCIENCE.

Histology. TEXTBOOKS.—The third part of the 'Manual of Histology,' edited by Stricker, has appeared, and contains the following articles:—The Blood-vessels of the Intestinal Canal, by Toldt; the Liver, by Hering; the Larynx, by Verson; the Lungs, by F. E. Schultze; the Kidney, by Ludwig; the Supra-renal Capsules, by Eberth; the Urinary Bladder, by Obersteiner; the Testicle, by La Valette St. George; the Ovaries, by Waldeyer; the Skin, &c., by Biesiadecki; the Serous Membranes, by Klein.

M. Robin has brought out a second edition of his 'Programme du Cours d'Histologie.'

THE BLOOD.—Geinitz (Pflüger's 'Archiv,' No. 1, 1870). has examined the action of hydrocyanic acid on the blood-corpuscles. If frogs were poisoned with cyanide of potassium or hydrocyanic acid, the blood-discs examined after death showed two remarkable changes. Their shape became rounder, or, in fact, very nearly circular, and their edges became crenulated and granular. The former change predominated when the acid was used, the latter when the salt. In another series of experiments a drop of blood was exposed to the action of hydrocyanic acid vapour in the "moist chamber" of Stricker. The blood-discs first underwent the changes above described, and then dissolved altogether, except their nuclei, the pigment becoming diffused through the liquid. Similar results were obtained with the blood of warm-blooded animals, the discs first assuming the characteristic "mulberry" shape, and then dissolving altogether. These changes are supposed to explain the brilliant colour which blood assumes when treated with hydrocyanic acid, the enlarged corpuscles reflecting more light than in the normal state.

Spontaneous Division of White Blood-cells.—Klein ('Med. Centralblatt,' January 8th) has observed the phenomena of division, especially in the white cells of the newt's blood. A drop is brought into the moist chamber under the microscope, and kept at a temperature of 25° to 30° C. Two masses may often be seen marked by a narrow neck of protoplasm, which either reunite to form one cell, or else divide into two, which remain permanently distinct. Another method of division is for a cell to flatten itself out into a disc, from the surface of which arises a kind of mound or elevation, containing a nucleus, which finally separates. The same phenomena were easily seen in the frog's blood, and also in human blood, when kept at a temperature of 35° to 40° C.

TENDONS. — Güterbock ('Med. Centralblatt.,' January 15th) has repeated the observations of Ranvier on the tendons of youug animals, and has come to the following conclusions: —The fissures which are found between the fibrillar bundles of the tendons contain chains of staff-shaped nucleated cells, which may be called connective tissue corpuscles, and which are less numerous in the tendons of adults than in those of young animals. The tendons are so rich in cells as to be little inferior to cartilage in this respect. These chains of cells pass uninterruptedly into rows of cartilage cells, where the tendons are inserted into cartilage.

MUSCULAR FIBRE.—Professor Krause has contributed to the 'Zeitschrift für Biologie' a paper on the structure of striated muscular fibre, in which he controverts the views of Hensen, which have lately attracted much attention. According to the latter observer, each transverse stria or disc of doubly refracting substance is divided by a disc of less highly refracting substance, which he calls the "median disc." This appearance is regarded by Krause as a misinterpretation of the appearances, and due, at least in part, to the action of water. The median disc he believes to be merely the central portion of the dark or doubly refracting substance, rendered paler by the action of water, and also brought into contrast by the greater distinctness given to the planes of contact of the dark and light substances. Krause has also seen the delicate line crossing the middle of the light substance, described by Hensen, and described also long ago by Dr. Carpenter, from Lealand's preparations ('Human Physiology,' 4th edition, p. 296). We are unable now to give a fuller account of this important paper.

LYMPHATICS OF THE EYE.—Schwalbe contributes to Schultze's 'Archiv' (vol. vi, part i, 1870) an elaborate paper on the lymphatic spaces connected with the eye. He confines himself in the present memoir to those of the posterior division of the eyeball. The posterior lymphatic system is defined as including the perivascular spaces of the retina, the perichoroid space with its efferent channels, and, finally, a lymphatic space between the outer and inner sheaths of the optic nerve, which, without communicating with the other two, opens directly into the arachnoidal sac of the brain.

Between the inner surface of the sclerotica and the outer layer of the choroid is a space, which was recognised by Arnold as a serous cavity, and to which the name arachnoidea oculi might be given. It is, however, occupied in part by some loose connective tissue with elastic fibres and stellate pigment-cells, which constitute the membrana supra-choroidea of Henle, and the lamina fusca of most anatomists. This is quite a distinct structure from the space or sac itself, and need not now be considered. The cavity itself, called by Schwalbe the perichoroid space, has the closest resemblance to the lymphatic sac of the frog, having inner surfaces, which even to the naked eye are sometimes smooth and shining, and are found, on careful examination, to show on both surfaces the epithelial markings characteristic of lymphatic spaces. The principal method made use of to demonstrate this strueture was the silver method of Recklinghausen, which had the effect of bringing into view a complete epithelial network, in the meshes of which were contained oval nuclei. No means was found of isolating the epithelial forms, but even without the silver treatment little shreds of membrane containing nuclei could be torn off. These appearances were best seen in the eyes of white rabbits, with more difficulty in pigmented eyes, and not at all in the human eye, since a specimen of this could never be obtained in a sufficiently fresh state. The perichoroid space being thus defined, injections were made into it with the view of determining its extent and connections. When completely filled with a coloured liquid the space was found to reach backward into the neighbourhood of the entrance of the optic nerve, and forward as far as just under the ciliary processes. The space thus forms a complete double envelope, surrounding nearly the whole of the eyeball. The injection never penetrated into the proper vascular tissue of the choroid, or into any of the chambers of the eye; it did, however, leave the space at four points, and these are points corresponding to the entrance of the venæ vorticosæ. It has been clearly made out that the injection passed through spaces encircling the veins, which, on a cross-section, were found completely surrounded by it. Through these perivascular canals the perichoroid space communicates with the space between the eyeball and the capsule or fascia of Tenon, which may be called the space of Tenon; and this was filled with injection from the perichoroid space. The space of Tenon further communicates backwards by a cylindrical channel surrounding the outer fascia of the optic nerve with the arachnoid space of the skull. The same methods of investigation as those spoken of above showed that this space is a true lymphatic sac, its internal surfaces being covered with a regular epithelium. The same conclusion has, as is well known, been arrived at by other observers with regard to the arachnoid itself, so that the continuity of the whole system is established. The final proof is given by throwing an injection into the arachnoid, which fills first the canal round the optic nerve and the space of Tenon, then the perichoroid space, and passes at the same time into the lymphatic vessels of the neck, showing the continuity of these cavities with the lymphatic system in general. Beside the canal just spoken of as surrounding the whole optic nerve, to which the author gives the name of *supra-vaginal* space, there is another included between the two fasciæ of this nerve, which he calls *subvaginal*. This is continuous with the arachnoid, from which it may be injected, and surrounds the whole optic nerve almost to its entrance into the eyeball; but here it stops short, not communicating with the proper lymphatic spaces of the eye. Its inner surfaces have the structure of a lymphatic sac, as distinctly as the perichoroid and other spaces.

Schwalbe makes some remarks on the method of investigation by impregnation with nitrate of silver. There can be no doubt that the markings produced on serous surfaces correspond to cell outlines; but he thinks they are caused rather by a precipitation of silver upon the edges of the cells by some albuminous substance adhering to them, than by a staining of any actual cement uniting the cells together. The term *endothelia* seems preferable, since they are not really identical with ordinary epithelial cells.

LUNG.—F. E. Schultze, in the third part of Stricker's handbook, gives a full discussion of the vexed question of the alveolar epithelium. According to his observations, made principally after injecting nitrate of silver into the air passages, the whole surface of the alveoli and bronchial terminations is in the fœtus covered with a continuous and homogeneous epithelium. After respiration is established this epithelium, though not ceasing to be continuous, becomes heterogeneous, and some of the cells, which are originally granular, polygonal, and clearly defined, become pale, transparent, and indistinct, thus producing the appearance of a *partial* epithelial covering, which has been described by many observers.

MIGRATORY CELLS IN THE SKIN.—Biesiadecki, in an excellent description of the skin, published in Stricker's 'Handbook,' gives an account of some important observations on migratory cells, which, though published before, have hardly attracted sufficient attention in this country. He describes as of normal occurrence in the deep layer (rete mucosum) of the epidermis, in the corium, and in the subcutaneous connective tissue certain cell forms having the following characters. They are round, oval, or irregular in shape, composed of soft, brilliant protoplasma, with a more or less distinct nucleus, readily stained by carmine, and agreeing generally with the lymph or white blood-cells in form, size, and properties. Being generally met with in the vicinity of blood-vessels, and differing in structure from the regular cells of the part, Biesiadecki regards them as identical with the "migratory cells" first described by Recklinghausen in the cornea, and since then often observed in diseased parts (being, according to other observers, extravasated white blood-cells). The number of these cells met with depends very greatly upon the vascularity or œdema of the skin, and they are enormously increased, especially in the deep layers of the epidermis, in certain pathological conditions. Such cells and their migrations seem to have a more important place in physiology and pathology than has yet been assigned them.

STOMACH.—K. Trütschel ('Med. Centralblatt,' Feb. 19th) has examined the terminations of nerve-fibres in the mucous membrane of the stomach of the frog. He finds in the submucous tissue a fine plexus of delicate fibres, in which cells are here and there embedded. This is in direct relation with nerve-branches, so that the whole must be regarded as a nervous structure. Other nervous filaments, which are not a part of this network, extend in the direction of the mucous membrane, and dividing into smaller branches reach the epithelial layer. In the mucous membrane itself is a layer of large multipolar cells, connected with one another by processes. These are also, without doubt, nervous structures. From these, moreover, nerve-fibres arise, which reach the epithelial layer, and there form a second plexus; while a third similar structure is described as lying immediately under the epithelium. Certain oval, clubbed bodies, which are met with between and among the cylindrical epithelial cells, are regarded as the ultimate terminations of the nerves. The evidence of the nervous character of these different structures rests principally on their coloration by chloride of gold and osmic acid.

LIVER.—Hering, in an account of the liver, published in the third part of Stricker's 'Handbook of Histology,' repeats his important and now well-known observations on the arrangement of the ultimate bile-ducts. His figures represent a complete network of intralobular capillary bile-ducts, which may be demonstrated by injection in the livers of the rabbit, amphibia, &c. This network has not yet been demonstrated in the human liver, as within a few hours after death it is no longer possible to inject the biliary system. Although it is impossible to avoid speaking of these finest bile-passages as capillaries, Hering does not regard them as at all similar in structure to blood-capillaries. The finest passages are without any true wall, being formed merely by the apposition of the grooves which run transversely across the edges of the liver cells (just so, if two bricks laid side by side in a wall had each of them a vertical groove in the corresponding position, they would form when placed together a vertical channel). These finest channels may be seen to pass into the smallest bile-ducts, which are furnished with distinct epithelium, and at the point of junction the epithelial cells lining the duct pass uninterruptedly into liver-cells. His observations have not confirmed those of Biesiadecki, Frey, and others, on the lymphatic sheaths said to surround the capillary blood-vessels of the liver, and he does not regard the existence of such sheaths as proved. Hering has also failed to trace the connection of nerve-fibres with liver cells described by Pflüger, though he has used the reagent (osmic acid) recommended by him.

Salivary Glands.—Mayer (Schultze's 'Archiv,' January, 1870) has repeated the observations of Pflüger on the nerve supply of the salivary glands. His very numerous observations do not permit him to confirm the results of Pflüger. He was able to discover very few undoubted nerve-fibres in the gland, the greater number of bands and fibres which are seen belonging to the vascular system, though he guards himself against the supposition of imputing to Pflüger that he has mistaken vessels for nerves. Mayer has never seen any medullated nerve-fibre penetrate an alveolus of the gland, and the structures which reach the outside of an alveolus are either vessels, or nerves which have lost their medullary sheath. He points out the great improbability of fibres resuming their medullary sheath after having lost it, and further, that it is quite contrary to the general rule that any nerve-fibres should preserve this structure up to their finest terminations. Mayer has seen fine filaments in connection with the nuclei or nuclear processes of cells, but entirely failed to obtain any proof that these filaments were nervous.

Development.—Cazalis ('Archives de Physiologie,' 1870,

No. 1) has examined the development of the muscular fibres in the diaphragm of the fœtus. He finds than these are among the earliest muscular fibres developed, and that at birth they have a higher development that the muscular elements connected with animal (as distinguished from organic) life. This he brings into relation with the immediate necessity to the infant of the function of respiration.

Embryology.—Schenck (Pflüger's 'Archiv,' Nos. 2 and 3, 1870) has observed a remarkable rotatory movement of the embryo of the common frog within its envelopes. The impregnated and growing ovum may be seen with a simple microscope to be continually rotating on an axis, which is perpendicular to the dorsal furrow of the embryo. The motion may be described as opposite to that of the hands of a clock, if the head of the embryo be regarded as the point of the clock-hand; or, in other words, an observer situated at the tail end, and looking towards the head, will see the latter turn to his left. The motion is incessant, and undergoes no change till the later stages of development, when the perfectly horizontal position is changed for an inclined one, the tail end being depressed from the preponderating right of the organs developed in it. The time of rotation varies in different specimens, but was generally from five to twelve minutes. Schenck believes ciliary motion to be the cause of the rotation, since it is increased by warmth, which also accelerates the movement of cilia, and is altogether stopped by dilute acids, which equally check ciliary motion. With the death of the embryo both ciliary movement and rotation are suspended.

Dublin Microscopical Club.

October 28th, 1869.

Dr. John Barker showed examples of the encysted or resting condition of *Ceratium cornutum*, which, unlike the corresponding smooth-walled state of the common species of Peridinium, has a double coat—the outer smooth—the inner densely covered by short, straight, blunt, projecting processes.— He also exhibited a fine specimen of the curious little infusorium lately drawn attention to by him, possessing the remarkable characteristic of inhabiting a hyaline barrel-shaped test, open at each extremity; a more particular account of this pretty and interesting form will hereafter be given by Dr. Barker.

Dr. Moore showed the pretty and curious aquatic cryptogamous plant, *Salvinia natans*, being the first time it had been exhibited in a living state at any scientific meeting in Ireland. The beautifully fluted sacs containing the antheridia, as well as the sporangia attached to the under side of the short rhizome of the plant, were ripe and in good condition for examination. The former constitute a pretty object as seen under a low power, the separate antheridia being globose in shape, each with a short stalk or point of attachment, and having a cellular reticulated appearance on the surface. The sporangia are much larger, pale coloured, and ovoid in form. During the conversation which followed the exhibition of these objects, Dr. Moore put forward a query whether the points of attachment visible on the antherida could be what Schleiden may have mistaken for tubes, which he states they emit and penetrate into the prothallia developed from the sporangia? In connection with the foregoing, Dr. Moore showed plants of two other genera of Rhizocarpods in fruit, namely, *Pilularia globulifera*, and *Marsilea salvatrix*, Braun. The plant of the latter he stated was raised from "receptacles" found in the pocket of the only survivor of that calamitous exploring expedition to the interior of Australia, on which our countryman, Burke, perished, after he and his companions had subsisted a considerable time on the "nardoo" of the natives, which consists of the "receptacles" of *Marsilea salvatrix*.

Rev. E. O'Meara showed specimens of a diatom coming close to *Pleurosigma Spenceri* which, however, he was disposed to identify as *P. Wormleyi;* he showed also *Amphora turgida* and *A. lyrata*.

Mr. Archer exhibited specimens of a minute and curious, seemingly novel, form of "trachelomonad" or "volvocine," remarkable for the possession of five prominent elongate hollow cornua, being in fact projections from the smooth and hyaline "lorica," four radiating equidistantly from the anterior portion of the organism, and directed obliquely backwards, and the fifth projecting posteriorly, all tapering and subacute at the apices; the "monad" green biciliated. A transverse view presented, thus, a very Staurastrum-like outline. Mr. Archer showed one or two other more or less closely related organisms, as yet hard to identify, if not novel; and trusted, perhaps, to be able to revert to the first of the foregoing on a future occasion, if, perchance, a search in the same spot, a small rock-pool at Greystones, whose water supply seemed to be about equally contributed by rain, by ooze from the bank above, and by sea spray, should reveal further examples.

Dr. E. Perceval Wright exhibited a series of preparations showing the form and arrangement of the spicules in *Aphrocallistes Bocagei*, a new species of this genus, which came from the Cape de Verd Islands. Dr. Wright's observations on this species will be found *in extenso* in a paper on some new sponges—*vide* antè, p. 1.

Professor A. Agassiz stated to the meeting that this interesting species had been found by Count Pourtales off the east coast of America.

November 18*th*, 1869.

Rev. E. O'Meara exhibited a pretty new diatom belonging to the genus Pinnularia, which he proposed to name *Pinnularia Collissii*, in memory of our late lamented Club member Dr. Maurice H. Collis. This he would presently bring forward in detail and figure in the Journal.

Mr. Archer brought forward an interesting condition of a minute Scytonematous plant. The slender filaments were interwoven in a variety of ways, and gradually tapered off until the extremities became a mere thread, the whole enveloped in a gelatinous matrix, the curious circumstance being that each ultimate extremity or "twig" bore a very minute elliptic cell attached at the extreme point. These cells increased by self-division and appeared to be enveloped by a special definitely bounded gelatinous covering, each presently borne on a stalk of its own, a bifurcation of the supporting filament taking place on each act of self-division. These elliptic cells were not ordinarily attached either by the middle or end, but obliquely, or as it were by the shoulder, so to speak. The whole thus presented somewhat of a tree-like appearance, bearing so many fruit-like structures on its leafless branches. These cells, when isolated, and as they occurred sometimes in the present material, were doubtless some of Kützing's many heterogenous "Palmoglœa," forms such as *P. micrococca* (Kütz.). It was interesting here to note the strictly genetic relationship of the cells to the filaments. It ought to be observed here that such forms as the present are completely distinct from the so-called "Palmoglœa macrococca," "P. crassa," "Brébissonii," &c. Regarding the Scytonemeæ, as Mr. Archer believed, as true, though aberrant or non-typical lichens, it might be assumed probably in this instance, that these egg-shaped cells at the apices of the filaments were "gonidia"—but what then of the central axis of green matter pervading these plants—likewise here present—which portion of the structure must be considered in Ephebe and others as the equivalent of the "gonidia" of typical lichens? No doubt these curious specimens were calculated to add to the puzzle, but every little observation as to structure, however partial or limited, might come in some day as useful in assisting to determine the true state of the case as regards these pretty plants, and hence the present examples, *quantum valeant*, were so far worthy an examination and record. Mr. Archer hoped some time to be able to revert to these and other specimens in this group more at large.

Mr. Robinson brought for exhibition "Reade's Prism," and very satisfactorily showed various *Pleurosigmata* by its aid.

Mr. Archer showed examples in considerable abundance of a new and very minute form of *Spirotænia*. This is exceedingly slender, being, however, long as compared with the diameter; the cells somewhat curved or arched, slightly tapering, ends truncate, the endochrome forming a single spiral reaching from end to end of the cavity, self-division transverse. As regards the width of the cells, this species is the most minute known, though *S. parvula* (Arch.) is shorter in length. The only form this could well be mistaken for would most likely be *Ankistrodesmus falcatus*, with which it pretty nearly agrees in size; but besides the spiral endochrome, the cells being blunt, not acute, at the apices, is a character striking the eye at first glance. Mr. Archer hoped to figure and describe this form more at large on a future occasion, naming it *S. tenerrima*.

Dr. E. Perceval Wright exhibited portions of the spicular skeleton of *Aphrocallistes Bocagei*, from deep-sea dredgings off the south-west coast of Ireland. For these he was indebted to the kindness of Professor Wyville Thomson, who informed him that some of the specimens were found living.

16*th December*, 1869.

Dr. John Barker showed various objects (*Conochilus volvox*, &c. &c.) viewed by aid of the new parabolic condenser, on the new principle contrived by him, by interposition of a stratum of water (or oil) between the *flat* top of the condenser and the slide. The objects now exhibited were beautifully seen on a very dark field, and well illuminated. Dr. Barker stated he was pursuing some investigations and experiments with a view to the adaptation of the principle to the higher powers, and hoped again to lay further results before the Club.

Dr. Moore showed a production sent to him from the County Wexford, found in a dried-up pool, and which, as seen in the hand, had all the appearance of so much cotton wadding, being quite white and hardly at all distinguishable therefrom; it had, indeed, occurred in such quantity, and was so like in appearance and feel to cotton wadding, that a mass of it had been employed to pack plate on its transit to Dublin. On being however placed under the microscope, it was found to be made up of the dried remains of a species of Œdogonium felted together in a very dense manner, and, in place of appearing white, as seen by the unassisted eye, it presented a certain amount of its natural green appearance, though the cell-contents were in most cases, of course, much contracted, and their natural arrangement necessarily destroyed. This indeed turned out to be the same Œdogonium exhibited in a recent condition by Mr. Archer, first at the Club meeting, July, 1865, and which he was still disposed to regard as most likely *Œdogonium setigerum* (Vaupell), though, as explained in the Minutes of date referred to, the plant might possibly be *Œdogonium apophysatum* (Prings.). The present dried examples showed, sparingly, the "dwarf males" with the "outer" antheridium, and the same series of mother-cells of androspores and the same shaped oogonia. This species, Mr. Archer thought, had not often presented itself; but the abundance in which it occurred in the present instance might perhaps indicate that it is not to be counted amongst the rarities.

Mr. Archer exhibited fine and beautiful specimens of *Vasicola ciliata* (Tatem, in 'Monthly Microscopical Journal,' No. III) new to Ireland, and interesting as being the second discovery only of this handsome infusorium. This, with its striking red "food-corpuscles," mingled with other colourless ones and vacuoles, its beautiful hyaline and transversely corrugated *vase* and its active ciliary motion, formed a remarkably pretty object. It must seemingly be accounted rare too, for though a comparatively conspicuous form, it is only the other day since its discovery by Tatem, nor does it appear to have been recorded elsewhere. The present gathering, made near Lake Belvidere, close to Mullingar, presented examples in the various conditions figured by Tatem, some *in sitû* in their cases, some undergoing division, and some freely swimming. Tatem's figures place the identity of this fine species beyond all doubt, though they seem to give rather too much prominence to the red granules, to the exclusion of the numerous more minute colourless ones which likewise pervade the body-mass, nor does he draw attention to the generally sufficiently striking feature presented by the marked pair of contractile vacuoles in the broad neck-like portion; nor, further, does he depict the surface of the body as faintly striate, after the manner of a Stentor. The outer marginal wreath of large anterior cilia appear to be comparatively rigid, whilst those within and on the sides are in constant wavy motion. This species sometimes takes in large objects as food, occasionally even longer than the body, such as a Pinnularia, which projects above and below, and bounded at each end seemingly by only the integument of the animal.

Dr. E. Perceval Wright exhibited and described a remarkable sponge from Greenland, for which he was indebted to the kindness of Mr. Edward Whymper. It belonged apparently to a genus near *Tethya*. It was apparently free, of the shape of the half of a small apple; the under surface was flat, the upper convex, and furnished with five to six star-like oscula. The investing sarcode layer was very thin, and the whole structure in external appearance much resembled the *Alcyonium mamillatum* of O. F. Müller. No such form was alluded to in Oscar Schmidt's 'List of Greenland Sponges.' A full description, accompained by figures of the sponge, will be given in a future number of this Journal.

Dr. Moss, R.N., exhibited some varieties of *Ceratium tripos*, collected during the voyage of H.M.S. Simoom to Ascension, in the months of October and November last. The specimens illustrated the peculiar fact that these cilio-flagellate Infusoria constantly occur connected together in chains composed of from two to eight or more individuals, generally, but not invariably, lessening in size from above downwards—that is to say, towards that end of the chain which presents the Ceratium with a free stock. Their attachment depends upon the reception of the extremity of the stalk of one into a little cup-shaped hollow in the flat surface of the body of the other, on the side opposite to that which presents the flagellum, and at the termination on that side of the cestoid ciliated groove. When a *Ceratium tripos* in a drop of water is placed upon a glass slide, it usually lies on its flat or slightly concave surface, with the upper angle of the carapace and the greater cornu springing from it to the left. In this position the ciliated groove, passing transversely across the convex surface of the body, divides it into two nearly equal portions, through the upper of which, close to the root of the right cornu, the attached stem of the next individual of the chain can be seen; and on the left side, in a somewhat corresponding position, a folding-in of the integument forms a canal for the flagellum, which is extended through an oral opening in the carapace, internal to the first serrations (in some varieties very prominent) on the base of the left cornu. It is, perhaps, worthy of remark that no exchange of endochrome between the connected infusoria has been observed, and that two varieties of Ceratium have never been found in the same chain.

Dr. R. M'Donnell exhibited preparations of nerves, which were much admired.

Dr. Archer exhibited a very pretty, seemingly new, Cosmarium, which most approached *C. cristatum*, Ralfs. Ralfs' description of that species, examples of which he had not himself seen, was made from drawings furnished to him by Sidebotham. Relying on their accuracy, the present appeared to be distinct from that rare species. In the present there are four lines of processes bordering the cell; not two only, but these prominences of but one kind. The form appears to be stouter and thicker than *C. cristatum*, whilst the little prominences ornamenting this pretty form are distinct from those belonging to any other in this genus—except, perhaps, *C. cristatum*—in that they are neither "pearly" granules nor spines, but short, thin, compressed, quadrate, vertical processes, the upper outer margin emarginate; thus, to a greater extent, resembling the similar ornaments of the Staurastrum exhibited by Mr. Archer at a former meeting of the club (*Staurastrum maamense*). A more detailed account of this pretty form, gathered by Dr. Wright at Glengariff, on a late visit thither, Mr. Archer hoped to be able to present on some future occasion.

Brighton and Sussex Natural History Society.

Feb. 10*th*. The President, Mr. T. H. Hennah, in the Chair. The receipt of 'Catalogue of Works on the Microscope,' by R. C. Roper, from the author; and 'Microscopic Objects Figured and Described,' by J. H. Martin, from the publisher, was acknowledged.

The Hon. Sec., Mr. T. W. Wonfor, exhibited a collection of galls found on British plants, made by Mr. W. H. Kidd; and read a description of each and of the insects producing them, as well as their parasites, drawn up by the same gentleman. This collection is intended for the Brighton Museum.

Mr. Wonfor then read a paper on "Seeds."

Commencing with the first appearance of the ovule, in the unexpanded flower-bud, as a pimple consisting of an aggregation of cells; its gradual development and impregnation by the pollen, together with its several parts, were traced; until the perfect seed, ready for dissemination and containing within it the embryo of the future plant, was fully formed. The various modes by which the seed is disseminated, the great number produced by some plants, the power possessed by some seeds of resisting extremes of heat and cold, and the wonderful property possessed by many seeds of preserving their vitality under apparently very adverse circumstances, for long periods of years, were next discussed.

On the subject of artificial selection, it was pointed out what might be done in increasing both the size and number of seeds and plants by carefully following the plan adopted with such success in the case of cereals by Mr. F. Hallett, of Brighton. Reference was next made to seeds as objects for the microscope. Having spent several years in the collection and examination of the seeds of wild and cultivated plants, as objects for the microscope, he considered few things in the vegetable kingdom presented such diversity of form, markings, and beauty. Although unwilling to lay down any law for classification, by means of the appearances of seeds he has often been able, in the case of unknown seeds, to determine the family to which they belonged, from certain peculiarities common to many plants of the same family. Among some of the most interesting families might be mentioned the Scrophulariaceæ, containing the mulleins, foxgloves, figworts, paulownias, &c; the Caryophyllaceæ, or pink family, containing a very great number of very beautiful seeds, not the least beautiful being the common chickweed and ragged robbin; and the Orchidaceæ, characterised by what had been termed the appearance of net-purses, each containing a single gold coin. The majority required no other preparation than that of mounting dry. Some, like the orchids, when mounted in balsam, formed good polariscope objects. For making out the several coats of the seed, the embryo, &c., sections cut on the plan recommended by Dr. Hallifax gave admirable results.

The paper was illustrated by a large collection of seeds and by microscopic preparations, including sections showing the several parts, made by Dr. Hallifax.

March 10*th*. The President, Mr. T. H. Hennah, in the Chair. The meeting was special, to receive a report of the committee on the subject of forming a Microscopical Section, a summary of which is appended.

A suggestion having been made that the usefulness of the Brighton and Sussex Natural History Society would be much extended if increased facilities could be afforded to its members for microscopical study, the committee recommended that as microscopical examination and the use of the microscope were almost indispensable to the pursuit of knowledge in natural history, it appeared necessary to form a section of the society to be called the "Microscopical Section," which should provide for the study of subjects connected with the use of the microscope, and for the more frequent intercourse of such members as were interested in microscopical study; that these objects could be attained by monthly meetings of the section, when papers on strictly microscopical subjects could be read; such reading to be restricted to twenty minutes, so that time might be afforded for the examination of objects and the comparison of observations; by the formation of a cabinet, to which members be invited to contribute slides, particularly of such objects as illustrate the natural history of Sussex—members to have specimens from the cabinet for home examination, under certain restrictions, and by the encouraging the exchange of slides among the members. The section to consist of all members of the society who signify their wish to the secretaries to join the section. The government to be under the present officers of the society until the annual meeting, when the committee shall suggest rules for its future government; the meetings to be held on the fourth Thursday in each month, at 8 o'clock, the chair to be taken by the president or a member of the committee, when, after the ordinary and special business of the evening, the meeting shall resolve itself into a conversazione, at which slides illustrative of the subject of meeting shall have precedence of other objects of interest and novelty. Before separating the subject of the next meeting shall be announced.

On the motion of Mr. Hazlewood, seconded by Mr. Wonfor, it was resolved: "That the committee's report be received, approved, entered on the minutes, and acted upon."

The meeting then became general; when a paper, by Mr. Clifton Ward, F.G.S., "A Sketch of the Geological History of England so far as it is at present known," was read by Mr. Wonfor, Hon. Sec., in which, from the earliest dawn of the Cambrian period down to the present day, the changes brought about by submersion, deposition, elevation, denudation, &c., together with a description of the animal and vegetable types of the various periods, were graphically described; while the amount of land above water in England at the different periods was represented by a series of fifteen charts.

It was announced that the 'Bryological Flora' of the County of Sussex would shortly be ready for distribution, the society having determined to publish it at once, instead of waiting the issue of the annual report.

Royal Microscopical Society.

January 12th, 1870.

The Rev. J. B. Reade, M.A., F.R.S., in the Chair.

Mr. J. Browing read a paper "On a Method of Measuring the Position of Absorption Bands with a Micro-spectroscope."

The Secretary gave an abstract of a paper by Mr. Alfred Sanders, "On an Undescribed Stage of Development of *Tetrarhyncus corollatus*."

Mr. Kent read a paper "On the Calcareous Spicula of the Gorgonaceæ."

February 9th.

The President in the Chair.

The President delivered the annual address. In addition to the usual *résumé* of the papers read at the Society's meetings he gave an account of the history of the Society and the reasons which had induced the Council to give up publishing their transactions in a separate form, and merge them in a popular monthly journal. He also stated that since this arrangement had been made, three members had been added to the ranks of the Society. The address was rendered unusually interesting by a communication of Professor Lister, of Edinburgh, giving an account of the scientific career of his late father, Joseph Jackson Lister, with especial reference to his labours in the improvement of the Achromatic Microscope.

It appears from the Treasurer's account that the Society has £1,059 in consols, and the Treasurer £28 in hand. It was resolved in future to stop the refreshments at the evening meetings.

MEMOIRS.

BIOPLASM, *and its* DEGRADATION; *with* OBSERVATIONS *on the* ORIGIN *of* CONTAGIOUS DISEASE. By LIONEL S. BEALE, M.B., F.R.S., Fellow of the Royal College of Physicians; Physician to King's College Hospital. With Plates XI, XII, XIII, XIV.

HITHERTO I have employed the simple term *germinal or living matter,* to denote that matter which takes part in the formation of all living beings and their tissues and organs; but the term is lengthy, and in some respects perhaps awkward and inconvenient. It cannot be used alone when speaking of a single particle, nor can it be employed adjectively. The word "*protoplasm*" has been much used for some years past, but the vagueness attached to it renders it unfitted for employment here. I require a word to denote *living, forming, growing, self-producing matter, as distinguished from matter in every other state or condition whatever.* Now "protoplasm" has been applied, both in this country and in Germany, to *lifeless* matter as well as to *living* matter, to *formed matter and tissue* as well as to the *formative* matter. And more recently Prof. Huxley and others have added to the confusion by giving it a still wider signification—so very wide, indeed, that almost anything that ever formed part of an organism may, according to their view, be denominated protoplasm. Dead matter and living matter, and roast mutton, boiled as well as unboiled white of egg, and a number of other things, moist and dry, having strueture and structureless, alive and dead, are said to be protoplasm, so that the word ceases to be distinctive of matter in any particular state. It becomes, in fact, useless.

The name I propose to give to the *living,* or *germinal self-increasing* matter of living beings, and to restrict to this, is *Bioplasm* (βιος, life; πλασμα, plasma). Now that the word *Biology* has come into common use, it seems desirable to employ the same root in designating the matter which it is the main purpose of biology to investigate. *Bioplasm* involves no theory as regards the nature or the origin of the matter. It simply distinguishes it as *living*. A living white blood-corpuscle is a mass of bioplasm, or it might be termed a *bioplast*. A very minute living particle is a bioplast, and we may speak of living matter as bioplasmic substance. A cell of epithelium consists of *bioplasm* or bioplasmic matter, surrounded by *formed non-living* matter, which was however once in the *bioplasmic state*. In the same way a germ of a fungus, as the yeast particle, consists of the *bioplasm* with an envelope of *formed material*, which last has resulted from changes occurring when the particles upon the surface of the bioplasm died. The bioplasm of the microscopic fungus or other organism may give off diverticula which may become free independent *bioplasts*. Each minute bioplast may grow, and in the same way give rise to multitudes of other bioplasts.

Progressive Change in Power of the Bioplasm.

Those marvellous progressive changes which occur during the development of the embryo, while the structures which characterise the organism are being evolved, are still but very imperfectly understood. We know, indeed, that all the complex tissues and organs of man and the higher animals are dependent for their production upon changes occurring in a minute mass of perfectly colourless living matter, in which no indications of form or structure can be discerned, but how these changes are brought about we have not yet been able to ascertain; nor is it conveying much information to the student if the teacher informs him that the perfect organism, with all its marvellous apparatus, existed "potentially" in the little colourless living embryonic particle; since it is impossible for anyone to distinguish the particle which is to develop a highly elaborate mechanism from that which is to produce a simple amœba as its highest developmental product. Hence, to say the structures evolved "existed" in the original mass of living matter is not true, and to qualify the assertion by the term "*potentially*" cannot make it more correct. All we *know* is that such and such structures will result, but we know this from previous experience, not from reasoning.

For the characters and composition of the living matter do not enable us to premise anything whatever concerning its formative properties. In the formation of man and the higher vertebrata the primary mass of bioplasm or living matter absorbs nutriment, and grows, and then divides and subdivides into numerous masses which are arranged in a definite manner, but what determines this is not known. From each of these in pre-ordained order, and with perfect regularity, more are produced, no doubt, according to "laws," but laws which we know nothing about, except that they are not physical. As this process of division goes on the resulting masses produce various substances, some having wonderful structure and properties. But the power of each series to produce these peculiar materials, which did not exist before and which cannot be extracted from the food supplied, differs from that of the series which preceded it, and so on until the complex structural basis of the organism is as it were laid down. There are masses of bioplasm to form nerve, others to produce muscle, others glands, and so on, all of which have been derived from one common mass; but the bioplasm destined to take part in the development of a gland will under no circumstances produce muscle or nerve. And yet with all this marvellous difference in power, which seems to be somehow acquired as development advances, there is, as far as is known, no difference in matter. The nerve- or muscle-producing bioplasm is, as far as can be ascertained, the exact counterpart of the gland or bone, forming bioplasm, and why one produces one tissue and the other a very different tissue cannot be explained.

All these different forms of bioplasm have descended from one, which may be regarded as the parental mass, but in regular definite and prearranged order; so that if from any circumstance the bioplasm which is to form a gland or other organ, or a member, is not produced, and does not occupy its proper place at the right period of developmental progress, that gland, organ, or member will be wanting in this particular organism. The manifestation of power or property to form special parts with special functions occurs in regular order, progressively in one direction only as the germ advances towards the particular perfect form it is to attain. The power once lost can never be regained, although life may continue to be manifested nevertheless, and perhaps more actively than before.—The particles of bioplasm which were to take part in the development say of the brain, do not receive at the proper period a supply of nourishment of the right kind or in the proper proportion. A well-developed healthy brain cannot in that case be formed. The particles may waste and die; or they may grow for a time and then cease to progress further; or they may grow, and live, and multiply, and form a great mass of matter, which however will never produce a brain or an organ capable of performing the functions which the brain was designed to discharge. They may multiply fast, and take up more nourishment than the brain cells, had they been formed, would have appropriated, but the brain with its marvellously complex intricate structure which involves gradually progressive changes, steadily proceeding during a length of time, will never be produced; and under no circumstances conceivable could any of these masses, or any of their descendants, develop one perfect brain cell. If progress towards the mature state be stopped at any point, the perfect state of development can never be reached, and the organism if developed must be imperfect. The development of other complex organs may have proceeded with perfect regularity, but the organism must ever remain incomplete in structure, and incapable of performing all the functions it might have discharged.

But although developmental power may be lost for ever, power of a different kind may be acquired *pari passu* during the rapid multiplication of bioplasm. Progressive advance in the capacity to form lasting structures and elaborate organs is characterised by the comparatively slow but regular and orderly growth and multiplication of bioplasm. Rapid multiplication of the bioplasm, on the other hand, involves degradation in formative power, which is at length entirely lost, never to be reacquired.

Degradation in power is commonly associated with increased rate of growth, increased faculty of resisting adverse conditions, and, in some cases, such is the vitality of the living matter that it takes up the nourishment which should be appropriated by healthy parts, and these are at length starved and deteriorate or are completed destroyed. The actively living degraded bioplasm may be capable of retaining its vitality although removed altogether and for some time from the living body, and, remarkable as it seems, it may grow and at length destroy other living organisms to which it gains access.

The poisonous "virus" of many contagious diseases is, I shall endeavour to show, living matter or bioplasm, which has been derived by direct descent from the bioplasm of a healthy organism, and I propose in this paper to give a sketch of some of the most important facts which have led me to adopt this view. The inquiry is of great interest, not only because it affects the question of the nature of the material concerned in the propagation of contagious diseases, but it will be found to bear upon matters of the greatest practical importance, such as the means of preventing the spread of contagious diseases, and of treating such disorders. In the first place I shall refer to the mode of multiplication of the bioplasm of man in health, and then endeavour to trace its degradation until a form of bioplasm destructive of healthy life and capable of infinite multiplication results.

Bioplasm of Animals and Man in Health.

Bioplasm of Amœba.

Among the lower, simplest living forms known are some very simple organisms consisting apparently of transparent structureless semifluid material. Seldom as much as the $\frac{1}{1000}$ of an inch in diameter, they vary much in size down to the most extreme minuteness and tenuity capable of being seen under the highest power yet made, equalling about 5000 diameters. These masses, apparently composed almost entirely of living matter, can move in any part, and in any direction. Portions of the semifluid material may protrude in advance of the rest of the mass, and coming in contact with protrusions from other parts, join these, and thus a ring or a series of rings may result. The protrusion may be withdrawn and the whole assume the appearance of a perfectly smooth globular mass. Such naked masses of bioplasm or living germinal matter may apply themselves to foreign bodies, and if these are small, completely invest them, so that they are at length seen in the interior of the mass embedded in its very substance. It is in this way that these simple forms of life are capable of effecting the solution of certain substances, and afterwards appropriating them. They increase in number in a very simple manner. If one of the protrusions above referred to be detached, artificially or by accident, a new and independent organism results. For never after a portion has completely separated does it again join, and form a part of the parent mass. So long as a pedicle remains between the two, though it be so thin as to be only just visible, the diverticulum may be withdrawn, and the whole form one single spherical mass of living, growing, moving matter. But if the communication be once completely severed two separate beings result, and these can never be incorporated so as to form but one.

Any one can study for himself the most important of the highly interesting phenomena which have been observed in these wonderful and simple organisms. Amœbæ can be readily obtained from water which has been left for a few days in a warm light room. Their growth can be watched from day to day, and their movements can be seen without difficulty. With the aid of high powers it will be found that the moving material is clear, transparent, and as far as we are able to discover, destitute of structure. It appears like matter of syrupy consistence which moves in all directions. No one has been able to offer anything like an explanation of these movements which every one can see. Authorities have expressed themselves as if they had been able to give a full and satisfactory explanation of the phenomenon, but there is nothing in their statements to justify the confidence which they seem to repose in the correctness of their own views. The cause of these movements is unknown, if not unknowable.

Bioplasm of Man.

But it must not be supposed that this wonderful capacity for movement and the power of taking up materials in the medium which surrounds them and converting these into the matter of their own bodies is a peculiarity of these very simple forms of existence. The movements are even now called *amœboid,* as if they were a peculiar characteristic of amœbæ, but so far from this being so, these phenomena are characteristic of the whole living world. They are, however, strictly confined to living beings, and nothing like them has been shown to occur in non-living matter. In man and the higher animals it is not always possible to see the movements of the bioplasm, for a very slight change in the circumstances under which life is carried on may cause its death; but in some cases, and these not a few, they may be seen in the living matter taken from man's organism, both in health and also in the diseased state.

The Living Matter or Bioplasm of Mucus.—If a little mucus which collects commonly enough upon the soft mucous membrane of the air passages be examined upon a warm glass slide, with the aid of a power magnifying 700 diameters, or upwards, little oval masses of germinal matter not unlike amœbæ will be seen in great numbers embedded in the viscid transparent material which gives to the mucus its properties, and which has been formed by the particles of the bioplasm.

By attentive examination movements will be observed in many of these masses, not unlike those above described in the case of the amœbæ. If the distribution of nutriment to the mucus be increased, the bioplasts enlarge, and divide and subdivide until vast numbers result. In some cases the entire mass appears to consist of the form of bioplasts ordinarily termed *pus corpuscles,* while the proportion of formed material which was abundant in ordinary mucus is exceedingly small. The bioplasm has multiplied so fast that there has not been time for the production even of the soft mucus.

Vital movements resembling those which have been described in the white or colourless blood corpuscles may be seen, but not so easily, in the bioplasm of young epithelial cells in that of cartilage, the cornea, connective tissue and other textures, and there can be no doubt whatever that all bioplasm possesses the power of movement, and that by this is to be explained the positions which the several masses respectively occupy in all the different tissues which they form, and in the preservation and maintenance of which, in a state of integrity, they play so highly important a part as long as life lasts.

Embryonic Bioplasm.—The growth and multiplication of bioplasm at an early period of development may be studied in an embryo, and many highly important observations may be made if the tissues of the chrysalis of the common blowfly be submitted to examination, especially when they have been successfully stained by the carmine fluid. A mass of formless bioplasm invariably represents the earliest stage of development of every tissue and organ. The bioplasm, which is concerned in the formation of the special tissues, emanates from this, and in many cases a sort of temporary structure is formed in the first instance in which the development of the higher tissue afterwards takes place. If one of the growing extremities of a fœtal tuft of the human placenta be examined, it will be found that the material which advances first, which grows away as it were from the tissue which is already formed, is a mass of bioplasm, which is dividing and subdividing into smaller portions, as represented in fig. 1, Plate XI. The loop of vessels gradually increases in the wake of this little collection of living matter which continues to move onwards as long as the organ continues to grow. These little collections of bioplasm bifurcate, and thus form branches into which vascular loops afterwards proceed. As in every other instance the first changes are effected by bioplasm, and upon this every kind of growth and development are entirely dependent.

White Blood Corpuscles, or Blood Bioplasts.—If a drop of blood be obtained from the finger by pricking it with a needle, allowed to fall upon a glass slide slightly warmed, covered with thin glass, carefully pressed and examined under a power of 700 diameters or upwards, here and there a colourless slightly granular, apparently spherical body will be seen amongst multitudes of the well-known red blood-corpuscles. These are the so-called white or colourless blood-corpuscles. They consist of living bioplasm or germinal matter, and exhibit movements like those referred to in the amœba and in the mucus corpuscle. The movements continue for some time after the blood has been withdrawn from the body. The colourless as well as the red blood-corpuscles vary much in size, although they are often represented as if they were of uniform diameter. These bioplasts multiply by giving off little diverticula, which become detached, and then grow into complete corpuscles. In the blood there are, besides the white blood-corpuscles, multitudes of minute masses of living matter, probably composed of the same material as the white blood-corpuscles. These were described and figured by me in 1863, and I showed that when the capillary walls became stretched by distension they would escape through little longitudinal rents or fissures into the spaces external to the vessels, where, being freely supplied with nutrient matter, they grew and multiplied, giving rise to the numerous corpuscles seen in this situation in inflammation. These minute particles are indeed the most important constituents of inflammatory exudation, and are the agents by which the important changes occurring in the exudation are effected.

Whenever the circulation is carried on slowly in any part of the body the colourless corpuscles grow and multiply, and at an early period of development, before the heart and lungs are fully formed, the only corpuscles are these white or colourless blood-corpuscles. This important fact maybe demonstrated by examining the blood in any of the small vessels of the embryo of a vertebrate animal. A very striking and beautiful example is represented in fig. 3, Plate XII, from the ovum of the turtle. The capillaries are seen to be filled with living growing blood bioplasts (white blood-corpuscles) every one of which was coloured by carmine fluid, and can be very distinctly seen in the specimen. Only here and there could an ordinary *red* blood-corpuscle be discovered.

In fig. 4, Plate XIII, I have given a drawing of part of a small vein, with a few capillaries opening into it, from a beautiful specimen of the *pia mater*, covering the hemispheres of the brain of a human embryo at the fifth month of intra-uterine life, to illustrate the same fact. The little veins were quite filled with blood bioplasts, very few of which had as yet become developed into red blood-corpuscles. In the capillaries represented in this drawing will be seen many very minute bioplasts which have been detached from larger ones and are growing. The bioplasts seen in the capillary interspaces are those which take part in the development of the other textures of which the *pia mater* is constituted.

In animals which hybernate, or which have been kept inactive in confinement for some time, and in man, under similar circumstances, many of the red blood-corpuscles in the blood-vessels are absorbed, just as they are from a clot formed in any of the smaller vessels, and in some instances from a clot situated external to the vessels, and the living bioplasts (white blood-corpuscles) grow and multiply at their expense. After a time such is the increase of the latter that the capillaries in many tissues are almost entirely occupied by them. This fact is illustrated by fig. 5, Plate XIII, which represents very small capillary vessels of the mesentery of the common frog in winter. The vessel is almost choked up with white blood-corpuscles, only one or two red ones remaining in the specimen from which the drawing was taken. Another illustration of this fact is given in fig. 2, Plate XI, which represents some of the capillaries from the bladder of a half-starved frog. The capillaries have much wasted, and contain no red blood-corpuscles whatever, their cavity being entirely occupied with fluid liquor sanguinis and masses of bioplasm, differing much in size, the largest particles having the ordinary dimensions of the white blood-corpuscles, while the smallest are so minute that they cannot be demonstrated under a power magnifying much less than 1000 diameters. It is remarkable that in this case the white blood-corpuscles are still growing and multiplying, and are, indeed, probably the active agents in the absorption of the tissues. In this specimen, from the most beautiful and delicate of all the tissues of the frog, may also be seen the very fine pale nerve-fibres, which I demonstrated some years ago. A fine bundle is seen at *a*, from which point it may be readily followed, as it divides into finer branches, ramifications of which are seen in every part of the drawing. The bundles of unstriped muscular fibres are marked *b*, while the bioplasm masses of the connective tissue corpuscles are represented here and there in the intervals.

So far I have endeavoured to show that the masses of germinal matter or bioplasm which are to be found in all parts of the tissues and organs of man and the higher animals at every period of life, and suspended in the nutrient fluids, notwithstanding such remarkable differences in power, exhibit the same general characters as those manifested by the living matter of the lowest animals and plants. In all cases it is the bioplasm only which lives and grows. Moreover, attention has been especially directed to the fact that the rate of growth of the bioplasm varies according to the scarcity or abundance of the nutrient material, and to the facility of its access. The bioplasts (white blood-corpuscles) of the blood increase in number, when the fluid in which they are suspended moves slowly as at an early period of life before the propelling apparatus is fully developed, or at any period of life when the circulation is retarded from any cause whatever. This remarkable growth and multiplication of the blood bioplasts seems to be determined by the altered condition under which life is carried on without necessarily any derangement of the health. The fact of the increase of the white blood-corpuscles in apparently opposite conditions of the system is thus very easily explained. A hybernating animal cannot be said to be suffering from disease, but nevertheless the blood in his capillary vessels contains a vastly increased number of bioplasts, and could hardly be distinguished from the blood stagnating in consequence of something impeding the circulation—a state of things which would be rightly regarded as disease. In this part of the inquiry we seem to be on the very confines of disease; in a sort of border land where the healthy process so gradually and imperceptibly shades into the morbid process that it would not be possible to draw a distinction in words, nor would the appearances which may be demonstrated to the eye enable us to define with greater exactness the special condition. In fact, up to this point there is no real difference. The state of things I have described if it continues, and if it leads to other changes, is disease. If, on the other hand, the circulation soon returns to its normal rate, the increased numbers of white blood-corpuscles soon pass into the circulation and are lost in the mass of the blood where they undergo further changes, and there is no further evidence of even a temporary disturbance of the healthy condition than is afforded by some slight disturbance of the nerves, giving rise, perhaps, in the case of man and the higher animals to slight pain, which soon passes off, and often escapes notice altogether.

From Health to Disease.

I have endeavoured to show that the only material in the organisms of living beings capable of growth and multiplication is that which has been termed *bioplasm, germinal, or living matter*. In fully formed tissues the proportion of this is very small. Still, all active change depends upon this living matter, however little there may be. If there be none, the tissue is as incapable of undergoing active changes as if it did not form a part of the body. The smallest particle of bioplasm possesses active powers, and if supplied with proper pabulum, soon grows. Each little bioplast grows, that is, increases, by taking up material differing entirely from it in composition, properties, and powers, and converts certain elements of this into matter identical with that of which it consists. After the bioplasm-particle has reached a certain size, division occurs. Instead of growing larger and larger, and forming a continuous mass of enormous size, as some have fancifully supposed to exist at the bottom of the ocean, portions are from time to time detached and separate themselves, moving away from the parent mass. Each of these little germs has properties in many respects like those of the parent mass. It lives and grows, attains a certain size, and may produce its kind in the same way.

Now, the whole human organism at a very early period of its development consists entirely of little masses of living or germinal matter like those above referred to. Each of these grows and divides and subdivides, so that multitudes at length result from the division of a few: and these are all the descendants of the first primitive germinal mass, which was derived from pre-existing germinal matter. After a time some of these cease to multiply, though they still live and take up food. The living matter of which they are composed undergoes change. It dies under certain conditions, and tissue results. In this way muscle, and nerve, and fibrous tissue, and bone, and hair, and horn, and nail, and all the other tissues, are formed. In the adult, however, there remain some masses of germinal matter which go on growing and dividing just as all of them grew and multiplied in the embryo. Among these are the white or colourless blood-corpuscles, which possess formative power even in old age in greater degree than any other kind of bioplasm in the adult. At the deep aspect of the cuticle, and below the fully-formed epithelium of mucous membranes and some glandular organs, are masses of germinal matter, which are dividing and subdividing in the same way throughout life. These, in the ordinary course, move towards the surface, and as they move, each, in the case of the cuticle, gradually forms upon its surface the hard cuticular matter (cell-wall) to which the properties of the epidermis are due.

It has been already said that the bioplastic masses of different organisms, and those in different parts of the same organism, possess very different endowments. From one kind of bioplasm is formed muscle, from another nerve, from another fat, and so forth, and yet all these kinds have directly descended from one. They could not be distinguished from one another, nor from their primary mass. Neither could one of these kinds of bioplasm in the adult develop a mass capable of producing the rest. Although no one could distinguish one particle from the other, each will produce its kind, and that alone. It would be as unreasonable to expect an amœba to result from a pus-corpuscle, or from a yeast particle, or to suppose that by any alteration in food or management a cabbage would spring from a mustard seed, or the modern white mouse from the descendant of an ancestral white rabbit, as it would be to suppose that muscle, nerve, brain, gland, or other special tissue might be produced indiscriminately by any mass of bioplasm of the adult, supposing that the conditions under which it lived were changed to any possible extent. Its powers, which are within, and upon which the capacity to develop depends, cannot be thus changed by any mere change in external circumstances.

The Production of Pus.—But it is very remarkable that the many kinds of germinal matter of the organism of man and the higher animals, though differing so much in power or property that one produces nerve, another muscle, a third bone, a fourth fat, and so on, will each under certain conditions give rise to a *common form of germinal matter or bioplasm differing in properties and powers from them all.* This is the form of bioplasm known as *pus,* which may go on multiplying for any length of time, giving rise to successive generations of pus bioplasts, which exhibit remarkable vital properties, although they cannot form tissue, nor produce tissue-forming bioplasts of any kind whatever.

It is evident from this that the power is manifested in one direction only—onwards. Embryonic living matter or bioplasm gives rise to several different kinds, not one of which can produce matter with the endowments of that which existed immediately before it, and from which it sprang. And yet every kind of germinal matter exhibits powers of infinite growth.[1]

When bioplasm or germinal matter lives very much faster than in health, in consequence of being supplied with an undue proportion of nutrient matter, a morbid bioplasm

[1] While, however, the process of division is proceeding, as has been described, in some cases a small portion of the germinal matter does not undergo division into masses of the next series, but retains its primitive powers. This remains in an embryonic condition after the tissue has been formed, and thus the development of new tissue, even in advanced life, is, in some cases, not only possible, but actually occurs. Many cancers and other morbid growths probably originate in these masses of embryo bioplasm which remain for a long time in a quiescent state embedded in some of the fully-formed textures of the adult.

results; and if the process continues for a short time, changes familiar to those conversant with pathological alterations occur upon a large scale.

In discussing questions of this kind, involving such minute details, we must, however, be most careful to avoid too hasty generalization, and proceed by very slow steps; and this is more particularly necessary if it so happens that our inferences in some measure accord with the views of speculative and excited persons, who are always fancying that we are on the eve of some grand discovery which is to revolutionise thought. Many might perhaps infer from the arguments advanced, that I incline to the view that the lowest living forms are capable of being produced by the retrograde development of higher forms, and that bioplasm even very high in the scale of organization, may produce forms of bioplasm approximating more and more closely to the lowest constant forms of life with which we are acquainted. A doctrine asserting that by continual retrogression through ages the descendants of the highest forms would gradually deteriorate until their only remaining representatives were monads, would not be very easily disproved, and might be supported by many ingenious arguments. It is a view that doubtless would recommend itself to some minds in the present day.

On the other hand, it is quite conceivable that cells and organisms may retrograde and produce various modified forms, without giving rise to any of those particular forms characteristic of the lower organisms with which we are acquainted. Nay, cells of different organisms might give rise to many different retrograde forms, and every one of these be very different from one another. It is obviously possible that there should be infinite advance and infinite retrogression in multitudes of parallel lines, as it were, without the resulting forms of any one line becoming identical with those of another. Just as it is possible to conceive infinite advance in the features of the dog, without any resemblance whatever to the human face resulting, and retrogression and deterioration of the latter proceeding to any degree, and continuing for any length of time without the development of the simian type of countenance.

Sufficient allowance is not made by many thinkers for the infinity of difference in structure and variety of change possible in living forms, without the production of two forms exactly alike, or any indication of the merging of one set of forms into another. It must not be forgotten for an instant that from such a marvellous storehouse of facts as has been placed at our disposal in nature, we may with very little ingenuity pick out series of facts in favour of many different general hypotheses; and however conflicting these may be with one another, it may not be possible to disprove any one of them in the present state of knowledge. The fact that masses of germinal matter, derived by direct descent from cells of one of the lower animals, may grow and multiply in man's organism, and *vice versâ*, might be adduced as an argument in favour of the original common parentage, countless ages back, of the predecessors of both; but there are, it need scarcely be said, facts and arguments tending to a different conclusion, and these must not be lost sight of in endeavouring to arrive at the truth.

I propose now to draw attention to the facts I have been able to observe in connection with the deterioration in power of bioplasm during that increased multiplication which results from the very free supply of pabulum.

Bioplasm of Epithelium.—When the germinal matter of the epithelial cells of certain mucous membranes, or that of other tissues of the body, or the germinal matter of the white blood-corpuscles, lives faster than in health, in consequence of being supplied with an undue proportion of nutrient material, it grows and multiplies to an enormous extent; so that one mass may perhaps be the parent of 500, in the time which, in a perfectly healthy state, would be occupied in the production of two or three cells. And in some ordinarily very slowly-growing tissues, the germinal matter may in disease divide and subdivide very quickly, although in the healthy state it would undergo scarcely any appreciable change in the course, perhaps, of several weeks or months. The increased rate of access of nutrient material to the living matter is the necessary condition of its increase. The living matter always *tends* to increase, but in the normal state of things it is only permitted to do so at a certain regular rate, which is determined by the even distribution and somewhat limited access of the nutrient material.

In certain cases in which an increased proportion of nutrient material is distributed to the epithelium of the mucous membrane,—as, for example, to that of the fauces,—the young epithelial cells grow and multiply so rapidly that the superficial layer of older and hardened structure becomes detached, and the free surface is formed of a thick layer of soft, spongy, epithelial elements, with, in many instances, but faint indications of division into individual epithelial particles. In fact, under the circumstances alluded to, growth is taking place too rapidly for the formation of the characteristic epithelial texture, though the changes are not so rapid as to lead to the production of actual pus. The spongy texture produced may be regarded as occupying the position midway between healthy epithelial tissue and the pathological germinal matter or pus. I have observed these facts in the young rapidly-growing, but as yet imperfectly-formed epithelial particles in specimens taken from the surface of the pharynx in a case of slight sore-throat coming on in a person enjoying ordinarily good health. The mode in which the masses divide and subdivide could be well seen, and the thick plastic character of the matter of which they are composed has been well given in drawings. The greater part of the material consists of living matter or bioplasm, some of which has probably undergone conversion into soft-formed material, which, however, still remains mingled with it. From any part of one of these masses diverticula might have been formed, and thus new bioplasts, each capable of undergoing conversion into an epithelial cell, result. Many epithelial formations exhibit much the same changes in disease, and the gradual transition from the healthy to the morbid state is beautifully indicated. Nay, we may almost conceive that it is by unremitting continuance of this very process, combined with irregularity in the rate of multiplication of contiguous particles, that the remarkable pathological formation, epithelial cancer, results.

If, then, the bioplasts of a tissue receive an unusually abundant supply of nutrient matter, they grow and multiply just like the amœba, the white blood-corpuscle, the mucus-corpuscle, and the pus-corpuscle, and they may give origin to pus. Masses of bioplasm which under ordinary circumstances would form cuticle, grow and live very fast, and lose their cuticle-forming property. The changes are well shown in fig. 6, Plate XIV, to the left of which, at *a b c d*, are represented separate cells, the bioplasm of which is growing and dividing and subdividing. The cells multiply faster than any cuticle cells, and the numerous descendants at last produced are pus corpuscles. From these pus bioplasts diverticula proceed, and particles are from time to time detached which are extremely minute, and by their movements may pass through very narrow chinks in tissues, and thus spread from the point where they were developed: not only so, but so minute are these particles of pus bioplasm, that, like the little germs detached from the yeast cells and other microscopic fungi, the amœba germ, and many others, the atmosphere will support them; they may thus be wafted long distances from the spot where they were produced. If exposed to great heat or cold, or to the action of certain gases or vapours, they will be killed, but in warm, moist air they will live; and if they fall in a favorable place, that is, where there is proper food for them, they will grow and multiply a thousandfold. But the yeast will not produce amœba, or the latter pus. The pabulum suitable for the first would kill the last.

Multiplication of White Blood-corpuscles in Disease.—Next, then, let us consider in what way the multiplication of the bioplasts (masses of germinal matter) of the blood in the capillary vessels in disease differs from the process which we have seen occurs in the vessels at an early period of development, and during the winter sleep at all periods of life in hybernating animals, and in man under physiological conditions which cause the blood to circulate very slowly, or to stagnate for a time in the smaller vessels of the body. As will be inferred from the remarks made in page 218, I do not think that any distinct line of demarcation can be drawn between the physiological and the pathological change. In *inflammation,* the phenomena above referred to proceed a stage further, which is unquestionably pathological. But even if this stage be reached, it by no means follows that the texture involved should not regain its normal condition and the previous healthy state be perfectly restored. On the other hand, it is quite certain that if the state of things now to be described proceeds to any great extent, destruction of tissue is inevitable and return to the original condition impossible. *Repair* may follow the injury, but this *repair* involves serious alteration in structure, with corresponding deterioration in action, without capacity for improvement and without the possibility, under any circumstances, of return to the former state.

When the circulation through the capillary vessels is impeded in cases of disease, the blood bioplasts (white blood-corpuscles) multiply, and the capillaries often appear to be filled with them, in which case they closely resemble the vessels of an animal during the early period of its development. This state of things always exists in inflammation, and the multiplication of the bioplasts proceeds to a wonderful extent. The appearances seen are not due simply to the *accumulation* of white blood-corpuscles, as some have held, but only partly to this, and mainly, as I pointed out many years ago, to their actual growth and increase. "If in any capillaries of the body the circulation is retarded from any cause, an increase in the white blood-corpuscles invariably takes place. In congestion and inflammation of the vessels of the frog's foot, the number of the white blood-corpuscles soon becomes so great as to impede and ultimately to stop the circulation through the vessel. Although the great majority are merely corpuscles that have been retarded in their passage, there can be little doubt that the corpuscles actually multiply in number in the clot that is formed."[1] This fact of the *increase* of the white blood-corpuscles has been overlooked in consequence of the examination not having been conducted with sufficient care, and with powers of low magnifying power.

The capillary vessel of an inflamed part being distended, its walls consequently become much reduced in thickness, and little longitudinal rents or fissures are here and there produced. Through these serum, holding in suspension very minute bioplasts detached from the larger ones growing and multiplying in the vessel, pass. Having thus extravasated, these particles, resulting directly from the subdivision of the white blood-corpuscles, make their way into the interstices of the surrounding tissues, and being nearly stationary, and abundantly supplied with nutrient pabulum, grow and multiply in the new locality, and at an increasing rate. The phenomena here described will be understood if fig. 7, Pl. XIV, be carefully examined. This has been copied from a preparation which was preserved in the year 1863. But the facts demonstrated were well known to me, and were described in my lectures before 1863, and were particularly referred to in a paper presented by me to the Royal Microscopical Society in that year. I did not come to the conclusion which has been recently advocated by Cohnheim, that an individual white blood-corpuscle passed through the wall of the vessel, and then changed its characters and became a pus corpuscle; but my observations led me to infer—and of the correctness of the conclusion I am fully satisfied—that the particles of germinal or living matter seen in such great numbers *outside* the vessels in many cases of inflammation at an early stage, result principally from the growth, division, and subdivision of minute particles of germinal matter which have passed through the vascular wall suspended in the fluid exudation. These masses of germinal matter (fig. 7, Pl. XIV) are the *descendants* of white blood-corpuscles, but they are not the white blood-corpuscles *themselves* which were previously in the blood, and which were circulating in that fluid. They may continue to grow and multiply like other kinds of germinal matter, until at last that rapidly-growing form of bioplasm, the common result of the greatly-increased growth and multiplication of every form of bioplasm in the living body, may be produced. Thus *the pus corpuscle* may be a descendant of the white blood-corpuscle,

[1] "On the Germinal Matter of the Blood, with Remarks upon the Formation of Fibrin," December 9th, 1863, 'Trans. of the Mic. Soc.'

as well as of the germinal matter of epithelium, and of other tissues, and we may trace back its parentage to the original embryonic bioplasmic mass, which must be regarded as the primitive ancestor of all.

Degradation of Bioplasm in Peritonitis, and the production of Contagious Virus.—In peritonitis we have an example of an imflammation which much more frequently proceeds to the formation of pus than inflammation of other serous membranes. The great vascularity of the peritoneum as compared with that of other serous membranes, may perhaps account for this fact. It is interesting to discuss briefly the characters of the different "inflammatory products," as they are called, resulting from peritoneal inflammation, varying in intensity.

In *slight inflammation* there is great vascular distension, as in other cases accompanied by the escape of exudation in which are suspended particles of bioplasm. The exudation coagulates upon the surfaces of the serous membrane, perhaps gluing them together. The fluid portion is gradually absorbed, and if the case progresses to recovery, much of the coagulated matter is also taken up, a little being transformed into fibrous tissue, resulting in a few "adhesions," or mere thickening of the serous membrane, as the case may be.

When, however, the intensity of the inflammation is greater, the little particles of bioplasm originally derived from the white blood-corpuscles, grow and multiply, and, with the fibrinous matter in which they are entangled, form transparent flocculi, which are suspended in the serous part of the exudation, or adhere here and there loosely to the peritoneal surface. Many of these flocculi are found to contain multitudes of bioplasm particles, and oftentimes a vast number of these are suspended in the fluid, and congregated here and there, forming little collections upon the surface of the delicate serous membrane, to which they adhere, and where they grow.

If the inflammatory process still continues, and increases in severity, the vascularity of the membrane becomes more marked, and the exudation is poured out from the blood more abundantly; the masses of bioplasm increase in number yet faster, and the exudation in consequence appears nearly opaque. The flocculi are of a yellowish colour, and look not unlike pieces of clotted cream which stick here and there to the peritoneum covering the intestines and the inner surface of the abdominal parietes. Not unfrequently the surface is smeared over in places with whitish pasty masses of soft cream-like matter, in the intervals between which the highly-injected vessels stand out with great distinctness. The masses of bioplasm would now be called *pus corpuscles.* Here then is an interesting example of the production of pus-corpuscles by the rapid growth and multiplication of particles of bioplasm which were once in the blood, and intimately related to the white blood-corpuscles.

But further: if, as is well known, a little of this material be introduced into the body, as sometimes unfortunately happens from a dissection-wound in the course of making a post-mortem examination, terrible inflammation may be excited in the person inoculated. The most tiny morsel of this virulent, rapidly-multiplying morbid bioplasm may give rise to a dreadful form of blood-poisoning, which may end fatally and in a very short time. And in some cases similar poisonous particles are so light that they are supported by the air, may find their way into the blood through the respiratory organs, or gain access to the circulating fluid even by traversing the narrow chinks between the epithelial cells of the cuticle.

Now, what is the nature of the matter inoculated, which produces these dreadful results? The virulent poison which sometimes destroys life in cases of dissection-wounds cannot be attributed to the presence of vegetable germs, for the period of its most virulent activity is shortly after death, but before the occurrence of putrefaction, when the vegetable fungus germs multiply.

It has been assumed that the poison in question is not developed until *after* death has occurred. But no one has shown that if inoculation were effected while the patient yet lived, the results would be in any way different. There is surely no more doubt that such poison is developed during life, than that small-pox and syphilitic poison, and many others, which, I have shown, are allied to it, and probably grow and multiply in the same manner, increase during life.

When putrefaction has actually set in, and bacteria germs are being developed in immense numbers, a punctured wound is not productive of the dire consequences which result if inoculation takes place within a few hours after death. In fact, the real virus loses its power when decomposition commenees. Before vegetable germs appear the virus is active; soon after these have been developed it is harmless. Its power cannot, therefore, be attributed to the germs but to something else which continues to live and remain active for a short time after death, and then becomes changed and disappears, the products resulting like those remaining after the death of other forms of living matter being appropriated by the growiug vegetable organisms which are seen in such immense numbers.

It would seem, therefore, reasoning from the facts in this one form of inflammation, that there can be little doubt as regards the nature of the poison. The minute particles of bioplasm or living matter produced in such great numbers as the inflammation advanced, are the actual agents, and it is by their rapid growth and multiplication in the lymph, and afterwards in the blood, particularly in the capillaries in various parts of the body, that the dreadful effects exerted by these particles in an organism into which some have been introduced by inoculation, must be attributed.

It seems, then, that every stage in the production of at least this particular virulent poison, may be watched and studied, and from the facts ascertained a connected history of its development may be compiled. We have seen that the first particles of bioplasm resulted directly from the white blood-corpuscles. In consequence of continued free access of nutrient material, series after series, generation after generation, of new particles was produced; each series degenerating in formative power, and acquiring new powers, but of mere growth and multiplication, and a capacity for living upon materials which would not have been appropriated by the bioplasm from which it originated. The following diagram may perhaps assist in rendering my meaning clearer.

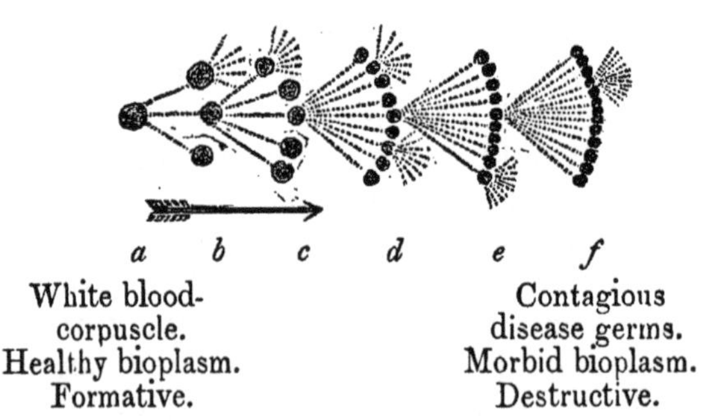

Successive series of living particles resulting from the growth and multiplication of a single white blood-corpuscle. Each series grows faster than the one from which it originated. In the plan, the process of multiplication is represented as if it only occurred in the case of one particle in each series, but, in order that an accurate conception of the process may be formed, similar radiating lines with bioplasts must be supposed to diverge from every part of the circumference of every particle. *a*, Is a white blood-corpuscle; *b*, *c*, *d*, and *e*, Successive series of particles which produce others, until at last contagious disease germs, *f*, result.

The contagious disease-carrying bioplast is then a descendant of the white blood corpuscle, or some other form of bioplasm or living germinal matter of the body which was once healthy. And these have descended from the original embryonic bioplasm mass. The death-producing disease germ, therefore, is not derived from the vegetable kingdom, and is not a parasite, but seems to have been derived from the living or germinal matter of man's own organism.

The evidence seems to prove that from normal white blood-corpuscles may be evolved after very rapid growth and multiplication, bioplasts or living particles, which possess the most virulent poisonous properties, destroying not only the organism which gave them birth, but able to kill any other organism into which they may be introduced. The higher formative life of the white blood-corpuscle is degraded, and its character for ever lost. But while this degradation has been going on, new powers of life have somehow been acquired. We cannot say there has been any loss of mere vitality, of appropriating pabulum, of growing and multiplying, for the new particles live faster than those from which they emanated, but there has been loss of constructive, loss of formative, power. Living matter derived from the blood, which performs a very active part in healthy changes in the organism, and without which healthy life cannot be sustained, escapes from the blood, grows and lives under new conditions, and from this results a living matter with properties or powers so very different that if it returns to the blood it may, by its increased rate of growth and multiplication, appropriate the nutriment destined for the healthy textures, and destroy the organism. It is a poison generated by the degradation of healthy living matter. From bioplasm which left the blood a formative constructive living substance, has been developed, a living matter so deleterious that if it re-enters the same blood, or that of another being, terrible derangement and irreparable injury are occasioned, if indeed death does not occur.

And there is reason to think that the generation of the poison of many contagious diseases, and all contagious fevers, occurs in the same way. It is certain that many cases of blood-poisoning, and various forms of idiopathic fever, depend upon the passage into the blood, and its dissemination through the system, of a poisonous bioplasm which has been generated in the body, the virulent bioplasm itself having resulted from the growth and multiplication of generations of particles derived by continuous succession from the normal bioplasm of the organism. These views have been further developed in the second part of my work on "Disease Germs," which will, I hope, be completed in the course of the present summer.

On the Minute Anatomy *of some of the* Parts *concerned in the* Functions *of* Accommodation *to* Distance, *with* Physiological Notes. By John Denis Macdonald, M.D., F.R.S., Staff-Surgeon H.M.S. "Fisgard."

(Communicated by the Director-General of the Medical Department of the Navy.)

No organ in the body exhibits more beauty of construction to unravel, or more interesting physiological problems to solve, than the eye; and it is no wonder, notwithstanding the great advances of modern science, that much room should still remain for the improvement of our knowledge of its minute anatomy, and the function subserved by every part in the economy of vision. It would be quite impossible within the limits of a few pages to deal with a twentieth part of the facts and views that have been advanced from time to time in relation to the subject of the present paper. Nor would that be the writer's intention, though it were possible, being merely desirous of making known some few anatomical particulars which do not appear to have been hitherto noticed, and of associating these with the function of accommodation to distance.

We commence, then, with the conjunctiva, though it is not immediately engaged in adaptation. As soon as we have traced this membrane to the margin of the cornea we may readily follow the epithelial coat over that structure, but the basement membrane in this locality is generally believed to be represented by the anterior elastic lamina of Bowman. The intimate union of the conjunctiva with the cornea is noticeable not only in dissection but in the pathological condition known as chemosis, in which, by intelligent design as obvious as any that can present itself in the eye, effusion of fluid is prevented from encroaching upon the domain of the cornea. The tying down of the deep surface of the anterior elastic laminæ to the cornea proper by oblique and decussating fibres, first described by Mr. Bowman, would, according to the present view, sufficiently explain the close union above noticed; but in the eye of the ox or of the sheep, in which an anterior elastic lamina is not demonstrable, are we to assume that there is no basement membrane between the cornea proper and the conjunctival epithelium? In a late series of investigations made with the view of settling this point, the writer has been quite satisfied of the existence of a delicate basement membrane in the locality indicated, doing away with the apparent anomaly of an epithelial pavement in direct contact with a fibrous structure.

In the human eye a thin structureless membrane supporting the epithelium is traceable all over the face of the cornea, and which, no doubt, from its very intimate adhesion to the anterior elastic lamina, has hitherto escaped detection. At *b*, fig. 1, Plate XV, this membrane is shown in vertical section, and at *b*, fig. 2, it is seen in face, and as it were ripped up from *c*, the anterior elastic lamina, near the border of the preparation, a portion of the conjunctival epithelium remaining at *a*. Fig. 3, Plate XV, represents a vertical section of the cornea of the pig, showing, faintly, the same membrane at *b*, with an anterior elastic lamina very similar in appearance to the corneal structure beneath, and very different from *e*, the posterior elastic lamina, though nearly of the same thickness.

In the ox and sheep, as above noticed, an anterior elastic lamina cannot be detected, though the tissue of the cornea proper is evidently more condensed and highly refracting immediately beneath the conjunctiva than elsewhere, as seen at *c*, fig. 4, Plate XV.

The substance of the cornea in the shark (Plate XVI, fig. 1, *h*) consists of a superimposed series of unbroken laminæ extending over its whole area, and seen in face; these layers exhibit delicate fibrous markings, taking different courses, though more or less uniform in each. At the border of the cornea they are seen to be directly continuous with the fibrous tissue of the sclerotic, into which the lamellation may be traced for some considerable distance, as shown at *c*. The alternately light and dark appearance of the plates would seem to be an optical effect due to the different directions taken by the component fibres; but if this explanation be not correct it is still a problem requiring solution. In my notes I have named the first and last of these plates respectively the anterior and posterior elastic laminæ, but I cannot now be sure whether they are quite structureless or not.

In the human eye the anterior elastic lamina, or Bowman's membrane, is considerably thicker than the posterior, or the membrana Descemetii, which appears to have been first observed in the eye of the horse. This latter membrane, fitting loosely to the cornea proper, is much more easily demonstrated than the former, and is lined posteriorly with a single pavement of transparent cells, which appear to rest upon a fine basement membrane continuous with the investment of the iris and ciliary processes and the lining of the choroid.

The pillars of the iris (Plate XVI, fig. 5, *g*), soon after they leave the margin of the cornea, break up into two principal sets or bundles of fibres, on the one hand diverging as they expand into the ciliary muscle (*o*), and on the other converging as they pass into the radiating tissue of the iris. The primary fasciculi are crossed at right angles superficially or externally, and thus bound down by a stout circlet of unstriped muscular fibres (*h*), corresponding with the inner wall of the circular sinus (*s*), or the space apparently contrived to permit of the play of this annular muscle.

The fibres are flat and band-like, and differ essentially from the tomentose or flaxy tissue composing the principal mass, if not the whole, of the radiating bundles, both of the iris and the ciliary muscle, which latter is decidedly a problematical structure, more especially when the light of comparative anatomy is brought to bear upon it. Thus, while it is present with the same histological characters in the eye of the owl (Surnia), for example, it is superimposed by veritable striped muscle, taking its origin from the inner fibrous surface of the cartilaginous and bony sclerotic. There may possibly be some organic muscular fibres associated with it, but doubtless the great mass of it is of a very different nature.

Immediately behind the circular muscle above described, some accessory fibrous bundles (*r*) pass outwards and backwards to the posterior boundary of the circular sinus, where they are incorporated with the sclerotic. These were first noticed by Mr. Bowman, and their office would appear to be to keep the ciliary ligament and fore part of the choroid in relation with the sclerotic during the action of the circular muscle. It is probable, also, that in man they only form the rudiment of that more important structure in the bird, with which the striped voluntary muscle is connected anteriorly.

Just within the attached margin of the iris the fibrous tissue is supplemented by a circular framework of stout-branched and reticulate yellow elastic element, essentially different from that occurring in other parts of the human body. This remarkable structure, acting through the medium of the radially disposed wavy tissue in connection with it, both sheathing the vessels and running in the intervening spaces, is the mechanical antagonist of the sphincter fibres of the iris. It thus fulfils an office, analogous to that of the ligamentum nuchæ in browsing animals, substituting and relieving muscular action, and obviating the necessity of so nice a proportion as should otherwise exist between the complementary contractions and relaxations of the radiating and circular fibres regulating the size of the pupil.

The strong fibrous tissue of the choroid and its appendages seems to be distributed, wherever it is required, beneath the delicate investing membrane, to use a familiar illustration, like the padding of a coat. Thus, it is to be found in the iris, surrounding its vessels, on the external and internal surface of the ciliary circle, in the tissue constituting the tapetum lucidum of the lower animals, and it is not absent, though it has not the brilliancy, in the same situation in man himself.

The ciliary processes are intimately adherent to the zonule of Zin, with an intervening pigmentary coating, the various little folds and corrugations of the one being received into corresponding recesses and depressions in the other. When the processes are forcibly severed from their attachment the pigment-cells appear, at least in many places, to be torn, as it were, into two layers, thus liberating the contained pigment-granules so as to stain the surrounding parts. Under a low power the general appearance of the ciliary circle is not unlike a beautiful though minute model of a mountainous country, and in the physiology of adaptation the little spurs and folds, taking a circular or transverse direction, play as important a part as the principal mountain chains, for thereby any tendency to dislocation is obviated when pressure is being exerted upon the fluid in the canal of Petit, and through this upon the free margin of the lens. Whether the old doctrine of the splitting of the hyaloid membrane at the outer angle of the canal be true or not, the transparent radiating tissue which is implanted anteriorly into the capsule of the lens near its border is covered with a distinct layer of homogeneous membrane bearing the impress, as above described, of the ciliary processes.

The intimate structure of the lens is such as to favour Helmholtz's view that adaptation to distance is principally effected by the alteration of the figure of this body, increasing or diminishing its convening power as the case may require.

It will be recollected that the band-like fibres passing from the limbs of the triradiate fissure on one side to the angles between them on the other are somewhat wedge-shaped towards their extremities, with interlocking margins. By this arrangement provision is made for the close approximation of the fibres by lateral pressure exerted through the medium of the fluid in the canal of Petit, ensuring also an equable convexity of the lens without dislocation of its structural elements; and, what is most important, the definition is not obscured by what would otherwise be equivalent to a flaw in the lens. Moreover, we find the lens gradually increasing in density from the anterior surface to the nucleus, and again, of course, decreasing in density from the nucleus to the posterior surface, determining a curvilinear course to the refracted ray, which we can but imperfectly imitate in our most scientific combinations of refracting media. It must be apparent from all this that a very gentle lateral pressure, as evenly applied as the plan of construction is calculated to permit, will produce such an alteration, both in the figure and density of the lens, as materially to affect the focal range. In reference to the calculation of Albers Dr. Kirkes remarks that "the change in the distance of the retina from the lens required for vision at all distances, supposing the cornea and lens to maintain the same form, would not be more than about one line;" and Dr. Young "estimated the necessary change at one sixth of the length of the axis of the eye;" but the former deduction is, perhaps, the more reliable.

Without multiplying authorities, it appears clear, however, that a very trifling alteration in the conditions of figure and density in the refracting media, and in their position with respect to the percipient surface, will suffice for all focal adaptation.

Though the remarks made in connection with the foregoing anatomical details have anticipated the conclusion to some extent, the view here taken of the mechanism of adaptation may be set forth as follows, taking for illustration a diagramatic section of the fore part of the eye of a pig (Plate XVI, fig. 5).

The pillars of the iris (*g*) being the common tendons, so to speak, of the radiating fibres and of the ciliary muscle (*o*), assuming these structures to be muscular, it stands to reason that their contraction can only produce on the one hand dilatation of the pupil, and on the other draw the ciliary processes (*i*) forwards and a little inwards, so as to stretch the ligament of the lens and draw that body forwards. Hand in hand with this change, the lenticular pit in the fore part of the hyaloid (*n*) would become shallower by proportionate pressure on the vitreous humour, while the pressure would be removed from the fluid in the canal of Petit (*m*), and thereby from the free margin of the lens (*l*). Now, as the sphincter surrounding the pupil antagonises the radiating fibres of the iris, so the annular muscular band (*h*) on the inner wall of the circular sinus, acting in concert with the sphincter pupillæ, would antagonise the radiating fibres of the ciliary muscle (*o*), and also draw backwards and inwards the pillars of the iris (*g*), so as to increase the convexity of the cornea. Moreover, the fore part of the sclerotic (*q*) would be also drawn inwards by means of those fibrous bundles (*r*) which enter it at the posterior boundary of the circular sinus (*s*). The effects of the changes here described upon focal adaptation are too obvious to require further elucidation.

Experimenters have found—1st. That the application of a direct stimulant, viz. Calabar bean or galvanism, will produce contraction of the sphincter fibres of the iris and thereby of the pupil. 2ndly. That a sedative influence, such as that of atropine, will induce relaxation of those fibres, and consequent dilatation of the pupil. 3rdly. That alternate stimulation of the third nerve and of the sympathetic respectively will cause contraction and dilatation of the pupil. 4thly. That if the sympathetic twigs are paralysed, contraction of the pupil will ensue, the passive antagonism, so to speak, of the sphincter overcoming that of the radiating fibres; and conversely, that paralysis of the third nerve will give rise to dilatation of the pupil, through relaxation of the sphincter. 5thly. As instanced by Dr. D. Argyle Robertson,[1] that after the galvanic stimulus has ceased to produce contraction of the pupil in the recently dead iris, actual dilatation takes place. Now, if it be admitted, as is most natural, that stimulants or sedatives will induce contraction or relaxation, as the case may be, in muscles supplied with nervous energy from any source, whether it be cerebral, spinal, or sympathetic, the following positions may be reasonably assumed to account for all the foregoing phenomena:—1st. That the sphincter fibres are more powerful than the radiating. 2ndly, That simple elastic force, supplementing the latter, would in some measure make up the difference. 3rdly, That in all probability agencies of a stimulating or depressing kind would exert their influence, primarily on the cerebral nerves, and secondarily upon the sympathetic, which is also known to retain its irritability for a longer period (Budge). 4thly. That, as in the normal play of the antagonistic radiating and circular muscular fibres, when one set is in a state of contraction the other is relaxed, and *vice versa*, so, when contraction or relaxation of one is induced by artificial means, the opposite condition is likely to arise in the other as an automatic consequence. The latter position is of great importance, for by taking it into account the often contradictory results of experiment may be reconciled.

However necessary it may be to seek a physiological

[1] 'Lancet,' Feb. 5th, 1870, p. 212.

indication of the existence of genuine muscle amongst the radiating tissue of the human iris, no such necessity can present itself in the eye of the pig, in which the sphincter pupillæ is antagonised by no less than five distinct sets of muscular fibres, leading to the inference that a well-conducted series of experiments upon the eye of that animal would develop important results.

On the BEST METHODS *of* STUDYING TRANSPARENT VASCULAR TISSUES *in* LIVING ANIMALS. By RICHARD CATON, M.B., Assistant Physician to the Liverpool Infirmary for Children.

DURING the last three years, the researches of Professor Cohnheim and other German observers have rendered the study of the relation between vessels and parenchyma particularly interesting to physiologists.

The amœboid movements of white blood cells; their relation to pus corpuscles; their spontaneous division, according to Klein; the entrance of pigment cells into the vessels, as recently described by Saviotti; along with other points, are just now deserving careful study by microscopic observers. The following enumeration of the chief transparent tissues suited for purposes of study in vertebrate animals, along with a description ot some improved methods of examination, may perhaps be useful to those who study the circulation from a physiological point of view; and also to those who are interested in it merely as a microscopic object.

I. *Circulation in Animals breathing by Lungs.*

1. *The Frog.*—In studying the circulation in any animal, the first essential is, that it be kept still on the stage of the microscope. This has generally been attempted by mechanical restraint, or sometimes by the use of chloroform.

I have found the subcutaneous injection of the hydrate of chloral answer much better. A solution of four grains to the drachm is a convenient strength, and the rule for its use for the frog is very simple: as many minims should be injected as the frog is drachms in weight. The injection is made under the skin of the back, with a morphia syringe. In the case of a painful operation a larger dose is needed, or woorali may be employed.

A grain of woorali in powder is added to an ounce of water; it dissolves to a slight extent. Two or three minims of this solution (previously shaken) injected, will render a frog of six or seven drachms' weight perfectly motionless for twelve hours. The drug does not produce its effect fully for one-and-a-half or two hours.

Woorali may be obtained from Messrs. Morson & Son, of Southampton Row.

The frog having been narcotised by one of these drugs, the circulation may be studied in any of the following structures:—

(*a*) The foot-web, which is very familiar to every one, but which, from its comparative opacity, is useless in observations requiring much nicety.

(*b*) *The Mesentery.*—A longitudinal incision $\frac{2}{3}$rds of an inch in length, is made through the skin and muscle of the abdominal wall, on the left of the middle line; bleeding is easily arrested by a gentle touch of the solid nitrate of silver. A loop of gut is carefully withdrawn by a pair of photographer's horn forceps, (if the ordinary metal forceps be used, it is most difficult to avoid injury to the delicate vessels on the surface of the gut).

Great care must be taken to avoid a tear of the mesentery.

A tin plate should be prepared, about the size of the stage of the microscope, and having a circular aperture in its centre half an inch in diameter. A ring of cork $\frac{1}{3}$ inch in thickness, is glued round the margin of this aperture. The frog is placed on its side, on that part of the plate which is remote from the pillar of the microscope, and the mesentery is laid on the perforated cork. It is a great convenience to have two little arms, one inch in breadth and two in length, projecting from the corners of the plate, on its farther side; these can be folded over the body of the frog: it is thus held quite steady, and there is no fear of the mesentery becoming displaced.

The plate may be easily made from an old biscuit box, by means of a stout pair of scissors. A piece of talc is more convenient than a thin glass, as a cover; it can be cut to any shape, and is less liable than glass to tear the vessels. The above is a modification of the mode recommended by Professor Cohnheim. Virchow's 'Archiv,' xl, i, p. 28.

Woorali is preferable to chloral in this instance.

(*c*) *The Tongue.*—The frog is placed prone or supine, the tongue withdrawn by means of the horn forceps, and stretched to the requisite degree of thinness, over a large perforated cork. The tongue is apt to become dry, and should be moistened with a natural or artificial serum: a mixture of water 100, common salt 1, and albumen 10 parts, has been recommended, and answers well.

(*d*) *The Lung.*—The frog being narcotised as before, is laid on its side, and a vertical incision is made from the axilla downwards for about an inch. The knife must be used carefully as the lung is approached; when the last layer is punctured, a small hernial protrusion is seen; the opening is now enlarged by a probe-pointed knife, and the whole lung springs out through the thoracic wall. With a low power, the circulation is seen very prettily by transmitted light.

2. *In the Toad* the tongue and mesentery may be exposed in the same manner as in the frog.

3. *In the Mouse and other small Mammals*, the circulation may be studied in the mesentery with advantage. The white mouse is preferable to the common one, being less liable to die of fright, and more easily handled. From six to ten minims of the chloral solution should be injected. Operation performed as in the frog.

II.—*Circulation in Animals breathing by Gills.*

1. *Fishes.*—The caudal fin in certain small species shows the circulation with great beauty. The difficulty of keeping the fish alive on the stage of the microscope for more than a few seconds, and its disposition to contract the fin when out of water, have hitherto prevented the general study of the fish's circulation. Being convinced of its value as a field for observation and experiment, I attempted the construction of an apparatus for supplying the fish with a stream of water while on the stage, and for keeping its tail fin still and stretched; after many unsuccessful efforts, the fish-trough shown in the sketch was produced, and has been found to answer perfectly.

A A is a rectangular piece of common brown gutta-percha, 3 in. by 2½, and ⅓ in. in thickness, having a circular aperture about 1 in. in diameter, in its centre; C is a trough of the same material, 1½ in. in length, 1 in. in breadth, and ½ in. depth, placed on the plate, sloping diagonally from the central opening in the plate to one corner; it is partly covered in, as shown in the drawing; D D is a little square glass stage, raised ¼ in. from the plate over the opening, and in a line with the trough; E is an inclined plane, connecting the surface of the glass stage with the floor of the trough; F is a small india-rubber tube, conveying water to the trough from a tap; G is a second tube, conveying the water away, and placed at such a level that the trough is always nearly full, but cannot overflow: H H′ is a loop of silk cord, which passes through two little holes, K K, in the plate.

The fish is placed in the trough (as shown in dotted outline), with its tail on the glass stage. The upper part of the loop of thread marked H′ is passed over the tail, and by pulling its lower end H, is drawn down over the narrow part

Outline Sketch of Fish-Trough on the Stage of the Microscope.

of the tail; by this means the tail is held still: L L are two small springs, which press gently on the glass stage; the tail is now stretched to its natural width, and each extreme edge placed under one of these springs.

It is well to have a tap on the supply pipe; one on the waste-pipe is also useful, as thereby an unpleasant bubbling may be prevented. If the trough is in order, not a drop of water escapes. The microscope should be inclined at an angle of about 45°.

A gentle stream of water being kept circulating, the fish may be left in this position some hours without apparent injury. Eight hours is perhaps the maximum.

In a suitable fish, free from parasites, the tail membranes are as clear as glass. They contain a close capillary mesh, The thinness and transparency are such as to allow of the use of dry objectives as high as the $\frac{1}{12}$th. In this case a film of talc answers better than a glass cover. The tail is held so still that the changes in progress in any given point where inflammation has been excited may be studied for hours with the greatest precision.

The amœboid movements of white corpuscles, also the changes of form of pigment cells, have, so far as I can judge, their maximum activity in the fish.

The parasites often met with are in themselves most curious objects, especially "Gyrodactylus elegans," which was described first in this journal (1861, p. 196).

The common stickleback, when full grown, is perhaps the best fish to use: the trough described just fits it. Small specimens of the gold fish may be used, if nothing more is desired than a beautiful object. For physiological purposes, fishes having little pigment, as the stickleback, &c., answer best; in addition, this fish is very hardy, and can always be readily obtained.

2. *The Tadpole.*—While losing its black pigment, and so long as the tail remains, the tadpole presents a beautiful view of the circulation, and of various tissues. Epithelium, nerve, connective tissue, striped muscular fibre (the last often in the act of contraction), are seen as they can nowhere else be seen. The changes in inflamed parts are shown to perfection, including phenomena of the greatest interest in relation to Professor Cohnheim's theory.

By putting the tadpole in warm water it may be kept motionless for a short time, as is well known; this method is, however, very unsatisfactory.

By a modification of the trough described above, the tail of the tadpole may be studied with the same ease as that of the fish, though for shorter periods, on account of the delicacy of organization of the tadpole. In consequence of the greater thickness of the tail, moreover, 600 diameters is about the limit of magnifying power practicable.

The tadpole-trough shown in the sketch will be seen to be on the same principle as that for the fish. AA is a similar glass stage, raised $\frac{1}{4}$ in. above the plate, and having beneath it an aperture in the plate. B is a trough entirely covered in, the cover of which (C) is here removed to show the interior. The trough is kept constantly full of water by the supply-pipe D; E is the waste-pipe; F is a little cage made of pieces of pin wire, and of such a size as to hold easily the head of a full-sized tadpole.

The apparatus being placed on the stage of a microscope inclined at an angle of 45°, a tadpole is deposited from a small teaspoon on the lower part of the glass stage (about the point G).

The head immediately falls into the cage F, the tadpole

turns on its side, the tail lying across the glass stage as shown in the dotted outline. The silk cord H is used exactly

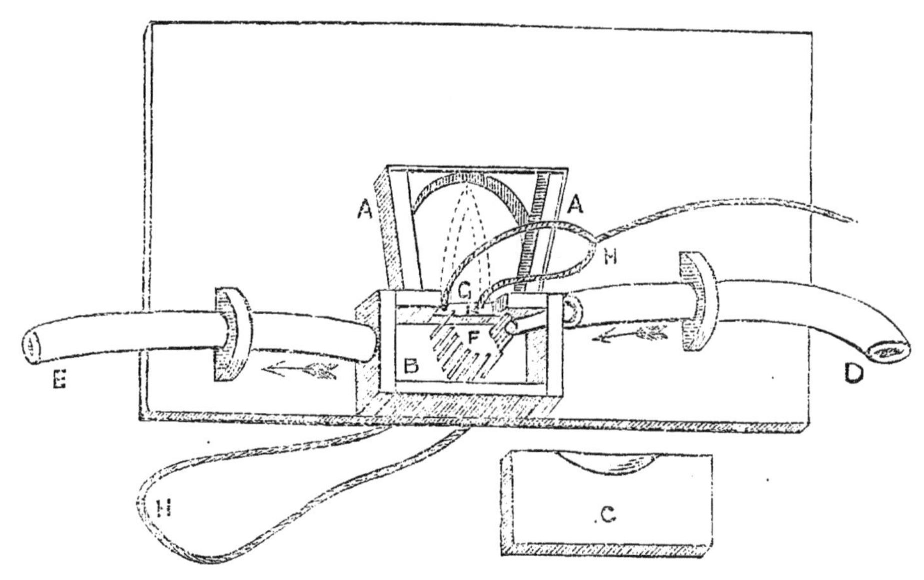

Outline Sketch of Tadpole Trough, actual size.

as in the former case, to hold the animal in its place. A cover glass is now laid on the tail. The tadpole should not be left in more than three hours at a time.

In the case of both fish and tadpole the circulation is apt to flag during the first five or ten minutes, then revives.

3. *Young Water-Newt.*—The circulation in the gills of the young newt is well known as a beautiful object; the great difficulty met with is that of keeping the animal still in the zoophyte trough. I have found hydrate of chloral answer well here. The newt is put into a solution of eight grains to the ounce, and left in until it becomes narcotised, which generally occurs in ten or fifteen minutes. Being now removed to fresh water, the sleep continues for some hours. The circulation continues active, and may now be examined with great ease.

The circulation in the frog's lung, and that last described, though beautiful objects, are of no great value for physiological purposes. The other examples are all suited for practical work.

I have employed all the methods described very frequently, and can speak with confidence as to their efficiency. The troughs are easily made from sheet gutta-percha, and are managed without difficulty after a little practice.

Studies on Inflammation.

By Professor Stricker, of Vienna.[1]

1.—On the present state of our knowledge of Inflammation.

"Inflammation is so interwoven with the theory and practice of medicine . . . that in all ages it has been made the pivot upon which the medical philosophy of the time has revolved."—J. H. Bennett, *Treatise on Inflammation.*

Microscopical inquiry has had the most important influence on the development of our knowledge of inflammation. Every step in the former department has advanced the latter, and each error there has made itself felt here. But beyond everything else it has been in accordance with the variations in our views as to the production and the life of the cell that the doctrine of inflammation has been modified and developed.

Rokitansky introduced into his valuable illustrations of inflammation the doctrine of cell-genesis of Schleiden and Schwann; Goodsir and Redfern made their fundamental studies, after Schwann's example, on cartilage, and the results they gained thereby came back to us across the sea.

The attention paid to the history of development and the extension of the connective-tissue theory became, through the labours of Schwann, Reichert, Henle, and Virchow, the prop of Virchow's school.

Finally, the last movement—the reforms in the cell-doctrine as initiated by Max Schultze, and Brücke—influence us still to this day, and the more so since Recklinghausen has directed one current of this movement into the path of the pathological investigation of tissues.

But in proportion as the excellence of instruments lighted us into the depths of the subject, the value assigned to general macroscopic appearances became weakened. In place of the idea "inflammation" formed at the bed-side and in the dead-house, and relating to a combination of appearances, we obtained definitions of more minute occurrences in the structural elements. So far as these definitions related to the life of diseased elements, a progress was thereby obtained which was satisfactory to the thinker. In an inflamed organ the

[1] 'Studien aus der Instituten fur Experimentelle Pathologie in Wien aus dem Jahre,' 1869.

living tissue was supposed to take on a heightened life, and to inquire into this life seemed to be the purpose of that line of thought. But at last another idea prevailed, an idea according to which the life of the inflamed tissue is held to be without any signification in the process of inflammation; but, on the other hand, the out-wandering of the colourless blood-corpuscles is supposed to form the essential feature.

Now absolutely nothing is left of the old idea of inflammation. The new movement has brought us, as will be proved, nothing but a doctrine of pus-formation, and has created the following alternatives—either we return to the older way of thinking, and hold the process of inflammation finished with occurrenees at the vessels, and cease to speak of an inflammation of non-vascular tissue; or, on the other hand, we do not do that, but reverse the clinical experience which teaches us that pus-formation is the consequence of inflammation, and derive inflammation from pus-formation.

I will explain this superficial historical sketch by a closer description of the different theories. It is not intended that this should be a history of inflammation, but an historical explanation of those questions which occupy us still at the present time, and to which we have to pay attention in new researches.

Source of Pus-corpuscles.

In the year 1846 Waller[1] published the results of his microscopical examination of a living frog's outstretched tongue, and considered that his observations confirmed the idea that pus-corpuscles are nothing but colourless blood-corpuscles. This author, in speaking of the resemblance of the latter to pus-corpuscles, expresses himself on the state of the point in question in the following manner:—"In consequence of this striking resemblance, observers have already supposed that the corpuscles of the blood give rise to those found in mucus and pus, and that these are simply corpuscles extravasated or filtered from the blood. An important observation has corroborated this theory, namely, that of the accumulation of the corpuscles at the inner sides of vessels which are subjected to prolonged irritation. On the other hand, it has appeared to other physiologists so improbable to suppose the perforation of capillary vessels by the corpuscles, that they have come to the conclusion that they are not derived from the blood, but, like semen or milk, are

[1] 'Philosophical Magazine,' 1846, vol. xxix, pp. 271 and 398. The knowledge of this essay I owe to the papers of Mr. Kosinsky.

formed on the secreting surface, in virtue of some plastic power of the fluids which are effused upon it."

Waller thought to have proved[1] by this experiment on the tongue the perforation of the walls of the vessels which seemed so improbable to the physiologist, and opposed, therefore, the doctrine, then in favour, of the formation of pus-cells in blastema. The opposition had no success, and Waller's statement fell into oblivion. The doctrine of cell formation in exudations was, indeed, soon given up in general, but this did not happen in consequence of Waller's inquiries. It was the researches of Goodsir,[2] Redfern,[3] and Virchow that produced

[1] It will certainly be of interest to the reader to learn in what way Waller contrived to prove this. I will satisfy this interest with the following note:—In the essay beginning with page 271 (loc. cit.) Waller represents a frog wrapped up in a piece of stuff, the tongue of which is pulled out and fastened down by pins. Such an experiment he had communicated previously to Donné, and when the latter made use of this experiment without naming the author, Waller brought a lawsuit against him for damages.

To the second treatise, beginning with page 398, a plate is added, from which it may easily be seen that Waller has done more than his contemporaries suspected. The drawings are so very much enlarged that it seems wonderful that Waller could see so much in so restless an animal. But the representations are so true to nature that the exactness of his observations cannot be doubted. Several vessels are stopped up with stagnant blood, and two are filled with only a few blood-corpuscles; in the former you recognise plainly the colourless blood-corpuscles, as the outlines of the red ones are not marked; in one of the two last-mentioned vessels you see two minute spheres separated by the outline representing the vascular wall; heaps of colourless blood-corpuscles adhere in different places to the outside of those vessels, filled with stagnant blood. Waller expresses himself, in reference to the best case in which this is seen, as follows:—"In some instances the manner in which the corpuscle escaped from the interior of the tube could be distinctly followed, that part of the tube in contact with the external side of the corpuscle gradually disappeared, and at nearly the same time might be seen the formation of a distinct line of demarcation between the inner segment of the corpuscle and the fluid parts of the blood in contact with it."

It is plainly to be seen that Waller was well aware what he was going to prove, and the priority of the discovery must doubtless be attributed to him. But in so far as I recognise the difficulty of a distinct assertion of the perforation of vessel-walls, I must insist upon the fact that neither Waller's drawings nor assertions were capable of inspiring confidence in this difficult question. I did not allow myself to form any definite conclusion until I had been able in the curarised animal to follow the very same corpuscles sticking in, and then passing through, the vessel-wall. As long as this is not the case it might always be suggested that one was dealing with two bodies, one adhering to the outside, the other to the inside, of the vascular wall; but such an engulfing (of the blood-cell in the vascular wall) has Waller neither described nor drawn, nor could I gather from his statements that he really followed the whole act of the squeezing through, though the passage just cited refers to such a process.

[2] 'Anatomical and Pathological Observations.'

[3] 'On Anormal Nutrition in Articular Cartilages.' Edinburgh, 1849.

this change. First, Goodsir and Redfern announced the growth and multiplication of cells in diseased cartilage. Virchow joined them, basing his support on his researches on connective issue.

In the following ten years, even up till 1860, observations in this province were nearly all carried out in the light of Virchow's doctrine, until Cohnheim,[1] in the year 1867, in a better position for making an attack than Waller, took up again the old idea of deriving the pus corpuscles from the blood.

Shortly before this I had introduced a method[2] which has become very important in this matter—that method, I mean, of bringing curarised animals under the microscope, and of examining the circulatory vessels, and the circulating blood itself, under high magnifying powers. In this way I could observe with leisure the vascular walls perforated by the blood-corpuscles, and so obtain a diagnosis which was beyond doubt.

Cohnheim has now taken up Waller's forgotten experiments, also by the use of curarised animals. He has recognised the exodus of blood-corpuscles in great masses as a constant occurrence, and has obtained for the theory founded upon this observation a recognition which it had never enjoyed before.

History of Experiments on Keratitis Traumatica.

As far as modern tissue-pathology is concerned, Bowman[3] was the first who brought the cornea into the domain of experiment. Whilst formerly the transparent organs only had served to ascertain, by experiments, the occurrences at the the blood-vessels, we find in Bowman, in accordance with Redfern,[4] the nutritional disturbances put into the foreground. "If we puncture or incise the cornea, the first effect is a change wrought in the natural actions of nutrition then existing in the wounded parts." With these words Bowman introduces his studies on keratitis traumatica, and explains, in an unmistakble way, that near the injured place the corneal corpuscles, which he still calls "nuclei or cytoblasts," multiply, and that these embryonal structures mixed up with old ones cause the milky dimness of the cornea.

The cornea has since been made by Strube,[5] under Vir-

[1] "Ueber Entzündung and Eiterung," 'Virchow's Archiv,' 1867, vol. xi.

[2] 'Sitzungsberichte,' lii, 1865.

[3] 'Lectures on the Parts concerned in Operations on the Eye,' p. 29. London, 1849.

[4] Loc. cit.

[5] 'Der normale Bau der Cornea und die pathologischen Abweichungen in derselben.' Inaug. Diss. Würzburg, 1854.

chow's direction, and then independently by His[1] and Langhans,[2] the subject of experimental studies.

The ideas prevailing at that time were supported by a number of fine observations and detailed and accurate statements; but on the whole, the view advanced by Bowman was but little changed.

Since then no remarkable experimental researches on inflammation of the cornea were published, until Recklinghausen's[3] incisive essay assigned a new path to tissue-pathology.

The doctrine of the derivation of pus from the corneal corpuscles was not altered by this publication, but Recklinghausen brought histological notions and methods,[4] which had only just then been devised, to bear upon the department of tissue pathology, and discovered the contractility and wandering power of the pus corpuscles.

Four years later Cohnheim[5] announced that he had investigated the occurrences in inflammation of the cornea afresh, and in the spirit of modern science and that he, with the assistance of a fine reaction discovered by him, had come to the conclusion that the corneal corpuscles take no share in pus formation. He was able, he said, to make structural elements plainly visible by chloride of gold, and in this manner demonstrated that they remained quite intact, however numerously the pus-corpuscles were formed.

After such a positive declaration all contrary observations that had been received until that time became doubtful, and so much the more when Cohnheim drew the attention of his fellow-observers to the point that cell-division in the inflamed tissue was in general admitted as true, but that, in reality, had not been observed by anybody.

In relation to this matter, Cohnheim pointed to the permeability of the uninjured vascular walls; he referred to the fact that in the process of inflammation colourless corpuscles really wander out, and in this way advocated the same law for the cornea which Waller had advanced for the frog's tongue without any success.

Hoffmann,[6] a pupil of Recklinghausen, tried then to defend the older idea conditionally in proving that, under certain

[1] 'Beitrag. zur normalen und Histologie der Cornea.' Basel, 1856.

[2] 'Zeitschr. f. Medicin,' 1861, 1 Heft.

[3] "Ueber Eiter-und Bindegewebskörperchen," 'Virch. Arch.,' vol. xxviii (1863).

[4] Compare 'Handbuch der Lehre von den Geweben, Methodic und Zelle.'

[5] Loc. cit.

[6] "Ueber Eiterbildung in der Cornea," 'Virchow's Archiv,' vol. xlii, p. 204.

circumstances, pus-corpuscles are, after all that had been said, formed out of the star-shaped corneal corpuscles. But Cohnheim[1] opposed this assertion, and, supported by new experiments, he rose with still greater boldness, and denied, without reserve, any participation of the corneal corpuscles, which he pronounced to be fixed, in the production of pus corpuscles.

Exudation and Disturbance of Nutrition.

If we want to find our way in the history of tissue-pathology, we must face quite independently of the question about the origin of pus, also the history of the knowledge of those changes in tissue which are peculiar to the process of inflammation. I mean, in the first instance, the saturation of the inflamed tissue by fluids, exudations, and the nutritional disturbances of the living elements. I have not yet been able to give myself up so deeply to historical researches as to find out to whom we have to ascribe the merit of having found out exudation.[2] The notion that exudation of liquid substances out of the blood into the tissue is a principal sign of inflammation has found, as far as I can see in the authors in my possession, its most important advocates in Bennett and Rokitansky.

Bennett[3] points to an abnormal exudation of blood-fluid, an effusion of serum, or an extravasation of blood, as a sure sign of inflammation.

The occurrences at the vessels themselves did not appear to him to influence the diagnosis of the process, and he expresses himself on the matter as follows :—" But it is only when the latter (exudation) takes place that we can state positively our conviction of the presence of inflammation " (p. 38). Rokitansky[4] states again that with exudation the process of inflammation is to be considered as complete (p. 178). But when the doctrine of cell formation in such exudations (blastema), based on Schleiden and Schwann, got put into the background, the doctrine of exudation itself soon shared the same fate. It was particularly the knowledge that the non-vascular tissues may exhibit the same disturbances, characterising the process of inflammation, as the vascular tissues, which produced this result.

In England, dating from the researches on cartilage of Goodsir and Redfern, the password " nutritional disturb-

[1] Loc. cit.

[2] According to a note by Virchow (' Geschwülste '), this knowledge took its origin in England.

[3] ' Treatise on Inflammation.' Edinburgh, 1844.

[4] 'Handbuch der allg. path. Anatomie,' 1846.

ances" has been spread abroad, and Virchow has given recognition and value to this password.

Although, during this whole period, the importance of exudation for nutritional disturbances was not entirely left out of consideration, yet people gradually accustomed themselves in studying the process of inflammation not to assign any value to the part played by the vessels.

But through Cohnheim's studies the attention of his colleagues has been directed again towards the blood-vessels, and even in so high a degree that nutritional disturbances are again put into the background. In the place of liquid exudation living-formed elements are introduced; but, nevertheless it is a matter originating in the blood that is pointed out here as the essential feature of incipient inflammation.

The Connective-Tissue Theory.

Another appearance in the sphere of our literature demands our attention, if we want to get clear about the principal questions of tissue-pathology; I mean the significance of connective tissue in the process of inflammation.

Virchow[1] has represented the connective substances as the source of all inflammatory new formations. These substances, the extensive distribution of which in the organism has become known by Reichert's studies, were supposed to be the real seat of all formative phenomena that accompany the process of inflammation. From the point of view of the inflammation theorist, you need only consider the organism as a framework of connective substances, in which nerves, muscles, and glands, not essential for new formations, are inserted. Virchow has recognised in this connective substance the persistent cells, and from these all-important results were supposed to arise.

From time to time great breaches were battered into this doctrine. The discovery of Remak and Buhl taught that the epithelial structures are among the sources of new formations, and no theory could be devised which should point out a relationship between epithelial and connective tissue.

Histologists agree, and Reichert[2] and Remak's works[3] have proved clearly by embryological investigation, that the epithelial structures belong to the category of glands.

The share of epithelium in new formations was nevertheless

[1] Compare 'Handbuch der Lehre von den Geweben,' Leipzig, 1868, p. 35.

[2] 'Das Entwickelungsleben im Wirbelthiere.' Berlin, 1840.

[3] 'Entwickelungsgeschichte.' Berlin, 1852, 1855.

acknowledged, and that of gland-cells decidedly denied. One single voice—that of Holm—which rose in favour of the new formation of liver-cells, had the fate which befalls a paradoxical statement contradicting the prevailing opinion. Virchow[1] himself made the breach wider by proving that muscle-nuclei multiply in the process of inflammation. This fact must weigh so much the heavier in opposition to his theory, seeing that the whole doctrine of the multiplication of the connective tissue-corpuscles was principally supported by the demonstration of the multiplication of their nuclei.

As soon as multinuclear cells were found, the expression that the *nuclei* had multiplied was allowed. It was different with the *cells*. The proof of a multiplication of these could only be given in the case of cartilage. In all other places one was only justified in conjectures, since the phenomenon was not observed directly.

Srelkow,[2] indeed, took refuge in considering muscle-corpuscles as analogous to connective-tissue-corpuscles—a view which has lost all possibility of holding its ground by O. Weber's[3] late researches.

In opposition to the above-named doctrine (Virchow's connective-tissue theory), were also Leidesdorf's[4] and my own demonstrations, that capillary vessels grow out in the process of inflammation, and that the capillary walls take a share in the parenchymatous inflammation. Our statements had at the time when they were made but little weight indeed in comparison with His's[5] emphatic declaration that in the cornea young vessels are formed from outgrowths of corneal cells. The discovery of Eberth, Aeby, and Auerbach, that by means of silver injection the cells out of which the capillary tube is built up could be demonstrated, was to our disadvantage, but in favour of the connective-tissue theory. But people are beginning to judge more soberly in this question. It is clear that our observation of genesis could not be purely imaginary. Without misjudging the important discovery mentioned above, people nevertheless admit the new formation of vessels according to Leidesdorf and Stricker's observations in inflammation.

In addition to all these attacks came Cohnheim's discovery that the inflamed tissues are inundated by colourless blood-corpuscles, and all the consequent statements as to new formations of tissue out of such elements.

[1] 'Virchow's Archiv,' vol. xix, p. 220.
[2] Ibid.
[3] Ibid., vol. xxxix, p. 216.
[4] 'Sitzungsberichte,' 1865.
[5] Loc. cit., page 95.

If we wish to be just in our review of the historical progress of our knowledge, we must point out that there exists up to the present time only one cell-structure of which the share in the inflammatory new formation of cells has not yet been spoken ; these are the nerve-cells.

The knowledge of the participation of gland-cells, muscle-corpuscles, and vessel-walls, in the process of new formation during inflammation, has scarcely lessened the great significance of the connective-tissue substance. Only a share has been claimed for those tissues in the process in which the connective-tissue corpuscles hold the first rank. It is only Cohnheim who excludes them altogether, and we see in his theory simply the extreme reaction from Virchow's older principle.

Virchow ascribed everything to the connective-tissue-corpuscles ; Cohnheim everything to the colourless blood-corpuscles. According to Virchow, form-elements that are not connective-tissue-corpuscles could only vanish in the process of inflammation. Cohnheim is already tending to a similar conclusion by assigning such a *rôle* to the corneal corpuscles.

Cohnheim's doctrines have found numerous followers. As a consequence of these it is denied that epithelial cells, liver-cells, and muscle-corpuscles, take any share in the new formations of inflammation. The proof on which these statements have been founded are, for the most part, not backed by careful consideration. But it would be equally inconsiderate to ignore such boldly made statements, and not to submit the matter to proof. So it has become necessary to examine anew, in all parts and places, the doctrines of inflammation. With this object I have myself taken up a series of the most important questions, or induced younger investigators to take them up.

The essays relating to them will be brought before you one after the other, and from these I will proceed to a consideration of the general result.

One of these essays will scarcely seem to find a place at first sight in this plan ; it is that concerning the yelk-cleavage and blastoderm-formation of the hen's egg ; but tissue-pathologists have, after Remak's and Thiersh's example, begun to appeal to the principles of the history of development in their theoretical studies. The hen's egg was selected for particular attention, simply because our knowledge of the development of this was regarded as the most advanced by the best informed of our fellow-workers. In fact, the quantitative proportions in the literature of the subject prove this too. For the purpose of tissue-pathology the history of the development of the chicken has been until now but little explored. The most important publications on this subject have been silent up till the present time as to what occurs in the oviduct. I allude to the yelk-cleavage. The only (and really very superficial) statements about it proceed from Coste, and these have not much reference to the formation of the layers of the blastoderm. Without any knowledge of yelk-cleavage, theories as to the elementary layers of the blastoderm are very vague, nor are the researches into the pathology of the tissues which are connected with it less so. It seemed advisable to me, therefore, to put the work in question in this volume.

The papers which follow this introductory essay by Professor Stricker are as follows:

1. RESEARCHES ON INFLAMMATION OF THE CORNEA. BY WILLIAM F. NORRIS AND S. STRICKER.

The cornea of the frog is inflamed either by boring its centre with a fine point of silver nitrate or by a needle and thread. The structural elements are rendered apparent by staining with gold chloride—five minutes in a half per cent. solution—after which the cornea is placed for twenty-four hours in very dilute acetic acid, and is mounted in glycerine. It may also be studied without staining. The authors demonstrate the formation of the numerous oblong pus-cells of the inflamed cornea, from the breaking up of the large stellate cells of the normal cornea, and give four admirably true figures drawn by Dr. C. Heitzmann, the artist of Stricker's 'Handbuch,' in illustration. They also treat of the changes of cells of the conjunctival and Descemetian epithelium, and of the active movements of the cells of the inflamed cornea.

2. ON CELL DIVISION IN INFLAMED TISSUES. BY S. STRICKER.

In this paper Stricker describes his method of "draining," by which he is able to keep tissues alive for some time under the microscope, and so observe changes due to living processes. Draining consists simply in passing in fresh blood-serum at one side of the glass cover, and sucking it out at the opposite side by means of a fine glass tube, so as to produce a circulation of nutritive oxygen-bearing fluid. It was suggested to Stricker by Ludwig's and Schmidt's experiments on the muscles. By this method Stricker has been able definitely to witness and to assert the division of cells. He has seen the division in inflammation of corneal cells, and also of certain cells in the tongue of the frog—a phenomenon which no one two years since, when challenged by Cohnheim at the meeting of German physicians and naturalists, could say he had seen.

3. On the Relations of Vessels and Nerves to the Process of Inflammation. By S. Stricker.
4. On Traumatic Encephalitis. By Dr. Friedrich Jolly, of Munich.
5. Researches on the Yelk-cleavage and Blastoderm-formation in the Hen's Egg. By Dr. Josef Oellacher, of Innsbruck.
6. On the Endogenous Formation of Pus-corpuscles in the Conjunctiva of the Rabbit. By Dr. L. Oser.
7. On the Inflammatory Changes of Muscular Fibres. By Dr. Janovitsch Tschainski, of St. Petersburg.
8. On the Tissue Changes in the Inflamed Liver. By Dr. Andrew V. Huttenbrenner.

The conclusions at which Holm arrived in his researches upon inflammation of the liver have since been disputed by Koster and Von Josef. Holm's description culminated in the positive assertion that on injury the liver-cells became fatty granular cells, which formed circles around the injuring body, and became finally converted into fibres, so that granular cells became granular fibres, which passed into the hepatic cicatrix. Furthermore, Holm taught that at places where the needle which is passed into the liver strikes the cells the inflammation is more intense than at places where it strikes only connective tissue. Although he afterwards spoke of nuclear disease in the liver-cells, and asserted the fact that the small cells (granulation-cells) were derived from the multinuclear liver-cells, he still added that he was not in a position to be able to follow out how the young cells could become free. Their appearance, he stated, in more or less compact heaps, as may sometimes be observed, probably favoured the above assumption. In assuming their origin from connective-tissue-cells, he could find no more sure holding point. To this last part of Holm's account Koster turned his attention, and this investigator, basing his ideas upon a more recent theory of inflammation, looked upon the pus-cells in the liver as colourless blood-corpuscles.

The statements of Holm and Koster concerning the formation of pus are, however, nothing more than the echoes of the then prevailing views. At the time of Holm's publication suppuration from parenchymatous cells was the order of the day, whilst at the time of Koster's migrating blood-corpuscles were coming into note. Thus pus was attributed by Holm to multinuclear liver-cells, and by Koster to cells which had wandered away from the blood-vessels.

If, again, the remaining portions of the remarks of Holm and Koster be considered, it will soon be seen that the latter observer had before him a process differing from that brought under the notice of Holm. In the course followed by Holm one meets with a scarcely appreciable suppuration; here the formation of fibres predominates. Koster, on the other hand, speaks of a process in which suppuration was the prominent fact.

The positive assertions made by Holm were opposed by as positive assertions from Josef, who maintained that the liver-cells were not converted into fibres. This observer held that the liver-cells fell away, and that the fibres of the subsequent matrix grew out from the connective tissue of remote acini.

Josef, in the course of numerous experiments upon animals, found, as he believed, that from a remote acinus a host of young elements pressed towards the injuring body, and on one fine day discovered that connective-tissue-fibres had been formed at a spot where some few days before only accumulated spindle-shaped cells could be made out. How it could have happened that connective tissue was observed under the microscope where for some days previously only spindle-shaped cells could be seen, the author does not tell us. Such an assertion presupposes that a certain part of the liver had been subjected to microscopic examination without having been torn from its connection with the living animal. This, we know, cannot be practised with our present means of investigation.

One other positive assertion made by Holm is opposed by Josef. The former observer maintained that the liver-cells arrange themselves in layers about the inserted needle; Josef, on the other hand, asserted that the cells about the needle undergo disintegration. A dog's liver was punctured by copper wire. This wire, after it had been allowed to remain in the liver for forty-eight hours, was withdrawn, and it was then found, on examining the adherent portions of liver, that the cells had been destroyed.

This assertion does not seem to be an improvement upon that relating to the formation of connective tissue. To convince oneself that the liver-cells about the needle disintegrate, it is necessary to make a section of the hardened organ at the particular spot, and to observe the detritus in this section. This, however, Josef did not do. He tore away fresh liver-tissue, and such tissue, indeed, as had been altered by the process of inflammation. What, then, are the characters upon which this conclusive "disintegration" is grounded? Probably that the cells were lacerated, or that fat-granules were found. Of little value would this discovery be to any one who had occasion to draw a wire through embryonic tissue, and then to examine what remained adherent. As we may see, the strict demonstration of the disintegration of liver-cells has been held no less established than the genesis of the matrix has been rendered certain.

Although the objections raised against Holm's view are unsupported by demonstration, still nothing remains proved for this observer. It would, therefore, be worth while to investigate again this interesting point. I have undertaken this work, and with regard to the changes in the liver-cells have arrived at certain trustworthy results. I thrust, as was done by Holm, a needle into the liver of a dog; at the end of twelve hours the animal was killed, and the liver placed in CrO_3. Sections were made, and the parts surrounding the puncture studied in the preparations thus procured. Already after twelve hours spindle-shaped cells could be seen arranged in layers around the needle. In consequence of the short duration of the animal's life after the puncture, it may also be almost concluded that the concentric elements are liver-cells. The demonstration of this statement may be made with irrefutable exactness.

The speedy appearance of the spindle-shaped cells struck me as remarkable, and I could not on that account shut out the idea that the primary cause of their extrusion was a mechanical one, and that the soft liver-cells were stretched to spindle-shaped bodies by the thrusting in of the needle. A simple experiment cleared up this conjecture. I opened the abdominal cavity of a living animal, and cut away a piece of the liver, in which I afterwards thrust a needle. After the piece of liver had been hardened in a solution of CrO_3 I made sections, and then observed the elements, namely, the tissue-cells, spindle shaped and arranged in layers around the punctured canal; other statements to the contrary must, therefore, yield to this experiment. In the excised portion of liver there could not exist colourless blood-corpuscles which had migrated to the parts about the needle with so much rapidity as to form, in the place of the speedily fading liver-cells, a concentric ring.

If one finds the concentric layer around the canal of puncture made in an excised portion of liver, also around the seat of puncture in livers which have no connection with the living animal for twelve hours, and, finally, a similar concentric layer on the third, fourth, or sixth day, it may at once be held as no longer doubtful that in all three cases we have to do with the cells of the liver.

Holm maintained that fibres proceeded from these spindle-shaped cells, and I am compelled, upon the basis of this set of observations, to support his assertion, and to bring into prominence that especially interesting circumstance, that the first irritation which acts upon the liver is a mechanical extension, which in the first place transforms the cells into spindle-shaped bodies, and that these then pass into fibres. Now, we have an explanation of the remarkable account given by Holm, that only at the place where the needle passes among the liver-cells do these bodies become much altered, and that the connective tissue is not so much affected. The compact connective tissue would not be mechanically affected by the puncturing needle in the manner that the soft liver-cells are, and thus the difference in the phenomena may be explained.

With injury of the liver by a needle the suppuration is very slight; there are but a few form-elements between the liver-cells that can be regarded as pus-corpuscles; but circumstances are different if the raw cut surfaces of an exposed liver be touched with diluted ammonia. There is then excessive suppuration, and in sections masses of cells resembling pus-corpuscles are found about the vessels, not only at the circumference, but also in the centres of the lobules. Here, then, we have to deal with the suppuration described by Koster. With the kind of irritation just mentioned there is no formation of spindle-shaped cells, except in parts where there has been much extravasation of blood, and here we have a fresh proof of the circumstances under which these spindle-shaped bodies are formed.

9. On the Conduct of the Fixed Cells of the Tadpole's Tail after Mechanical Irritation of this Organ. By E. Klein and H. Kundrat.

In order to prove the amount of stability of the ramified cells found in the tail of the tadpole, we have examined this again when in a state of inflammation.

Before giving the results of our observations, we will describe the methods by which our investigations were carried out.

Vigorous tadpoles were placed in a small quantity of a slightly coloured solution of curare, and were allowed to remain in this fluid until they were fully paralysed. Irritation was then set up; this consisted in cutting off the tip of the tail, and afterwards pricking the portion of this organ retained in connection with the body with a fine needle, or passing over it a firm brush furnished with fine hairs.

Simple excision of the tip of the tail in most instances led to no good result, since, except in a small sore lying at the margin of the wound, in which stasis and extravasation had been produced, no changes could be observed. In like manner, puncture of the tail in relatively few instances only was attended with any success; stasis and extravasation were similarly established, and so prevented examination. On the other hand, the application of the small brush after excision of the tail proved very successful.

By an occasional application of water to the whole of the creature the tadpole was kept moist. From time to time, usually at intervals of from one half to one quarter of an hour, the application of the brush was renewed.

Immediately after the application of the irritant masses of colourless blood-corpuscles accumulate in some of the vessels near the injured part, and through the walls of the vessels several of the corpuscles may be seen to make their way. There are presented to observation near the vessels cellular elements, which, although branched, differ very much from the branched cells situated at a distance from the injured vessels, or in unirritated portions of the tail. These are not, as in the normal condition, marked by sharp angles, nor are they so smooth and large, but appear contracted and studded with small prominences, from which processes are given off. Moreover, they are considerably smaller, and are often drawn together so as to form a small rounded lump which sends forth on all sides long single or branched processes.

One may best attain conviction concerning these changes by delineating, before the application of the irritant, at a readily accessible spot, several branched cells lying near vessels. The tail may be then brushed, and observation again directed to the spots in question.

When the irritation is slight the cells will soon regain their previous normal form; they again become smooth, acquire their sharp angles, from which processes set out. With a sufficiently strong stimulus, however, further changes in form supervene on the above-mentioned alterations; at one spot a prominence is extruded, which is gradually elongated into a process, or the body of the cell becomes acuminate at one spot, and a process extruded at this point; the extruded processes are afterwards slowly withdrawn, whilst at other spots similar processes are again put out. During these occurrences the cell mass also changes its form.

These changes in form generally proceed gradually. In the course of from two to four hours single cells have already undergone considerable alteration; they appear to be swollen and clouded by granules; the number and length of their processes thereby become less on the whole, since individual cells are now possessed of only one or two more short thick processes.

Our examination could not be extended until the complete retraction of the processes, for as soon as granular stasis came on the apex rapidly became clouded and the creature died.

We have obtained preparations of tadpoles' tails which. in about two hours after they have been irritated in the manner above described, were cut off and coloured in the usual way by chloride of gold. In the successful preparations we missed at once in the neighbourhood of the vessels the characteristic forms of branched cells, as observed in the gold preparations of normal tails. We found at numerous spots, especially near vessels in the walls of which colourless blood-cells were included, more irregular, large, and granular cells set with small bosses, and furnished with two or three short and relatively thick processes. Besides these could be observed transition forms from the cells just described to the symmetrical non-granulated cells furnished with numerous branched processes.

10. On Inflammation and Suppuration. By S. Stricker. (*Concluding Essay.*)

We distinguish, as is well known, two ingredients of pus, namely, formed elements and a fluid in which these are suspended. Concerning the source of the fluid there is no dispute. It is held as a general opinion that this proceeds directly or indirectly from the blood. We discuss here the genesis of the formed elements.

The resemblance of these formed elements to the colourless corpuscles of the blood long ago favoured the view that the pus-corpuscles had their origin in the blood. Advancing knowledge, however, has much diminished the value of this opinion.

We have learnt that the corpuscles of pus, like those of the blood, are contractile; we recognise in both some which are more and some which are less granular; we allow, finally, that their absolute size, and also the relative proportion of the cell itself and of the nucleus, may present manifold variations. But so far we have enumerated all the hitherto known characters of young cells. It may then be assumed that pus-corpuscles are similar to blood-corpuscles, for the reason that both present an embryonic form, and hence that the former are derived from the blood.

Any disputation, however, is opposed by the recently demonstrated fact that at the commencement of the process of inflammation hosts of colourless blood-corpuscles leave the circulatory system to be distributed in the tissue. The question whether pus-corpuscles be derived from the blood need not, then, be kept open any longer. We must now enquire from what *other* sources pus may be derived. To this question several answers have been given.

It has been shown that pus-corpuscles themselves may undergo division, and it has been rendered probable that in profuse suppuration the chief mode of origin of pus is to be sought for in this division process. It has, morever, been proved that epithelial cells, partly by division, partly by endogenous growths, may produce a young generation, and also that connective-tissue-corpuscles in the course of the inflammatory process undergo peculiar changes, which render the probability of the division of these bodies no longer open to doubt, and upon the results of which peculiar changes in the character of pus-corpuscles depends. Finally, it has been proved that the corpuscles of muscle may multiply.

We can thus in the first place express, as quite a general proposition, "*that pus-corpuscles take their origin from various sources.*"

The physiological mother tissues which have in this manner given rise to pus-corpuscles must, in respect of their altered function, have been subjected to a quantitatively or qualitatively altered chemical process. Let us say, in deference to a conception by which so much has been gained for pathology, that some disturbance of nutrition must have taken place.

It has, moreover, been now proved that the process of inflammation may result in an outgrowth of capillary vessels, that liver-cells may be transformed into filaments, and also that, under certain conditions, the nuclei of nerve-cells may multiply. All these occurrences likewise indicate disturbance of nutrition, which must be interpreted in the sense of an elevated vital process. For increase in size, change of form, and the appearance of fresh nuclei, are vital phenomena; and when such are observed where formerly they did not exist, some portion of the vital processes must have been augmented.

We see, then, *that the process of inflammation is accompanied by an elevation of certain functions of the cellular elements involved in the process.* This may be expressed as a general proposition, since the phenomena mentioned have been observed in all cell types.

Since we are thus brought back to the definition given by Virchow, and must regard local disturbance of nutrition as an important sign of inflammation, we may put the question whether by this definition the conception "inflammation" is exhausted, or its nature determined. This question must no longer be answered in the affirmative.

The escape of colourless blood-corpuscles always takes a part among the phenomena of inflammation. That this process cannot be associated with disturbed nutrition only, is obvious enough.

Disturbed nutrition, moreover, is not an exclusive sign of inflammation, since it necessarily claims a part in all processes in which new growths, using the term in a restricted sense, arise. If we add to the disturbance of nutrition the escape of blood-corpuscles, we have not even then exhausted the process of inflammation, since it has not been proved, and even is not probable, that simple new growths are free from emigrated blood-corpuscles.

If we chose to stand by an exclusively clinical conception of inflammation, we must then define it in another manner, or rather we become unable to define it; we must describe it after the manner of Celsus. We must enumerate the signs of the process in order to present the clinical process, and among these signs "*disturbed nutrition*" must occupy a prominent place.

A not less important place must be ascribed to the "*exudation.*"

Virchow has underrated the importance, not of the exuded fluid, but of the process of exudation. "The exudation," he says, "always proceeds from the blood; it is not, however, driven out by the action of the heart, but is attracted by the action of the tissue elements."

When I look to recent literature, I find that there is but little need for me to contend much against this method of viewing the matter. The importance of exudation in the process of inflammation, again, stands with practitioners and experimenters in much the same position as it did at the time when it was so strongly maintained by Bennett and Rokitansky. A majority of the younger generation now believes that inflammation consists in the migration of the colourless corpuscles of the blood.

We must, then, arrange afresh the signs of inflammation, and, passing from experiment to the region of history, collect from the chief works of our ancestors that which is able to hold its ground against recent views.

With this object, I lay stress upon exudation. It is an

important, perhaps next to disturbed nutrition, the most important, sign of inflammation.

I have shown in another essay how important is the current of blood-serum for the activity of movable cells. I believe that the importance of exudation must be regarded in a double sense. *In the first place, the current acts as a mechanical stimulus; and secondly, the fluid of the blood has an influence as a material for nutrition.*

It did not lie in the plan of my present work to inquire into the more minute processes associated with exudation, and therefore I will not discuss this important question and the opinions regarding it without further special investigations.

Moreover, the knowledge of these processes is not indispensably necessary for deciding the characteristics of inflammation. For this purpose the expression "circulation disturbance" suffices, and this we must enrol among the signs of inflammation. It would be better for us provisionally to make use of this general sign, than to bring forward the signs "*heat*" and "*redness*." "Heat" is not always associated with inflammation. It has been proved that in cold-blood animals no increase of temperature is manifested. We shall avoid contention if we say, that in inflammation local disturbance of circulation, increased exudation of the fluid and formed elements of the blood, and disturbed nutrition, together with increase of cellular elements, follow one upon the other. The last link in the chain is the transition of normal and relatively fixed cells into active cells, and the multiplication of the latter by division or endogenous generation.

If we now ask what precedes the disturbance of circulation, the answer to this question is very soon given. It must follow violence to the vessels. We have seen that in traumatic keratitis the vascular nerves must intervene; in other localities and under other conditions direct influences may be exerted directly upon the walls of the vessels. Whether those influences are derived from causes lying within or without the organism does not affect the general question of inflammation. We say that somewhere a lesion sets up changes in the local circulation, and thereupon follow the processes previously alluded to.

One of the chief reasons that induced Virchow to ascribe so little importance to disturbance of circulation in the process of inflammation, was the fact that hyperæmia does not always result in inflammation. But I assert that, among those disturbances of circulation which precede inflammation, we have not to consider hyperæmia alone. Besides hyperæmia, other lesions have to be regarded. To these attention has been directed by Ludwig, Loven, and Samuel.

The point is, not that there are hyperæmiæ that are not followed by inflammation, but that every inflammation is preceded by disturbance of the circulation. The process may attack both a tissue that is directly vascular, and also a tissue which derives its nutritive material from vessels lying at a distance; in the latter case the lesion is to be referred to the distant vessels.

It was by the help of this negation that Virchow maintained the direct action of the lesion upon the tissue itself to be the primary cause of inflammation.

That cells are susceptible to the action of stimuli has been clearly proved within the last ten years. But that a stimulus applied to one cell should in the course of hours and days be transmitted to other cells, is a view that at the present day is hardly tenable. Such a mode of action might have been assumed so long as the connective-tissue-corpuscles were regarded as forming a system of anastomosing cavities. But this is no longer considered to be the case; we know now that the connective-tissue-corpuscles are independent organisms, which under certain conditions only become connected together. That irritation does extend from one irritated body to another has been disproved by recently acquired experience.

If the objection should now be made that we can explain in no other way how the extension of inflammation takes place, the remark must be regarded as unscientific. That I do not know anything concerning the nature of an occurrence does not justify me in having recourse to baseless hypotheses.

I assert once more that the direct irritability of cells has been proved by evidence. All increased activity of cells must result from the application of a stimulus. What we do not know, however, is whether the lesion gives rise to irritation in parts remote from its attacking point. The flow of the exudation might here be taken into consideration. And if any one should remind me of the fact that in the neighbourhood of the injured spot the greatest amount of cell growth takes place, even when, as in the cornea, the vessels be distant, I would not attempt to explain this phenomenon by views inferior to that of Virchow in their probability. I could, for example, say that injury to the tissues, produced either by the knife or by caustic, causes much movement of animal fluids, just as a cavity in the course of a stream may give rise to a movement independent of the general current

This explanation would be much more conformable to the logic of facts than the view of a prolonged and continuous operation of mechanical action.

As matters now stand, the continual propagation of injury from cell to cell cannot be proved. The remote action of injury upon the vessels, however, is proved; and so we have well-based right for stating the following succession of phenomena in inflammation set up experimentally:—

Injury, circulation disturbance, exudation of fluid and formed ingredients, nutrition disturbance, and new growth.

Not one of these signs taken alone is decisive. An injury may be applied; a disturbance of circulation may be manifested under our eyes, without inflammation being the result, and without our being able to decide, with our present means of investigation, whether this disturbance of circulation essentially differs from that of an inflammatory character.

That exudation alone does not constitute inflammation is proved by œdema. Again, we have no means at our disposal by which we can recognise the finer signs distinguishing inflammatory exudation.

Finally, in some cases, we may meet with other instances of disturbed nutrition and new growth, where the clinical observer cannot diagnose inflammation.

I have in my writings shown how probable it is that this succession of inflammatory phenomena includes a causal nexus. But for scientific determination this nexus is not needed to be fully followed out. When I define a body by its coloration and its specific weight, these signs may or may not stand in a causal connection. The two peculiarities are associated merely as distinguishing signs.

Since we have, through the reform in our knowledge concerning cells, attained the position of being able to distinguish old from young cells, the view that young cells proceed from old cells has been rendered doubtful. So long as cells were considered as vesicles matters were not so clear as they are at the present day; the notions concerning division of cells then represented the extent of our knowledge.

Remak, in his views upon cell division, imagined that processes grew from the cell-membrane into the interior of the cell, and caused division of its contents. How to conceive of the cell-membrane and its mode of growth, at that time there was no need to inquire. The membrane then was not regarded, as at the present day, as a passive enveloping strueture; it was an integral part of the living cell, and concerning the finer processes of life one would not, and was not required to, speculate.

At the present day things are changed. Recklinghausen has pointed out that fixed cells exist in the cornea. According to Max Schultz, yelk-cleavage corresponds to a contraction of the active cell mass. How, then, would the fixed non-contractile cells divide? Recklinghausen attempted the solution of this question, and pointed out that fixed cells in the process of inflammation acquire a certain degree of activity. At the present day it has been proved by evidence that they may become quite active. The question, however, has not been much simplified. The fixed are the older cells, the active the younger cells. Do the old cells, then, we inquire, become young again?

If facts do not assist us we need not undertake the discussion of this question. But facts do exist, and we must, at least, sift these if we cannot give an explanation.

In certain epithelial cells we have learnt to recognise an endogenous generation. The young brood is slowly extruded from the vesicular membranous envelope. Here the cell has not thus become young, but a portion of the cell mass has divided itself into segments, and has deserted the mother mass.

Whether the segment lying in the interior of an epithelial cell of this kind—that is to say, the portion which serves as the origin of pus-corpuscles—does not in its physiological condition perform any movements, for the reason that it is enclosed, or because in this state it is wanting in capability, is an open question. It is only certain that a portion of a formed element, intended for certain functional ends, has become converted into an active body capable of acquiring another shape.

I have followed directly, in the tongue of the frog, a transition of this kind. With care the metamorphosis can also be made out in the ramifying corneal cells, and capillary vessels may be observed to throw out processes.

We see thus that form-elements that serve distinct functional ends, and which under certain physiological conditions present very slight if any changes of external form, may under special influences become active.

The amœboid movements recently investigated do not distinctly correspond to the physiological ends of simple form-elements.

By putting these facts together, we see that several structures, in consequence of the process of inflammation, are deprived of their functional ends, and are reduced to a condition in which there is a tendency to procreation, that is to say, they become active, accumulate in masses, and divide, either partially or wholly, or not at all.

From the preceding remarks it is evident that the quality of being brought to such a condition is not limited to one or two groups of form-elements. It has, moreover, been rendered sufficiently prominent that the stability of the molecular equipoise varies in different forms of tissue.

Horny matter and connective substance (the cellular elements being excepted) have, without doubt, the greatest stability. It has, indeed, been much questioned whether in these sufficient traces of a substance of unstable equipoise can be found to characterise them as living organisms. We can scarcely expect a new growth to consist of them.

On the other hand, there are the amœboid cells, the molecular equipoise of which is the most unstable. Investigations on the irritated cornea have, in fact, shown that these bodies are the first to undergo change, and, as examination of the frog's tongue has proved, they are the first to divide.

In what gradation we must arrange all the tissues with regard to their irritability, will have to be decided by special work. Such gradation, however, does exist. In the process of inflammation we have no exclusive position to assign to the connective-tissue-cells or epithelial cells; we have cells possessed of slight, and cells possessed of much, stability; cells such as, under certain influences, soon fall away from their physiological aims, and others which are less ready to do so.

It has been also shown that variations may occur in the same form of tissue, since the ramified cells of the cornea are not all affected in the same way. This experience compels us to imagine a gradation of age in the adjacent corneal bodies. It is not improbable that there are some of these bodies which do not multiply—matrons which no longer bear children.

On some MIGRATIONS *of* CELLS.

By E. RAY LANKESTER.

IT has been remarked that the lower animals often furnish us with valuable evidence as to the signification of structures and appearances in higher organisms, or man himself, by placing before us those structures in an exaggerated or schematic way, enabling us, as it were by caricature, to catch the true meaning of an arrangement of parts not clearly comprehensible in the more complex form. Again, a valuable class of evidence is furnished by the study of comparative physiology in the kind of natural experiment which certain animals may exhibit in the absence or diminution or increase of organs present in man, the function of which is explained by the habits or condition of life of such animals. Even the invertebrata are capable of furnishing evidence of this kind—as, for instance, with regard to hæmoglobin, which is so curiously distributed among Mollusca, Crustacea, and Annelida.[1] It is, perhaps, even more largely in the department of histology or minute anatomy that our knowledge of human structure is in intimate relation with the study of the lower animals, and, indeed, of plants, for the doctrine of Schwann was based on that of Schleiden. Modern ideas as to protoplasm and the character of cells in the tissues of the highest animals have proceeded from the study of the naked Rhizopoda, and increased knowledge of the histology of invertebrata is continually throwing fresh light on that of man himself.

A most important fact relating to the function of the minute ultimate masses of living matter or cells of the tissues, which has gradually been established during the last year or two, is that certain of them move, not only in the way of sending out processes and retracting them, manifesting contractility, but actually have the power of locomotion, and pass through what appear to be solid, but are rather to be considered as viscid masses which oppose their course, or through yielding points. Thus the out-wandering of the white and red blood-corpuscles through the capillary walls, first observed by Dr. Augustus Waller in 1846, has been established. A still more curious phenomenon is announced by Saviotti, who has observed the pigment-cells of the frog's web to pass into the capillaries, and become carried along in the blood-stream. A third form of locomotion is that noticed by Ecker and Stricker in the

[1] See on this subject, 'Quart. Journ. of Micro. Sci.,' 1869, and 'Journal of Anat. and Phys.,' Nov. 1869.

large cells of the frog's ovum at an early stage of development of the blastoderm; these large cells, having the power of movement, as may be witnessed on the stage of the microscope, advance from the floor of the cleavage cavity, and pass to the roof to form part of a new layer of the blastoderm, as similarly His has observed the large cells in the hen's ovum to move upwards and pass between the two primitive layers of the blastoderm. Again, Besiadeki has drawn attention to cells occurring in the fibrous tissue of the skin, which have in all probability advanced into this situation from some other part, making their way as if by an instinct to the position which they finally occupy.

As belonging to the same group of phenomena, may we not also signalise the fertilization of the ovum by the spermatozoa, as suggested by Dr. Wyman, and of the plant ovule by the pollen-tube or antherozooid? Though there is to some extent and in some cases a direct passage prepared for the moving element to use, yet there is almost invariably either direct penetration of the ovum-cell itself or of surrounding tissues. The out-wanderings and in-wanderings of corpuscles through the walls of capillaries and small veins may very well be coupled with and help us to understand the penetrating movements of male reproductive corpuscles, carried to such an extent as to affect the whole of the tissues surrounding an ovule or even neighbouring flowers in some plants;[1] and, on the other hand, in animals being so far-reaching as to bring the spermatozoid through the tunics of the ovary and other seeming obstructions as inferred from cases of extra-uterine fœtation.

The ubiquitousness and the power of many parasites of passing through membranes,—*e. g.* of Gregarinæ, Sarcocystis, Miescheri, Trichina, Bacteria, &c.—is also worth mention in reference to the penetration of tissues and migrations of cells.

The annelids offer very beautiful subjects for histological study, and in them is to be observed a constant and normal migration of cells, which may have some interest in relation to the migrations of cells above mentioned. The perivisceral cavity of the Oligochæt and Polychæt annelids is a large space between the muscular body wall and the axial alimentary canal, lined with a more or less complete endothelium, in some Polychæta ciliated, traversed by thin muscular septa communicating by segmental openings with the exterior, and containing a colourless liquid of oxidizing properties similar to those possessed by solutions of hæmoglobin, and in which

[1] The fertility of the closed flowers of *Viola odorata*, some *Campanulæ*, &c., is perhaps thus explained.

float a great number of corpuscles. In the Oligochæta, the terrestrial and fresh-water forms, the perivisceral cavity communicates with the exterior by large convoluted ciliated ducts, and a closed vascular system is constantly present in addition, containing a solution of hæmoglobin. In many Polychæta there is no shutting off from the perivisceral cavity of a closed vascular system.

In leeches the cavity is reduced to a system of fine vessels by the blocking up of its chief bulk with cellular tissue; it is as though the corpuscles in the perivisceral fluid of Chætopods became connected with one another, and aggregated into a continuous mass. The properties of the perivisceral liquid of Chætopods, and its relation to the parenchyma of leeches and trematods, must cause us to regard it as something more than a great excretory cavity, and make it rather comparable, but only by way of analogy, to such an organ as the blood-lymph system or Hæmochyle of Vertebrata. This being the case, the origin of the corpuscles which float in this perivisceral liquid becomes important in relation to the origin of corpuscles, which pass in or out of the vertebrate hæmochyle system, and their analogy may suggest explanations of facts observed in the one or the other. That there is any homology, in the sense of homogeny, *i. e.* genetic community of origin, between these vascular systems or those of any of the invertebrate groups and the Vertebrata, must not be supposed.

In several genera of the Oligochæta I have carefully examined the corpuscles of the perivisceral fluid, and endeavoured to trace their origin. There is no doubt that the majority of them are detached from the layer of large cells containing yellow granules, which surround the alimentary tube and blood-vessels. (Pl. XVII, fig. 5.) This layer of cells varies very much in its form in different genera, and is directly continuous through the layers of cells which enclose the segment organs, or, as in *Lumbriculus*, through those which enclose the lateral blood-vessels also, with the endothelial cells of the body-wall. In *Lumbriculus* the cells on the lateral blood-vessels have the same coarse yellow granules as those on the alimentary canal and on the dorsal vessel, seen in all Oligochæta without exception. In Limnodrilus the cells on the intestine are exceedingly long and large, and have the character of Becker-zellen—goblet-cells—very well marked. (Pl. XVII, fig. 4.) The cells of the hepatic tunic of the intestine, as it is sometimes termed, exhibit this form at one period of their development in all Oligochæta; and it is probable that they discharge their contents or a part into the intestine to assist digestion, as described originally by d'Udekem, being, in fact, unicellular glands; but they are also thrown off into the perivisceral cavity and form a great number of its corpuscles. The endothelial cells of Limnodrilus seen on its segment organs, and incompletely lining the muscular body-walls, are large, clear, and globose, and these too are thrown off to contribute to the perivisceral corpuscles. In *Enchytræus* the cells of the hepatic membrane have a very definite oat shape, and so have those of the perivisceral fluid, and they are very numerous. (Pl. XVII, fig. 6.) Their form is so well marked that they are quite characteristic of the genus. The perivisceral corpuscles differ from the hepatic only in having lost their yellow granular contents.

In *Lumbriculus*, besides elongate corpuscles from the hepatic membrane, are also to be found somewhat round corpuscles with large amœboid pseudopodia, which apparently are thrown off—that is to say migrated—from the endothelium. (Pl. XVII, fig. 1.) In addition to these, in *Tubifex* and *Lumbriculus* there are very long, band-like corpuscles, which are nothing else but migrated muscular-fibre cells. (Plate XVII, fig. 3.) In Tubifex the endothelium of the perivisceral cavity is very incomplete, being chiefly developed in a narrow band on each side of the cavity between the dorsal and lateral rows of setæ; the muscular tissue consequently presents a free surface to the perivisceral fluid, and the cellular elements of the muscular tissue may be *seen in the act of separating*, to pass a floating existence in the perivisceral fluid. Besides these various contributions to the perivisceral corpuscles, there is a more remarkable lot, which are derived from the generative organs.

The Oligochæta do not expend all the generative elements which they develope. The large masses of sperm-cells and of ova which commence development do not all attain maturity and become extruded; but a considerable number, varying in individuals, undergo a degeneration, and become corpuscles of the perivisceral fluid, where they are especially to be observed in the autumn months. (Pl. XVII, fig. 2.) The dimorphism of the ovary in Tubifex observed last year by Dr. Fritz Ratzel, is, I believe, to be explained by the fact that after a certain number of ova have been deposited, or in accordance with other conditions as to copulation, &c., the ova do not develop to their full size, but form closely packed masses of cells of a smaller size (his Enchytræus form of ovary), which gradually lose the character of ova altogether, and are at length separated, and float in the perivisceral cavity as corpuscles. Such a series of changes I have carefully traced in Tubifex. The cells of the large accessory gland (which probably secretes the cementing material of the spermatophors—*Pachydermon* of Claparède),[1] situate on the efferent canal of the male organs in Tubifex and Limnodrilus also, at the end of the season of sexual reproduction, become broken up and discharged into the perivisceral cavity, where they float.

These facts of the separation of cells from various organs to float in the fluid of a large cavity, would not be worthy of particular attention if we regarded them as simply thrown off to be got rid of, and as having no further function when once floating in the perivisceral fluid. Such a view cannot, however, be held, for the perivisceral fluid, as before observed, is a highly-organized liquid, like the blood; and the cells, after becoming detached, and in fact *migrating* into it, are not found to be dead, but remain there for some time, slightly changing their form, and performing some part or function which has importance in the worm's economy. What that may be it is not easy to determine, but I think there can be no doubt that there is gradual absorption of such corpuscles as those from the generative organs—perhaps of all. Very characteristic large corpuscles, with excessively fine radiating processes which originate in the exuberant male generative glands after reproduction in Tubifex, are still to be found in the perivisceral fluid in October, in some numbers, though slightly shrunken in size, so that the absorption in that case does not appear to be very rapid. (Pl. XVII, fig. 2.) If one of the functions of the perivisceral fluid be the reassimilation of living matter which has attained a condition unfitted for the work of the part in which it originally developed, and this be the explanation of the migration of hepatic, muscular, endothelial, and generative gland-cells into a common reservoir, can we find any parallel to such a process in the breaking up of blood-corpuscles in the spleen or elsewhere? If, on the other hand, the migrated cellular elements of the perivisceral fluid have more active functions, and are not destined to be simply absorbed, may we not see in their change of function and position a parallel to some of the cases already noted in Vertebrata, and mentioned at the beginning of these remarks?

Remarkable stellate pigment cells forming bands of colour, and belonging to the Endothelium, are sketched in Pl. XVII, fig. 7. They are from an unidentified species of Limnodrilus occurring in the Thames, near Barking.

[1] See April number, page 148. I have this spring studied the spermatophors carefully, and have ascertained their mode of formation, &c.

NOTES *on* DIATOMACEÆ.

By Professor ARTHUR MEAD EDWARDS.

I AM one of those who have always strongly advocated the keeping of written and drawn notes by observers of nature. However crude and imperfect the drawings may be, however incomplete the written descriptions, yet, if made conscientiously and with due regard to facts, stating what the observer *thinks* he sees, they always possess the value of truth, and at the same time serve to place upon record and impress upon the mind many things that would otherwise pass unheeded, and those often of great value. So by following out such a plan, the mind of the student is drilled in system, the great secret of success in all scientific observations, as well as in other matters. For a long time I have kept a book in which, from day to day, and immediately as observations are made, memoranda are jotted down, often accompanied by sketches, coloured or not, as the subject requires. And on looking back, I frequently find in my older notes the key to some puzzling phenomenon undergoing investigation at a later time. Let not the observer plead the excuse that he cannot draw; I believe that everybody can learn to draw sufficiently well to give a truthful, if not artistic representation of what appears before his eyes. Every one can write well enough to say what he sees when required, and drawing is but a short-hand system of writing.

I believe, also, that when a student of nature has recorded anything that he thinks will be of value or interest to others, he is in duty bound to make such observations public. To illustrate my belief thus expressed, I thus communicate some brief extracts from my note-book, and if they prove acceptable, will from time to time do the same again.

My notes are of observations made by means of the microscope, and the first is relative to one of those curious atomies of the vegetable kingdom, the Diatomaceæ. A few days since (Sept., 1869) I made a gathering in a ditch communicating with the salt water of the Hudson River, opposite the city of New York, at Weehawken, N. J. Of course the water in the ditch was salt, and, in fact, in it last spring I had caught specimens of Stickleback (*Gasterosteus*) which had come up there from the river to spawn, as is their wont to do. The Ten-spined Stickleback (*G. pungitius*) I had found very plentiful, and mixed with it a few individuals of the Three-spined (*G. aculeatus*); in fact these fish occurred in such numbers that when the water became foul, as it did by evaporation, the bottom of the ditch was literally covered with their dead bodies. The gathering, however, I have to speak of at the present time was made for the purpose of procuring Diatomaceæ, and consisted of specimens of an alga belonging to the genus *Enteromorpha,* having attached to it more or less firmly numerous Diatomaceæ and animals. The commonest form of Diatom was a *Cyclotella,* and seemingly fixed in some manner to the *Enteromorpha,* for it was not shaken off by pretty rough usage. How it was fixed I could not detect; most likely by means of a mucous envelope of such tenuity that it is not readily seen.

The next most common form is the truly wonderful, inexplicable *Bacillaria paradoxa,* the paradoxical bundle of sticks. Often and often have I spent hours looking at this marvel of nature, the motion without apparent cause or mode, an invisible joint which, as a friend of mine, an engineer, once remarked, would be a fortune to any one who would discover it, for here we have several sticks forming the bundle, moving over each other without separating, and yet the use of the highest powers of the microscope has failed to detect the means of their union into one mass or composite group of individuals. This grouping of individuals together, which we so commonly find among the Diatomaceæ, as in *Schizonema, Achnanthes, Melosira,* and a host of other genera, appears to me to have its analogue in the animal kingdom in the Polyzoa; which, although generally fixed, yet at certain periods throw off motile forms by means of which the species is distributed. Do not the Diatomaceæ do likewise? I am of opinion that they do, and I shall produce evidence on that point further on. As to the *Bacillaria paradoxa,* the oftener I watch it the more it puzzles me. Not long since I saw one specimen (of course I mean one bundle of individuals) slide out to its utmost limit across the field of view, and then, becoming entangled with two others, which likewise were made up of many individuals, some eight or ten of its frustules (as the complete individuals are called) were twisted round almost off from the rest, so as to lie at right angles to them, and when the group containing the largest number of frustules receded to their former position, which they soon did, the eight or ten seeming by the act of twisting to lose their power of motion among themselves for the time being, were dragged along in a helpless condition, and twisted completely around one revolution, so as thereafter to fall back again into their places, when all went on as usual. That is to say, the regular motion of all the frustules over each other succeeded.

Now what kind of a joint can it be that permits of such eccentric movement? As I have already said, I am more puzzled than ever.

For some time back a discussion has been taking plaçe in some of the European journals as to whether this plant be an inhabitant of fresh or brackish water. What I have observed points to the fact that it will live in either. I have collected it in brackish water at Hoboken, N. J.; my Weehawken collection was from a ditch connecting directly with the salt water of the Hudson River at its mouth, and some years since I gathered it in the sweet fresh water of the Fishkill creek, along with Desmids and other truly fresh-water plants, which, as far as we know, will not live in water containing any appreciable amount of salt, and then, also, in winter and under the ice, but nevertheless in an active condition. And I have taken my salt-water Weehawken gathering and diluted it with several times its volume of fresh water, and yet it seems to flourish after many days, and the *Bacillaria* is apparently more active than when first procured. So, also, the other Diatoms which are present along with it evidently profit by the change, for they have increased rapidly and are in vigorous motion.

Along with the *Bacillaria* in the brackish water at Hoboken, I found numerous individuals of an *Amphora*, which I have known in this neighbourhood for many years, and which I considered unnamed as yet. To it I have given the provisional name of *A. lanceolata*, on account of the form of its outline. This genus has always been considered an epiphytaceous one; that is to say, one which grows attached to other plants or submerged substances, yet this form was free and in active motion. In fact I think it was one of the most lively Diatoms I ever saw. So another smaller species of *Amphora* which is common near here, is always, as far as I have noticed, free. Here we have species appearing both in the free and attached conditions, and this is even more strikingly illustrated in *Schizonema*.

Bacillaria paradoxa is usually set down as the most rapid in motion of the Diatomaceæ, its velocity being recorded by Smith, as he measured it, at over one two-hundreth of an inch in a second. This is certainly pretty quick when we consider that the length of the frustule is only ·0025 of an inch. But my experience has been that its velocity varies in every degree from that mentioned to perfect rest; at times some individuals will be in rapid movement, while others are motionless; and also I have remarked that from sunrise to noon seems to be the period during which, under ordinary conditions, the movement is most active, while during the afternoon it is very sluggish, and at night almost *nil.* This *Amphora*, as I saw it at the time mentioned, was moving even more rapidly than I ever saw a *Bacillaria* move, and that with a steady onward progression very different from that of most naviculiform diatoms.

It appears to me that in *Schizonema* and similar genera, which consist of siliceous loricated naviculiform frustules enclosed in membranous tubes, as soon as a rupture of the investing membrane takes place, by fracture or tearing asunder, almost immediately a knowledge of the fact is in some way communicated from the point at which the opening occurs to all other points of the tube, as at once the contained frustules which hitherto have been at perfect rest or, at most, only moving to a very slight extent, and even then in an extremely sluggish manner, become animated in their motion, and the most of them move towards, and attempt to escape from, the opening made. And this evidently does not result, as might at first have been supposed, from any pressure exerted upon them from the closed end of the tube, and which, therefore, only shows itself when the obstacle in the shape of the investing membrane is suddenly removed. For the motion is the true lively action peculiar to the living individual in the naviculiform Diatomaceæ, and is not in all cases towards the opening made, but often many, or, as in some cases which have come under my observation, most of the frustules begin to move in an opposite direction at first, while at the same time many escape by the opening in the tube, and thereafter assume vigorous motion in the surrounding liquid. Again, usually some of the frustules being, as at first appears, carried along by the stream constituting the mass of those moving towards the opening, all of a sudden seem to change their minds, or are struck with an idea, if I may so express myself, and here and there will be seen individuals which at once alter the dircetion of their course and move in exactly the opposite direction, or backwards, as we may say. The individual frustules as they escape from the ruptured end of the investing tube and enter the surrounding water, do so with the peculiar trembling and apparently uncertain movement so characteristic of many of these organisms.

It will be well to note that these observations have been mainly made on *Schizonema Grevillei*, a species occurring very commonly in New York harbour, although I have noticed the same thing to happen with other species of the same genus, and, if I am not mistaken, in the allied one, *Homœocladia.*

After a time it would seem that the broken end of the tube becomes closed again; perhaps by the deposition of new matter, or it may possibly be by the action of the surrounding water upon the fluid within the tube, if it be of a different composition (which would seem to be extremely doubtful, however), as the frustules no longer attempt to escape, and resume their quiescent state from which they have been startled by the accident of the rupture, or they move over each other up and down with the same irregularity which is commonly the habit of these forms.

I am strongly of opinion that certainly in some of the cases in which I have seen this escape of frustules take place from the investing tube, it has not resulted from any rupture caused by my manipulation, but would seem to be a normal occurrence. In fact, at such times the diatom is taking upon itself the active or free condition by means of which the species is to be distributed. And we must believe that such is the habit of all so-called epiphytaceous forms, otherwise it is not easy to comprehend how the species become so widespread as many of them are, for we have not at present any authentic notice of the formation of free swimming spores in this family. It is hard when making such observations as those I have here recorded, to believe that these organisms are not endowed with sentient capacities, especially when one sees, as I have, a free frustule of such a *Schizonema* apparently perseveringly attempt to regain a lodgment within the tube from which it had some time before escaped, by means of repeated dives towards the hitherto open end, which has since become closed. I have observed such struggles continue for a minute or more, but never with the success apparently desired.

Many months since I mentioned at one of the meetings of the Lyceum of Natural History in New York, that I had seen two apparently different genera of Diatoms existing within the same investing tube, and now I wish to place that fact upon record, and state one or two more instances of the same mode of growth. During the month of March, 1868, I found in the harbour of New York specimens of *Schizonema Grevillei* in active motion within their investing tubes, but accompanied by a much smaller form possessing a totally different outline from *S. Grevillei*, being blunter at the ends, and with parallel sides on S. V. During the same month, and also in April, I found this mode of occurrence very common, and also *Schizonema Grevillei* and a *Homœo-*

cladia in the same tube, and *Schizonema cruciger* and the small form mentioned above, both in the same tube, and *S. cruciger* and *Grevillei* in the same tube. In all these cases the frustules were in lively motion, passing over each other from one end to the other of the tube. In May of the present year, 1869, I found growing in the salt water of the "Mill pond" at Salem, Mass., *Schizonema cruciger* and *Nitzschia closterium*, W. S. (*Ceratoneis closterium*, C. G. E., and *Nitzschiella closterium*, L. R.), both in the same tube. And here it will be necessary to say something in regard to the form I have called *Nitzschia closterium*, as I shall thereby, I hope, be enabled to clear away a little fog of synonyms. Neither Smith, Kützing, nor Rabenhorst describes or figures any species living within a tube like *Schizonema*, the frustules of which have an outline and markings similar to *Nitzschia closterium*, so that it is not likely that they ever saw anything but the free form or condition of this species. However, Ehrenberg figures and describes, under the designation of *Schizonema ? Agardhii* (*Die Infusionsthierchen*, 1838, p. 343, t. xx, fig. xvi.), a form agreeing with this, but the structure of the frustule is that of *Nitzschiella* of Rabenhorst, so that the specific name of this species should be *Agardhii*, whatever its genus be decided to be hereafter. For the present, as it is nearest allied to the forms grouped under *Homœcladia*, it had better be placed in that genus, so that the synonomy would stand thus :

Homœcladia Agardhii, C. G. E. (sp.). Abhand. K. Akad. Berlin. p. 311. 1833.

Ceratoneis closterium, C. G. E. 1840.

Nitzschia closterium, W. S. 1853.

Nitzschiella closterium, L. R. 1864.

What are we to say to such facts as these I record, as well as that of which I sent an account and illustrating specimens to the late Dr. Walker-Arnott,—and which has been noticed by Mr. F. Kitton, who examined my specimens, in Hardwicke's 'Science Gossip' for May, 1869, vol. v, p. 109,—of the occurrence of what are usually considered two distinct species of *Gomphonema*, viz., *G. capitatum* and *G. constrictum*, both growing upon the same stipes or stalk! But this is not all. Since then I have made gatherings at the same place, and still find the above two forms growing upon the same stalk, and two others of totally different outline which appear also upon the same stipes. So that here we would have four hitherto considered distinct species arising from the same individual. I do not name the two last mentioned forms, as I am in some doubt with regard to the names that have been applied to them. The question of what is the individual in the Diatomaceæ is again raised by the observance of these facts, as well as those I described in my "Note on a point in the Habits of the Diatomaceæ and Desmidiaceæ," read before the Boston Society of Natural History, January 8, 1868, and published in their 'Proceedings,' vol. xi, p. 361. The specimens illustrating the remarkable mode of occurrence of the two forms of *Gomphonema* which I sent to Dr. Arnott unfortunately did not arrive until after his death; but, speaking of my having so found them, he wrote to me in the last letter I received from him as follows. I feel that I am justified in publishing this extract as it is of such importance; and I also know, from what he wrote to me, that he himself would not object to my doing so were he still living.

"Your discovery of *Gomphonema constrictum* and *capitatum* growing on the same stalk is interesting, if you are not deceived. When a *Gomphonema* spore grows on a weed, the stalk (which is merely the external mucus collected at the one end) is formed by the growing frustule. It is not the stalk (or in *Schizonema*, the tube) which produces the frustule, but the frustule which produces the stalk or tube. Then when the frustule self-divides, several are formed, either side by side, or each may project a new stalk; but seldom with much regularity. Now every frustule and valve arising from the same spore *must be precisely alike*, being all formed from the original frustule by repeated self-division; and as self-division merely repeats the same identical form or variety, it is not easy to understand how it is possible to have two varieties of form on the same stipes. If there be no mistake on your part, you will overturn all the present views of the production of new frustules and valves. It is more easy to suppose that a frustule from another stipes had become agglutinated to the stipes. But as you say you have sent some in a bottle, I will examine it with care when it arrives. Every spore may produce a different variety, but it is not easy to understand that the same spore, or stipes, can give rise to different varieties. As for the two species (*G. constrictum* and *capitatum*) I have never been satisfied that they are distinct, and probably *G. herculaneum* is another variety."

For my part, from the mode in which the two new valves are formed within and between the two old ones, when self-division takes place, I can readily understand how a great variation in both outline and sculpture can occur. In this case the two forms have exactly the same sculpture, and the difference between them merely lies in the outline of the valve.

From my knowledge of how greatly this character varies in the Diatomaceæ, I, from an early period in my studies, considered these two supposed species to be but forms of one, and this discovery proves that my surmises were correct; at some future time I may have something to say with regard to the genus *Gomphonema*, and what, in my judgment, constitutes a species in it. I am now engaged, and have been for some years, working up several genera, with the express purpose of determining the true lines of specific distinction in them. And I must be permitted to here enter my earnest protest against the custom which has become so wofully common, in England more especially, of manufacturing species where they do not exist.

The labours of such self-supposed students of nature are more than thrown away. Our books become crowded with worthless synonyms, and this branch of biology has, in consequence, fallen into disrepute among scientific observers generally. If those who have the opportunity of securing and examining specimens of Diatomaceæ, would only study them a little more carefully, and if they must publish, do so only after properly maturing their knowledge, we might hope to learn something of the life history of these strange atomies. Better that really new species should for ever remain unnamed, than that such contributions to the literature of the Diatomaceæ, as appear from time to time in foreign journals, should ever see the light. It is a curious fact that almost every one who becomes possessed of a microscope of sufficiently high magnifying power, at once imagines that he is abundantly armed and equipped, as well as qualified, to attack and overcome the most difficult problems in biology. Hence we find the most startling discoveries put forth by very immature observers of nature who suppose themselves to be students, but who have really given little time or thought to study. No branch of biology, perhaps, has been more cursed with supposed discoverers of this class than the Diatomaceæ, until a man comes to be appreciated by the number of species he can manufacture. By far the largest number of observers who are attracted to these beautiful and wonderful atomies forget that we have in them presented to us for investigation one of the most puzzling problems in the whole group of phenomena, illustrating that which we call life, but on the contrary appear to consider them as "simple organisms," whose morphology and life history, as well as classification, are therefore proportionally easy of comprehension. I have devoted many years to the earnest study, under varying conditions, of these examples of complex simplicity, and pity it is that others who have not spent so much time over this branch of organic existence should not have been so fortunate as I was in possessing a wise and patient counsellor in the late Dr. Walker-Arnott. I can truly say that had it not been for his invaluable friendly advice, I, too, would have doubtless ranged myself with the manufacturers of species and synonym accumulators. Often have the kindly words he has written me made me pause ere I, as he pithily remarked, "rushed into print" with supposed discoveries, which I would have been ashamed of thereafter. Dr. Arnott says "a microscopist looks on everything as subservient to the microscope, and that whatever he sees, and which appears distinct to the eye, he thinks ought to be described or figured as distinct. I am, on the other hand, a naturalist, a botanist in particular, and use the microscope, simple or compound, as a necessary evil, merely to enable my eyes to see better minute structures, but whether these differences amount to specific or generic importance, or are only peculiar forms of one species, is the result of analogy, a mental process which can only be attained by a training in botany in all its branches, for many years." Natural objects, like the Diatomaceæ, which can only be seen after they are magnified several thousand times, and then only under peculiar circumstances of illumination, must be difficult of comprehension, even if their life history were much more simple and more easily studied than it is. I cannot too strongly caution the intending student of this enticing branch against trusting to a few and hasty observations made upon the dead skeleton of the plant. It is only when they are studied in the living state that the Diatomaceæ can be understood, and even then only with difficulty.

But one more abstract from my note-book and I must draw these remarks to a close. In the early part of November, 1868, I made a collection of *Colletonema vulgare*, and for some time have been able to keep it alive in a bottle so as to study its peculiarities. And here let me say that many minute forms of both animal and vegetable life which I have been unable to rear otherwise, I have found to flourish in phials with small necks, or those with large ones, which have the aperture partly stopped with a loose cover of some kind. It would seem that the gases given off from the human body, and accumulating in dwelling rooms, in which I have kept specimens, are deleterious to these small forms, and the partial closing of the vessel prevents, to a great extent, their entrance. My specimens of *Colletonema* flourished finely and grew considerably. I have been thus enabled to watch them, as I may say, building their tubes; this species, consisting of Naviculæform frustules enclosed and freely swimming about in tubes, after the manner of *Schizonema*. In fact there is nothing to separate these genera, except that the first inhabits fresh water, whilst the latter is an inhabitant of the sea, where it is to be found generally in profusion, covering larger algæ and rocks. The extension of the tube takes place after the following manner. As the frustules increase by the process of subdivision common to all of the Diatomaceæ, of course the two frustules thus formed occupy double the space of one, and as the cell division is continually going on, after a time the tube must become choked with individuals. At this period in their existence they appear to be extremely active, moving with increased rapidity up and down the tube as freely as their crowded condition will permit. Whether the end of the tube is never closed, or opens at certain seasons, I have been unable to determine; at all events it is now found to be open, and the frustules slip over each other until they reach this opening, and one or two will project outside as if prospecting, and will occasionally return within the general envelope. When a frustule thus projects from the open end of the tube, it never, as far as I have seen, rushes onward with the vigorous motion with which it moves within the envelope, but this is doubtless only so when the tube is being lengthened. It can be easily understood that if the species be disseminated by the distribution of perfect frustules, as seems to be most likely, that they must then escape from the tube after the manner I have recorded above as taking place in the allied genus, *Schizonema*. When one or two frustules have projected from the open end of the tube, they often immediately come to a rest just beyond the tube, or do so after moving over each slowly outside of, but in a line with, the tube. While at rest there appears to form around them a transparent mucous sheath, which, so that it may not fix them in their position, is kept in a tube form by the frustules again moving over each other, and thus, as it were, fashioning and smoothing the inside of the tube. This sheath becomes more and more dense, until it is plainly visible as forming an elongation of the tube, when the frustules again project from the end, and a new portion is added. I have in this way seen a tube grow across the field of the microscope, and the closely packed frustules extend themselves in single file, each just overlapping those in front and behind it. The membrane constituting the tube, although dense and strong, is somewhat elastic, but not very much so, for I have seen three or four frustules become wedged together by one attempting to pass backwards, whilst the others were moving forwards, and at such times the tube does not stretch to accommodate the crowding, but yet is often bent by the force of the moving frustules. In fact this force must be considerable, as is evidenced by the size of the obstacles, as grains of sand, which a small Diatom will move; and in *Colletonema* I have seen the tough tube membrane bent inwards so as almost to collapse by such a crowding as I have mentioned.

As I have mentioned Mr. Kitton's paper in 'Science Gossip,' I must here take the opportunity of saying something in regard to that article, and I feel sure that he will not take amiss what I shall say when he understands the spirit in which it is written.

First, I wish it to be understood that the specimens and the letter accompanying them were sent to Dr. Arnott for his opinion; thereafter I intended to publish the facts treated of myself. However, as Mr. Kitton has made public his opinions on the specimens, I will now give mine; but the fact of its being a private letter of mine from which he quotes, and one never meant to be made public, must explain what I there say. He has considered the "queer form" to be a new *Fragillaria*, and has named it *crotonensis*. Evidently he does not agree with that portion of my letter which he quotes, when I say: "I am not in favour of naming forms after places or persons, but strongly incline to distinctive and descriptive names." If the form were a new species I should have named it myself; but Dr. Arnott at once said it was likely to be *Fragillaria capucina*, var. γ, and such I see Mr. Roper points it out to be in the July number of the same periodical.

As it may be of interest in connection with this point, I may say that in a previous specimen of the sediment from the Croton water which I had sent him, Dr. Arnott informed me he had found the following species:—*Cymatopleura elliptica*, *Navicula trinodis*, and *gibberula*, *Surirella craticula*, *Denticula obtusa*, *Epithemia zebrina*, *Cocconeis Thwaitesii*, *Achnanthes ventricosa*, *Cyclotella rotula* and *operculata*, *Orthosira orichalcea*, *Gomphonema tenellum*, and another intermediate between *dichotomum* and *intricatum*, most likely the latter.

On a NEW SPONGE, *TETHYOPSIS COLUMNIFER*. By C. STEWART, ESQ., F.L.S., Curator of St. Thomas's Hospital Museum, with Plate XVIII.

For the opportunity of examining and describing this interesting sponge I am indebted to the kindness of my friend Prof. W. H. Flower, of the Royal College of Surgeons.

I found it in a jar containing marine animals collected on the coast of tbe Philippine Islands by the late Mr. Cuming; in the absence of further evidence this may be taken as its probable habitat.

The sponge consists of an irregular hemispherical basal portion, a little less than an inch-and-a-half in diameter, thickly incrusted with pieces of shells and small stones; its flat surface has either been torn off from a rock or from a continuation of its substance completing the sphere; this surface shows long spicula radiating from the centre imbedded in firm sarcode, and supporting on their distal extremities a white superficial layer, usually about the fortieth of an inch thick, but frequently greatly increased as the interstices between the extraneous substances are filled by it; the torn orifices of four canals may also be seen.

From the convex surface of this, the basal portion, arises a stem-like prolongation three-tenths of an inch in diameter and one inch and two-thirds in length, it is free from any foreign particles except a few small grains of sand at its extremity, but is roughened by numerous elevations of the dermal membrane arranged in a right-handed spiral of half a turn, which is not always strongly marked.

On section it shows a dense axis of spicula occupying a third of its diameter, from which laterally flattened processes extend to support the dermal membrane, which is raised by them into the elevations already alluded to; the very large intermarginal cavities thus formed communicate freely throughout the entire stem, and with the canals of the basal portion of the sponge.

The large spicula of the basal mass are of the form termed "expando-ternate" by Dr. Bowerbank; the long shaft of the spicula reaches from the dermal membrane to near or quite to the centre of the sponge, where it terminates in a sharp point, its ternate distal extremity supporting the dermal membrane; mixed with these are a few of the same size but pointed at both ends. The sarcode spicula are stellate and about the 1200th of an inch in diameter. The dermal membrane of the base readily tears into fine fibres; it is densely crowded with stellate spicula of about a fourth the size of those of the sarcode, their radii being very short.

The dermal membrane of the stem is very thin but similar in structure. The axis of the stem is formed by a direct continuation of the large spicula of the basal part of the sponge immediately beneath the origin of the stem; they are arranged in a similar manner with their long pointed shafts directed downwards, their triradiate distal extremities assuming an asymmetrical character, one of the three terminal radii of each spiculum being greatly prolonged to form with other similar prolongations derived from its neighbouring spicula a flattened band which supports the dermal membrane.

The principal interest attached to this sponge is, that it consists of two parts extremely different in external appearance, and also in the aspect which they present on section, the spicula, however, being of the same type throughout the entire sponge, and only slightly altered to perform a special purpose.

I have been unable to detect either pores or oscula, probably owing to their rapid closure on removal of the sponge from the water.

REVIEWS.

Forms of Animal Life. By GEORGE ROLLESTON, D.M., F.R.S., Linacre Professor of Anatomy and Physiology in the University of Oxford. (Macmillan & Co.)

THE book which Professor Rolleston has produced is remarkable in many ways, and has special points of excellence which raise it very far above any general work on comparative anatomy, published in this country for many long years.

We would first of all draw attention to the great care and pains which have been bestowed on his work by the author. No one knows so well, as one who has personally watched the progress of this book, the infinite trouble which Dr. Rolleston has taken to assure himself of the trustworthiness of every statement made therein. Many long and careful dissections have been made, solely for the purposes of the book, so that the author might state with confidence, and of his own knowledge, what he does say. Again, the bibliographical research which he has made is extended far beyond what are the usual limits in such matters, and the student will find the reference to authorities for a host of observations and doctrines, the origin of which he will have previously ignored, besides quotations from the latest and most important contributions to the science. When we consider the amount of time which has been given to this work, and remember also the energy and care with which Dr. Rolleston carries on his teaching at the museum of Oxford, we feel that the less unselfish men of science owe a debt of gratitude to one who so far foregoes the pleasure of working in the field of original exploration, and gives himself so fully to the noble work of teaching. Men who have worked in the museum at Oxford know well the great power which the Linacre professor there exerts; how he has drawn many unemployed minds into the current of work, and by his example of untiring energy encouraged all as a great teacher can. His book will extend his sphere of influence beyond the limits of Oxford, and may we hope attract new pupils to his laboratories.

This book is not a book for the very youngest beginners, but it is a book for all who wish to push well into the study of zoology. A special feature in its plan is its practical character. It is not a book to read quietly through with easy reference to figures and diagrams, but it is a book to grapple with and to master, and when this is done the student will have obtained a sounder and more adhesive knowledge of comparative anatomy than he can from any other book we know of. The first part gives descriptions of the large groups of the animal kingdom, which may be read as easily as the author's style will permit; the second and third being detailed descriptions of actual dissections or of drawings, require careful comparison with preparations and specimens to which it is *absolutely essential* for the student to gain access, either in some museum, or by his own dissections, as far as possible. We allude above to the author's style, which no doubt will appear a difficulty to many in reading the book, on account of the dense packing of words and allusions into a single sentence. But let us not be understood as objecting to this style; its very difficulty has advantages, it arrests the attention and demands the thought of the student, and it is greatly preferable to the mystifying and wearisome verbosity of some writers on biological subjects. Let us take an example from a part of the work relating to microscopic organisms. "The Gregarinæ would by most writers be considered, as they are here, to be the lowest of the Protozoa. Their ento-parasitic habits, however, which will account for much of the simplicity or degradation of their organism, must not cause us to overlook their close affinity to certain forms of Rhizopoda, especially the *amœbina;* and it has been rather from considerations of convenience, which, in the absence of any actual demonstration of genetic affinity, have weight in classification, that they have been here separated from that class. The Rhizopoda are by some writers placed higher, by others lower, in the scale of life than the Infusoria; but the 'polymorphismus' of their more complex forms, amongst which the Radiolaria are usually included, may be considered in some sense to counterbalance the higher grade of specialisation to which the Infusoria in virtue of their digestive, reproductive, and motor organs must be allowed to have attained. The Spongiadæ should, for the same reason, and in the same sense as the Rhizopoda, be placed in co-ordinate rank with the Infusoria." This quotation gives an idea of the suggestive manner in which the questions of classification and interpretation of structures are discussed by Dr. Rolleston. There is a vast deal more, than a mere expression of judgment as to the position which the Gregarinæ should occupy, in the above sentence. The correlation of parasitism and degradation of structure, the claims of convenience and of the principle of genealogy in classification, the weight to be attached to polymorphism as against individual development, are all incidentally touched upon with much advantage to the student. The constant reference to and enunciation of principles deduced from general study, forms one of the peculiarities of Professor Rolleston's teaching, and where it may not be possible to agree with some of these axioms, yet it is excellent for the pupil to have such briefly stated conclusions on which to thread his facts and exercise his own powers of thought. We cannot commend too highly the categorical way in which the most prominent facts relating to the larger groups of the animal kingdom are set forth. Dr. Rolleston follows Gegenbauer mainly in his classification, and in describing the chief classes of vertebrata, of mollusca, vermes, &c., gives such a body of knowledge as is to be found in no other book—in the English tongue certainly —besides copious references to recent and ancient authorities, which will be valuable to the teacher and observer as well as the class student.

A further peculiarity of this book, which we would point out, and which in our judgment gives it a special value, and accounts for much of the freshness of its style, is that Dr. Rolleston looks at forms of animal life as probably no other distinguished anatomist of the day can do; he looks at these forms *not* as an artist; he does not draw; and he does not accordingly treat morphology at all from the same point of view as does the observer, who instinctively apprehends and generalises a structure in a forcible sketch, as for example Gegenbauer does. Though we have to regret on this account the rather hard and unreal appearance of some of the woodcut figures in this book, over which we know both draughtsman and author spared no pains; yet it is due to this same cause that Professor Rolleston is so careful in his descriptions, and so accurate in the use of language, an immense advantage for the student. The style of treatment in this work is rather by close and careful definition and detailed description, appealing to the logical faculty, than by diagrammatic outlines and word-sketches, dependent on an artist's perception of form. Though we think it impossible to overestimate the value of artistic power in the morphologist, it yet is exceedingly valuable to have so cultivated a thinker as the author treating morphology apart from this power, for so far he here stands almost alone as the representative of unbiassed thought.

There are two points in which we may venture to express dissent from the author's classificatory views, and there are probably many others in the book which are open to discussion, but make their enunciation by Professor Rolleston none the less valuable and interesting. Above we have quoted a sentence in which Dr. Rolleston says he would assign a rank to Spongiadæ among the Protozoa equal to that held by Rhizopoda, and for the same reason, viz., the polymorphism exhibited by some of the forms of these groups. We cannot call to mind any polymorphism in the Radiolaria to which Dr. Rolleston alludes, for simple aggregation does not constitute polymorphism, and the association of units to form a secondary aggregate is carried no further in these creatures. In the sponges, on the other hand, there is most complete polymorphism amongst the primary units, that is to say, "histological differentiation"; and not only that, but the secondary aggregates formed of these differentiated elements with endoderm and ectoderm, antimera, and central osculum, exhibit polymorphism in their aggregation to form tertiary aggregates in some cases, so that certain abortive "persons," as Haeckel terms them, share the mouth of a central "person." These characters of the sponges seem to us to separate them by a huge gap from Rhizopoda, among which we never see a trace of "division of labour" structurally expressed, or individuation, in connection with the aggregation of units in such forms as are compound, and they go far to justify Haeckel's and Leuckart's placing of Spongiadæ as Cœlenterata. Our second point is as to the dissociation of Trematods and Leeches, and the arrangement of the latter with the Chætopods. We are strongly persuaded that the digestive, vascular, reproductive, muscular, tegumentary, and locomotive system of the Leeches are but slight modifications of the Trematod's, and suspect that in placing Discophora with Chætopods, too much weight is given to a physiological phenomenon of great variability, viz., the presence of the Hæmoglobin in the vascular fluids (not in homologous vessels) of the two groups of worms, which have but very remote genetic affinities.

To the private student, and indeed to many teachers, a more exact account of methods of and apparatus for dissection in different cases would have been very valuable, and we feel sure when it is known that this is a strongly felt want, Dr. Rolleston will not delay to supply it.

To students attending classes in our universities and elsewhere, to those working in their own studies, to all interested in any branch of comparative anatomy, we most earnestly, and with the confidence which comes of experience, commend "Forms of Animal Life" as a thorough piece of work, and certainly the best book on comparative anatomy in our language.

New Method of Fixing the Objective.—We have received the following from Professor Claparède, of Geneva.—"I am anxious to recommend to all savants or amateurs who occupy themselves with the microscope a very convenient addition due to the invention of Professor Thury, engineer of the Genevese Society for the Construction of Physical Instruments, and employed by the same Society in the construction of its excellent compound microscopes. This addition consists of a spring clip, which is fixed at the extremity of the tube to which the objective is screwed. This clip has a part turned in steel, with the greatest care, under which it is only necessary to slide the objective, which can be thus easily put on and taken off instantaneously. For this purpose a small intermediate piece is screwed on to the objective, which is made to fit exactly into the clip.

"This invention, which is very much to be preferred to the so-called revolving system, presents besides the advantage of a great economy of time in changing the objective, that of permitting a more exact mechanical centering of the objective—than that which one could obtain by the screw; and it also allows the employment in the observation of an object, of that part of the objective which gives the best images. This system can be adapted directly to all microscopes provided with the English screw, and by the modification of the intermediate piece as required, can be adapted to the objectives of all the principal English, French, and German makers."

English versus French Objectives.—One of the editors of this Journal has now had considerable experience of the objectives of Hartnack, of Paris, and his pupil, Verick, the former of the Place Dauphine, the latter of 2, Rue de la Parcheminnerie, Paris, and he is anxious that all who use the microscope should know of the excellence and cheapness of these makers' glasses. At Vienna, in the laboratories of Rokitansky and Stricker, the glasses of Hartnack are alone used, and their excellence well proved by the work done: they are also in use in most of the German laboratories. Hartnack's

No. 10 *à immersion*, is a glass which in its working power is considerably better than the English $\frac{1}{12}$th, and has the great advantage of admitting of the use of covering glass of some thickness, besides giving ample light. We make this statement after careful comparison in various kinds of work with the best English glasses, and with the assistance of experienced manipulators. This glass costs eight pounds, whereas for its equivalent in this country we should certainly pay twice that sum. Microscopists, and especially young students who have not money to throw away, should remember this; and most confidently can we assure them that they may rely on Dr. Hartnack's work. Still more remarkable is the difference between the prices of the dry French glasses and the equivalent glasses of English makers. We possess a No. 8 of M. Verick (nearly the same as Dr. Hartnack's No. 8), which, after careful comparison with first-rate English glasses, we consider most satisfactory in performance. A very celebrated English microscopist—who should know better than any man the value of a high power—declared to us after comparison, that this No. 8, which cost two pounds, was quite as good a glass as his $\frac{1}{12}$th, which cost eleven guineas. Well, these are facts for the consideration of English students, especially of those who are thinking of buying microscopes or high powers. We strongly advise them to purchase the foreign glasses, and to thus induce the English makers to offer their work at a reasonable price. The main cause of the great difference of price is, we believe, in the difficulty of working Lister's system, which is that to which the English makers adhere; but English prejudice and a close market have more to do with it. In the matter of distance between cover-glass and lens, and amount of light, the foreign glasses have a clear advantage over English ones, independent of price.

Another subject, which is of the same nature, is the relative value of the English and foreign microscopes themselves. We protest against the long-tubed, many-wheeled, awkward English model, and much prefer for *work* the small French body. It is a great deal cheaper, which is one important thing for the student. One English maker has recently brought out a quite small microscope-body, which we should highly approve with one or two alterations. Such a body with a French No. 4, and 8, or Nos. 3, 6, and 8, should not cost the student more than six guineas or seven, and would be actually a first-class instrument.

We may here add one word as to the No. 15 (immersion) of Hartnack—a glass equivalent to a $\frac{1}{33}$rd on the English scheme—and of which as yet not more than six have been made. Dr. Woodward, in America, has not yet tried this glass on Nobert's lines, and we should be much interested to know what it can do on that test. It is an admirable glass to work with, very nearly as easy as the No. 10, admitting of a glass cover of No. 2 thickness, and not requiring a condenser. We had recently an opportunity of working with this glass at living muscular tissue in the laboratory of Professor Stricker, of Vienna, and hope soon to try one against the English glasses. The No. 15 costs sixteen pounds.

Development of Gregarinæ.—In a private letter our distinguished contributor, Dr. Edouard Van Beneden, informs us that he has recently made some highly important observations on the development of the Gregarinæ of the Lobster, which he described in detail in the January number of our Journal for this year. He has observed the formation of the pseudo-navicells, and finds that these give rise each to a protoplasmic body with two long processes, one of which is very mobile like the appendage of *Noctiluca*; the other is motionless, but elongates, growing at a great rate, and at length separates from the rest of the mass, becoming the adult Gregarine; thus in its separation from the original form by division recalling the metagenesis of Echinoderms. Dr. Van Beneden has also carefully traced the formation of the anterior chamber of this species. The importance of these observations, as far as Protozoa are concerned, cannot be overestimated. Lieberkuhn's observations on the development of the monocystic form of the earthworm are at length succeeded by a knowledge of the development of a *true* Gregarine, with two chambers, which before was totally wanting.

Freshwater Radiolarians.—We urgently draw our readers' attention to Mr. Archer's notices in the report of the Dublin Microscopical Club in our present issue. His detection of the central capsule in one of the freshwater Heliozoa is of the greatest importance, and we are glad that so valuable a piece of evidence of Radiolarian affinities has turned up. Meanwhile how is it we do not hear of any of Mr. Archer's beautiful forms from English localities. Will not the microscopists of Yorkshire hunt up their moor-pools—and add to our knowledge of this interesting class of forms—for the discovery of which science owe so much to Mr. Archer? There are moor-pools, too, in Devonshire, in Wales, in Cumberland, and in Scotland as well, which, no doubt, contain further evidence of these Protozoa and new forms, such as Mr. Archer's Labyrinthulean, the discovery of which must be of the greatest interest.

New Facts as to Bathybius, Coccoliths, and Coccospheres.—Ernst Haeckel, in a paper published too late for extended notice on this occasion, describes Huxley's *Bathybius Haeckelii*—first made known in this Journal in 1868. He gives figures of the network of protoplasm, which is the essential part of Bathybius, and also numerous figures of the Coccoliths, Cyatholiths, Discoliths, and Coccospheres. In a remarkable new Radiolarian from Lanzarote (Canaries), Haeckel found masses of concretions exactly like the Coccospheres, in fact, indistinguishable from them; and he raises the questions whether these were taken in by the Radiolarian, or whether the enormous quantities of Coccospheres found in the sea-ooze have come from such Radiolarians (to which he gives the name *Myxobrachia*), or whether the resemblance is in fact a proof of identity. He refers to Wyville Thomson's statement that the oceanic ooze was actually alive with a sticky, glairy protoplasm, and is rather inclined to accept Bathybius as one of his Monera, and a most important one, and to leave the question of the source of the Coccoliths still open, than to suppose that the protoplasmic matter and concretions are simply offsets from the siliceous sponges or any other of the sarcodic organisms of the sea bottom.

Dr. Carpenter, we believe, has not been able to satisfy himself of definite movements in the glairy matter of the Atlantic ooze, such as would be expected from a specific organism as *Bathybius* is supposed to be. He thinks the protoplasm *may* be only that of the various sponges, foraminifera, and radiolarians, whose hard parts are there in abundance, as well. He has, however, seen little balls of sand or arenaceous matter, held together by protoplasmic matter, which have so definite a shape that they may owe their form to organic origin. Possibly they represent the very simplest form of those sarcode-organisms which avail themselves largely of arenaceous matters in forming a skeleton.

A new Moneron.—On the coast of Norway, Haeckel observed last year the life-history of a new form belonging to his Monera. He proposes to call it *Magosphæra*, and will shortly describe it in full. It is an aggregation of long flagellate cells, something like one of the Volvocinea in this respect. At a certain stage of growth, the flagellate cells separate, and seeking the bottom of the vessel in which they are kept, draw in their flagellum, which assumes the characters of ordinary protoplasm, and they then move as *Amœbæ*. The conversion of ciliate into pseudopodial locomotion is very interesting and important when we find that Haeckel has observed the reverse change in the development of the cilia of the embryos of Siphonophora, and in the cellular elements of the Sponges.[1] When the cells of Magosphœra have become amœboid, they take on an encysted stage, and undergo a cleavage, exactly like yelk-division ; rupturing the cyst, the cleft mass comes out, consisting of an aggregate of cells which develope flagella, and thus the original form is reproduced.

Glycerine-jelly.—This appears to be an excellent medium for preserving objects which will admit of a little 'clearing' action on the part of the glycerine. It is not so powerful in this way as glycerine itself, and is far easier to use, setting firmly. It is more 'clearing' than Deane's jelly, and melts at a somewhat lower temperature. We have found it very useful for mounting worms of various kinds, especially Annelids with the view of demonstrating the form of the setæ. That which we have used was obtained from Mr. Baker, of Holborn. We believe it will be found very useful for a host of objects. The easiest way to use it, is to put a small bit on the glass-slip, melt it over a lamp or near the fire, then place in the object to be mounted and carefully lay on the thin glass-cover. All is then over.

A Method of Mounting delicate Tissues.—Dr. Macintosh, of Murthley, whose beautiful book on Nemertean worms we elsewhere notice has used the following method in preparing and preserving sections of these worms for the purpose of studying the tissues. He hardens in alcohol, and after making sections carefully, washes in water. Superfluous water being got rid of, a drop or two of a concentrated solution of the chloride of calcium is added and the cover-glass immediately cemented down. No cell of any kind is used, and some preparations have been kept thus for four years. Those who know the delicacy of the tissues of Invertebrata will highly value this method, which enables the observer to retain his preparation unchanged by the too great clarifying power of glycerine or resins.

Method of Staining and Mounting in use at Vienna.—The section of tissue to be mounted is cut with a simple broad razor (the large knives specially made are not to be recommended) well covered with alcohol or water, as the case may be—the tissue being embedded in a mixture of wax and oil poured into a small paper tray if necessary. The section is then placed in

[1] In the development of the spermatozoa of the oligochæt Annelid, *Tubifex*, I have observed a similar phenomenon, and what is even more remarkable, very large, active, fusiform masses, exhibiting very rapid movement like a cilium, and possessing at the same time the contractility of a pseudopodium.—E. R. L.

a solution of carmine in ammonia, from which the smell of ammonia has disappeared, till sufficiently stained, *i.e.*, three to twenty minutes. Then it is well washed in distilled water, agitated by blowing into it through a glass tube; thence for half a minute in a watchglassful of water with one drop of acetic acid added thereto; thence into absolute alcohol. After ten minutes in this, all the water being extracted, it is placed in oil of cloves, which completely clears it in a minute or two; and then is mounted in solution of gum Damar in turpentine—such as is sold by artists' colourmen. This method is used by Meynert, the great student of brain-structure at Vienna, and in the laboratories of Stricker, Klein, Brücke, Wedl, and Rokitansky. Dr. Klein cuts, stains, and mounts sections of the hen's blastoderm in twelve minutes by this process. Glycerine is often used to mount in after carmine staining, but is found not to keep so well. Glycerine is used for mounting gold and silver-stained preparations, but does not preserve them for more than a few weeks or months—the staining becoming at length diffused and uniform throughout. Any one who would suggest a means of permanently preserving the beautiful silver preparations of cornea and lymph-capillaries, &c., would perform great service, and confer a boon on those who use these admirable reagents.

Nitrate of Silver Staining.—The weight of authority and experience is now decidedly in favour of Von Recklinghausen's interpretation of results obtained from the use of nitrate of silver. Perfectly fresh tissues are to be placed in a half per cent. solution of silver, left for ten minutes, washed and mounted in glycerine, and exposed to sunlight for half an hour or more. Epithelium is to be brushed gently away before excision if underlying tissues are to be stained. The frog's cornea is admirably stained by scraping off the conjunctival epithelium and rubbing with lunar caustic. In half an hour the most marvellously beautiful demonstration of stellate lacunæ is obtained. Professor Stricker and Dr. Klein are clearly of opinion that the cavities demonstrated in fibrous tissue, and in the cornea, are not artificial products; Schweigger Seidel still thinks they are; Robinsky, having utterly failed in using the method, tries to make out that Von Recklinghausen has made the most childish misinterpretations. We have a gold preparation which shows exactly the same stellate lacunæ in the cornea as do the silver preparations; hence they are not due to any special precipitation by silver. But we have not yet heard of anybody who has patiently used the silver-method and succeeded in making preparations, who is not convinced of the accuracy of Von Recklinghausen's views on the subject, for which we refer the reader to 'Stricker's Handbuch.' We caution our readers that those who discard the silver-method are those who have not had the *skill* to obtain the results which it can afford.

Microtomes.—We have to notice the microtome of His (Schultze's Archiv, 1870, 2nd part); of Ranvier, made by Verick, of 2, Rue de la Parcheminnerie, Rue St. Jacques, Paris; of Stirling, of Edinburgh, described in the last number of the 'Journal of Anatomy and Physiology;' and of that of M. G. Rivet, also made by M. Verick. That of Prof. His is by far the most complex and important, and is a modification of one devised by Prof. Hensen. We cannot now describe it in detail, but we do not think any results can be obtained by its use which the educated hand will not give. Ranvier's and Stirling's microtomes are both adaptations of the old section machine described long ago in Carpenter's 'Microscope.' They are very useful, as Dr. Cleland testifies in our April number. The chief point in Dr. Ranvier's little instrument is the use of elder-pith to surround the object to be cut; this is then squeezed into a cylinder with a broad flat top, and is worked up from below by a screw; the razor is drawn steadily across the flat top, and good sections of such objects as the spinal cord or a worm are obtained at once, even by the beginner. M. Verick supplies this instrument for ten francs. Mr. Stirling's differs only in being larger and fixed to the table. The microtome of Rivet is a very pretty thing admirably adapted, we would suggest, for cutting cucumbers, if constructed on a larger scale; and certainly useful where it is wished to take many microscopic sections. Two grooves in a block of wood run parallel to one another, but one forms a slight incline; along this groove runs a clip holding the object to be cut, along the other groove runs the knife fixed at a suitable angle for cutting. The movement of these two bits is managed by the hand, the grooves being deep and the clip and knife being attached to large bits of wood which run firmly in the grooves. The knife having been made to cut a piece of the object held by the clip, it is clear that a slight movement of the latter up its inclined plane will raise it a very little, and when the knife is again drawn along, a very thin section, depending for thickness on the amount which the clip was pushed along, is obtained.

QUARTERLY CHRONICLE OF MICROSCOPICAL SCIENCE.

Histology.—*Teeth.—Development of the Milk and Permanent Teeth in Man.* By J. Kollman, with two plates, eighty-five pages, Köll. u. Sieb. Zeitschrift, 2nd part, 1870. The author of this lengthy paper gives certain conclusions at which he arrives. Every embryonic tooth, he says, possesses a tooth sac. From the string or cord of connection between the original epithelial organ of the primary tooth-germ and the enamel-organ of the secondary tooth germ, arise certain club-like branches with round cells; these are quite free from any connection with vascular loops, but each bud, or epithelial-branch, as the author calls it, can give origin to a tooth. Upon this arrangement depends the abnormal increase of the number of the teeth. The second teeth always take their first origin in the median line, never at the side of the milk tooth. Cell-metamorphosis in the enamel-germ of the second tooth proceeds with extreme slowness. The cell-brood of the oral mucous membrane persists in the remnant of the epithelial organ which is spoken of above as the string of connection (verbindungstrang), and the processes and buds from this cord for years retain the power of starting the development of teeth. Dentes accessorii and dentes proliferi are entirely different in their origin. The *membrana præformativa* is really as such an artificial product. It is, in fact, the young condition of the enamel cuticle, of which some persons have erroneously denied the existence. The cells of the enamel-germ become variously modified in the course of development into (*a*) stellate cells of the enamel pulp; (*b*) later they form the innermost layer of the tooth sacculus, after the transformation of the enamel cells, and they take on the character of young connective tissue cells which finally become changed in all those animals with a cement layer on the crown of the tooth into (*c*) bone-cells or osteoblasts. The membrana adamantinæ and membrana eboris are best comprehended as lamellar epithelium. The one is produced from the cells of the embryonal Malpighian rete mucosum, the other from connective tissue cells. The enamel arises not by mineralization of the enamel-cells, but by mineralization of a substance exuded from those cells. The dentinal processes or threads extend through the whole length of the dentinal tubes, and through their branches. The dentine-cells do not calcify, nor is the dentine produced by the calcification of such cells. Interglobular spaces may produce the appearance of a lamellar disposition of the dentine, but are not, as was once thought, the result of a true lamellification. The splitting up of the dentine by weathering into lamellæ does not depend on interglobular spaces. The ivory of the elephant's incisor, in which the concentric lamellæ are so often seen, possesses no interglobular spaces.

Microzoology.—*On the Structure of the British Nemerteans, and some new British Annelids.* By W. Carmichael McIntosh, M.D., F.L.S.—This magnificent work, which the Royal Society of Edinburgh has done honour to itself in producing, is the result of the employment of holiday time and the few hours of leisure of a busy medical man in charge of a county lunatic asylum. Thirteen plates illustrate the memoir, which occupies more than a hundred quarto pages, and is of great value throughout. Dr. McIntosh is well known as a writer on the anatomy of invertebrata; he is the author of a work which the Ray Society is to publish on the British annelids and nemerteans, in which the species will be carefully discussed and figured. To judge by the drawings he has exhibited at meetings of the British Association, and by those in the present work, Dr. McIntosh's volume will be one of the most beautiful ever produced on zoological matters. In the paper published by the Royal Society of Edinburgh the anatomy and histology of species of Ommatoplea, of Borlasia, of Cephalothrix, of Tetrastemma, and Lineus, is given with the greatest minuteness, the worms having been studied very closely under the microscope by means of transverse sections. The writings of Rathke, Max Schultze, de Quatrefages, Claparède and Keferstein, and of Harry Goodsir, Johnston, and Dr. Williams, are thoroughly examined by Dr. McIntosh and most minutely criticised; he differs very considerably from all of them on many important points of the anatomy—especially as to the structure of the proboscis and its sheath; and from the great care he has bestowed on the investigation he is in a position to really decide the questions which he has taken up. The development of some forms is also given. We observe *Pylidium* of Müller quoted without a reference to Prof. Van Beneden's recent declaration that the supposed metagenesis in this case is a parasitism, the worm being quite distinct really from the pylidium. Dr. McIntosh does not say whether he would accept this view of the matter or not. It would be impossible for us here to even give a summary of the contents of Dr. McIntosh's wonderful paper without entering into a vast number of controversial topics as to anatomical details in these worms. One little thing we can allude to is a very curious ciliated parasite which drills the body of Borlasia through and through, and is very like an *Opalina;* but Dr. McIntosh does not describe contractile vesicles, nor nucleus in it. The interesting point is the segmentation of this ciliated creature, recalling the *Opalina prolifera* of Claparède, and adding another to the list of segmented infusors, and tending to upset the view that segmentation can be held to furnish any criterion of genetic affinity in groups of organisms. Keferstein has observed a similar parasite in the planarian *Leptoplana.* Dr. McIntosh draws a wide line between the group of which Ommatoplea is the centre (Enopla), and that of which Borlasia is (Anopla). With the exception of *Cephalothrix* and another all the British Nemerteans can be grouped round these two types. We congratulate Dr. McIntosh on having secured Mr. Ford to execute his plates. We would simply say that they are the very best illustrations of microscopic structure that we have seen drawn by an English artist, and equal to the best of German work. We hope Mr. Ford will find time to undertake more microscopic plates: we must not, at the same time, forget that the author made the original drawings himself, the excellence or roughness of which makes all the difference in the engraver's work.

Zoologists and microscopists may point to Dr. McIntosh's elaborate paper with much pride as a specimen of British scientific research, which we heartily wish did not stand so much alone in its excellence.

If Dr. McIntosh had given a brief résumé of the points which he chiefly desired to establish at the end of his paper, it would have been a convenience.

Protohydra Leuckarti, a marine stock-form of the Cœlenterata. By Dr. Richard Greef, of Bonn. 2 plates. Koll. v. Sieb. Zeitschrift. 1st part, 1870. The interesting form here described by Dr. Greef was observed by him in 1868 at Ostend in a quantity of diatom-scum and algæ, which he brought home for observation from the oyster park. It is remarkable as presenting very much the histological characters of the common fresh-water hydra, but its gross structure is far simpler, for it has no tentacles whatever. Moreover, it reproduces by a process of transverse fission, very much like

Stentor; and Dr. Greef did not observe any sexual condition. It is probable, as he argues, that if sexual organs do develop, the same simple form is maintained as in hydra; and that a medusoid shape is not attained by this animal. The interest of this creature is great in relation to Darwinian views—and especially since Haeckel's proposition to associate the Sponges in one group with the Cœlenterata. As Greef justly observes, this is the simplest Cœlenterate known, having no tentacles, although it possesses nematophors and pigment cells. Is it not rather through this form than through the Anthozoa—as Haeckel wishes—that we must approach the sponges? In any case this Protohydra gives a very conclusive answer to one of the objections raised against the assimilation of Sponges and Cœlenterata, viz., that all the Cœlenterata take in their food by tentacles, whilst no Sponges do so. Here is an undeniable Cœlenterate with no tentacles whatever. Whilst we are very much disposed to accept Haeckel's proposal to group Sponges and what are now called Cœlenterata together, as Leuckart originally proposed, we would be very careful not to claim too close an affinity for the Sponges with the particular group of Corals. The development of radial septa and the calcareous skeletons of the Calcispongiæ present remarkable points of agreement with the Anthozoa, but the point at which the ancestry of Sponges and Corals meets probably includes within it the stock of the Hydroid polyps as well, which may be supposed to have presented closer resemblances to corals in the structure of their earlier representatives than what we see now.

Embryology.—*Researches on the Structure and Development of the Arthropoda. 4. Praniza (Anceus) maxillaris. 5. Paranthura Costana. 6. The Decapoda Loricata.* With five coloured plates. Koll. v. Sieb. Zeitschrift, 1st and 2nd parts, 1870.—Dr. Anton Dohrn, formerly of Jena—an ardent entomologist and skilful observer—has now been devoting his energies for some time to the study of the development of Arthropoda, with the object of bringing facts to light bearing upon the descent theory and the genetic classification of these animals. He has recently been working at Naples, where he intends to stay for the purpose of prosecuting these studies. The contributions above named are admirable pieces of work, in which the details of the embryogeny and the later development of the nervous, vascular, and digestive systems in these crustaceans is discussed. Dr. Dohrn's figures of embryos are drawn with brilliant colours, to give greater distinctness to the complex disposition of parts, which he sketches diagrammatically. With such admirable workers as Dr. Dohrn and Dr.

Edouard Van Beneden in the field, the knowledge of the embryonic history of Arthropods is being immensely increased. There is, nevertheless, plenty to do for those who have the time and patience to enter upon these very difficult inquiries.

The Genetic Relationship between Ascidians and Veterbrates. By Professor Kupffer, of Kiel, three plates, sixty pages. 'Max Schultze's Archiv,' 2nd part, 1870.—In the January number of this Journal we gave some account of the very remarkable researches of Kowalewsky on the development of Ascidians, and some confirmatory remarks of Kupffer. In the present paper Kupffer gives his observations on the development of *Ascidia canina* in detail, very beautifully illustrated. The ova and young stages of this species were obtained at Kiel at the end of July and during August, so that those of our readers, who care to do so, can this year follow out these observations on allied species. The eggs are placed in a watchglassful of sea-water for observation under the microscope—a low power being used, and the chief difficulty being to catch the required view of the ovum at the particular stage in its early development which may be desired, since the changes proceed rapidly. Professor Kupffer gives full details on all points; we here would mention this, that he has thoroughly satisfied himself as to the mode in which the nervous system first originates—a matter concerning which Kowalewsky was a little uncertain. Kupffer has clearly seen a groove form on the outer surface of the ovum, extending from the opening of the alimentary cavity, and spreading round the egg like a meridian. This deepens and widens, and finally the cells close in above it, leaving it as the primitive nerve-cavity within the embryo. The nerve-cord originates clearly as an open canal which becomes covered in by the growing together of its walls above, as in Vertebrata.

The History of the Development of the Siphonophora. By Ernst Haeckel, 1869.—Professor Haeckel's researches were made two years since in the Canaries; they relate to the genera Physophora, Crystallodes, and Athorybia of oceanic Hydrozoa; they are published in quarto, and illustrated by fourteen beautiful plates, by the Utrecht Society of Art and Science, by which body they have been crowned, and from whom Professor Haeckel has received a prize. "All appearances which accompany the individual developmental history of the Siphonophora are to be explained solely by the palæontological development of their forefathers." This was the motto which Haeckel adopted, and in his work he has brought forward some most interesting facts to establish its truth. The work consists of historical introduction, remarks on the fundamental form and topography of the Siphonophora larvæ, individual development of *Physophora*, varieties and monstrosities of *Physophora* larvæ, systematic remarks on the new Algamidean genus *Crystallodes*, individual development of *Crystallodes*, experiments on the multiplication of *Crystallodes* larvæ by artificial division, varieties and monstrosities of *Crystallodes* larvæ; individual development of *Athorybia*, reflections on the individual development of the Siphonophora and on the signification of this in the elucidation of their palæontological development. The plates illustrating the work are admirable, and are, as Professor Haeckel observed to us, among the best executed which he has ever had done. Among the more general points of interest in this very important memoir, we will here point out a few. The cells which result from the yelk cleavage exhibit amœbiform movements, as has been observed in the eggs of fish and batrachia. It is important to note that the single egg cell divides itself into two, as observed in some Molluscs, Insects, and recently by Dr. Van Beneden in Trematod worms, this being the first stage of yelk division. How this is to be reconciled with the total disappearance of the germinal vesicle in Vertebrate ova, and in those of Annelids after fecundation and before yelk division has become apparent, is a matter of considerable importance. The amœbiform processes to which the cells give rise become *cilia*, thus establishing a connection between these two kinds of processes. In some of the young forms described and figured, Haeckel recognises a rudimentary representative of the axial canal of the hydroid polyp which he supposes appeared as one of the developmental phases of the ancestors of the Siphonophora. *Physophora* developes directly by conversion of the whole yelk mass, being in fact holoblastic. *Crystallodes*, on the other hand, buds off from the primitive yelk bag so as to develop an embryo and attached yelk bag, being meroblastic; the two genera are at the same time closely related as members of the Siphonophora. Professor Haeckel made some very curious experiments with *Crystallodes*, cutting the young ovum just after yelk division into two, three, four, or even five pieces. Each part thus cut off continued to develop, but varied in the extent of its subsequent development according to the degree of division which had been practised. These observations are really worthy of most earnest attention from all physiologists; they seem to demonstrate that the earliest formed cells of the organism have a common *potential* as well as an undifferentiated *immediate* function. None of the cells of the blastoderm at the stage when the ova were cut could have had a function in the immediately subsequent development distinct from that of its fellows. The cessation of development after a point was due to the diminution of the 'matter of life.' Professor Haeckel shows that the formation of the colony of which these composite Siphonophora are made up, takes place by budding from the stomach of the original polyp developed from the egg. An interesting observation which he copiously illustrates is the remarkably potent effect of disturbances, such as light, jarring of the vessel in which the eggs are kept, &c., in modifying the course of development of the larva and producing deformities. It is just one of those pieces of evidence which Mr. Herbert Spencer will be glad of to help him in establishing his view that the direct action of external agencies on the organism (direct equilibration) as well as the indirect—natural selection—is efficient in producing species. Here we have a case in which the slightest abnormality of condition produces, not arrest of development, not death (phenomena which sometime occur thus), but strange, seemingly irrelevant deformity; hypertrophy of one part and atrophy of another, of a most curious kind.

The difference between the nematophors of the larvæ and of the adult Siphonophors is another excellent point which Haeckel makes. He points out very carefully and figures the structure of the two, the larval being so much simpler as to belong to another type. This interesting fact may well be grouped with other cases of the presence in larvæ of the same organs as are present in the adult, but of another type, *e.g.*, the external gills of young ganoids and sharks, compared with the gills of the adults, the bivalve shell of the larval Anodon, and the shell of the adult, the six hooks of young Tænia and the circlet of the hydatid; most nearly parallel of all, the very strange bristles of many larval Annelids, and the totally different bristles of their adult forms.

The new genus *Crystallodes* is remarkable for its dense and firm character, contrasting so strongly with the other Siphonophora. It has a series of hard masses or joints of curious angular form which fit together supporting the polypites. On account of these bodies Haeckel terms it *Crystallodes*.

PROCEEDINGS OF SOCIETIES.

Dublin Microscopical Club.

20th January, 1870.

Dr. John Barker exhibited the dotted structure of *Pleurosigma formosum* under his dark-ground illumination, as also some fine rotatoria, which were shown in a very beautiful manner and with great comfort to the vision on a completely dark field, the effect, viewed binocularly, being very pleasing. Some were shown very satisfactorily under higher powers ; more detailed results of his experiments in this direction Dr. Barker would prepare for a future occasion.

Dr. Moore showed exceedingly pretty stellate hairs from a Japanese fern, a species of Niphobolus, well adapted as a polariscope object.

Rev. E. O'Meara exhibited a slide containing several interesting diatomaceous forms, for the possession of which he was indebted to the kindness of G. M. Browne, Esq., of Liverpool. The material was obtained from the careful washing of oyster shells picked up in Dublin Bay. Among the species of rare occurrence found in this slide were *Nitzschia spectabilis*, *Coscinodiscus omphalanthus*, and *Cos. concinnus ;* the last-named was but a fragment, but large enough to leave no room for doubt as to its identity. Special attention was invited to two species of great rarity and interest: *Actinocyclus triradiatus*, Roper, and *Navicula bicuneata*, Grunow, and Cleve. The former species, though found in England, Mr. O'Meara had never before obtained from an Irish locality. The latter, so far as he could ascertain, has not been hitherto recorded as having been discovered in the United Kingdom. The only localities in which it had hitherto been found are the Adriatic Sea at Porto Piccolo, near Castel Muschio (Grunow), and Gullmaren in Sweden or Norway (Cleve). Grunow's description is as follows:—*Navicula bicuneata*, nov. spec.—" On the secondary side linear, slightly constricted in the middle, ends obtusely cuneate, median line direct, central nodule slightly oblong, transverse striæ, fine, longitudinal striæ obsolete, two not very conspicuous submarginal sulei, length 0·0048″, breadth 0·0010″;" colour of the dried frustule violet-brown. Grunow, " Ueber neue oder ungenügend gekannte Algen." 'Verhandl.

der k. k. Zool. bot. Gesellsch. Wien,' band x, 1860, p. 546, t. i, fig. 4. This author expresses an opinion that the form so described may be only a variety of *Nav. liber*, W. Sm., but Mr. O'Meara, after a careful comparison of the two forms, was not disposed to consider the affinity between them so close, and thought he was sustained in this view by the following remarks of Cleve:—"Grunow has not described the front view,[1] which in the specimen I had the opportunity of examining was cuneate in outline (kilformig), as in the case of Gomphonema and Novilla, for which reason this species ought to be referred to a new genus distinguished from Navicula by the cuneate outline of the front view." Cleve, "om Svenska och Norska Diatomacéer." 'Ofversigt af Köngl Vetenskaps Acadamiens Förhandlingar,' Stockholm, 1868, taf. iv, figs. 3, 4. Only two examples of this interesting species occurred upon the slide, one perfect, the other a fragment; in neither was the front view exhibited.

Mr. Archer presented numerous and fine living examples of a remarkable form, from the fresh water, which for the present he must relegate to the Rhizopoda, though presenting such an extraordinary resemblance to Cienkowski's lately established "family" (or "class"?), Labryinthulea, as to render it exceedingly probable that in that group (be their actual nature and affinities what they may) the present form should find a place. In Schultze's 'Archiv für Mikroskopische Anatomie,' bd. iii, p. 274, t. XV, XVI, XVII, in a memoir entitled 'Ueber den Bau und die Entwickelung der Labryinthuleen,' Cienkowski has given an account of a type or group of sarcodic beings named as above, and founded on two forms discovered by him, amongst algæ, on piles in the harbour of Odessa. These organisms, as stated by that author, are characterised by the possession of three principal portions or constituents, the *central mass*, the *spindles*, and the *filamentary tracks* (which last term Mr. Archer thought might be a convenient and suitable translation of the word "Fadenbahn," employed by Cienkowski). It would be out of place here, inasmuch as to do it satisfactorily would take up too much space in making this for the present but fugitive record, to give a résumé of Cienkowski's account of these organisms; a brief reference has already appeared at the time in the pages of this Journal ('Quart. Jour. Micr. Sci., vol. vii, p. 277); those who may take an interest in the matter will refer to Cienkowski's paper itself. Suffice it here, for the present, to state that the organism now shown by Mr. Archer presented the above three characteristics, with some minor differences in detail. We have here, then, the "central body-mass," the "spindles," and the "filamentary tracks," the differences alluded to being, 1st, that whereas the *spindles* in Cienkowski's two marine species are described as nucleated, these bodies, in the present form, are not nucleated; 2nd, that the total body-mass presents the remarkable

[1] "Sidoytan" (the *side view*) of Cleve = to the *front view* in our phraseology.

tendency to becoming repeatedly encysted or coated with a thick hyaline many-layered covering, the long and densely arborescent body-mass becoming now and again protruded through an irregular, seemingly torn-like, aperture; 3rd, that immersed in the body-mass occur often numerous irregularly figured deep crimson-coloured pigment-granules—giving to these organisms when viewed under moderate powers a decidedly red colour. Cienkowski's forms, too, did not show incepted food—these did, of varied kinds, and considerable quantity. Nothing could be more beautiful or striking in the way of beings of this type than the noble specimen now under view; streaming out from the opening in the coat or cyst emanated a large trunk-like projection of the body-mass densely charged with orbicular bluish-coloured granules, and the accompanying variously figured reddish ones, this trunk-like portion becoming by degrees ramified in a highly compound and extensive manner into numerous and wide-spreading branches; these branches giving off the linear and numerous *filamentary tracks*, the whole reticulated in the most intricate and varied manner, and the ultimate *tracks* showing the *spindles* (the orbicular bluish bodies adverted to, now spindle-shaped, seemingly, by pressure) slowly executing their strange progression up and down the tracks, often in long files or rows, at other times crowded or isolated. Another difference—one of habit merely—(in addition to the bluish, not orange or colourless, spindles) is, that whereas one of Cienkowski's species would take many hours (as many as 24) to spread out and show their arborescent appearance on a slide, as many minutes would suffice for such a grand *oak-tree* as that now under view to grow up, as it were, before one's eyes. As to the supposed mode of progression of these spindles, and other points in connection with this remarkable form and with those of Cienkowski, as regards that author's conclusions, Mr. Archer ventured upon some observations at the meeting. To discuss these would take too long in brief minutes like the present, and must be deferred to a more fitting opportunity; but even should this curious organism be found on the Continent ere that opportunity could be had by Mr. Archer, still it would be something to have secured in our pages a record of priority of its detection. He would not venture here, then, to give any further *description* for the present, but should anything further of interest connected with it reveal itself, he would take an opportunity to bring it once more before the Club. It must be at least regarded as interesting to meet with, in the fresh water, a form at least so *close* to the marine organisms forming, according to the views of so acute and distinguished an observer as Cienkowski, a new type of sarcodic beings. It is true that some might hold that the non-nucleated spindles would keep the present organism out of the Labyrinthulea, but even were it so, nothing could more closely or more surprisingly simulate his forms than that now exhibited. Let not those who may peruse this crude note think we have here to do simply with a Gromia; if not a Labyrinthulean, then we have, at all events, a perfectly distinct and novel fresh-water rhizopod, and in that respect sufficiently interesting.

Prof. E. Perceval Wright recorded the occurrence in a new situation of the new rhizopod, *Cystophrys Haeckeliana* (Arch.), the first specimens met with by Mr. Archer being from Callery, these from near Carrig mountain. The groups were often fewer than in the original specimens, but even the smaller examples were sometimes met with, having encompassed a comparatively large diatom with a thin stratum of its delicate sarcode body, the cellular structures, however, retained the while, giving the diatom the appearance, viewed under a low power, of being surrounded by merely a thin stratum of independent globular cells.

24th February, 1870.

Mr. Archer showed the seemingly rare and very unique looking infusorian, *Cœnomorpha medusula.* This is a very singular little form, somewhat medusa-like in figure, but asymmetrical and with a long posterior tail-like process. Its agile and rapid movements render it difficult to follow and examine, hence the figures in Pritchard—it may be assumed taken "flying,"—are, perhaps, contradictory and not very graphic. This pretty form would well deserve to be worked out more accurately as regards its form and structure; it does not appear to be known (or at least well known) to Stein.

Rev. E. O. Meara showed Epithemia globifera, Heiberg, "De danske Diatomaceæ" (p. 103, fig. 22). Heiberg assigns this species to the brackish-water forms, but adds the following remark, "Hitherto found only in Hasmark Moor in Northern Fyen; inasmuch as this moor contains both salt and fresh-water forms, it remains uncertain where the peculiar habitat of the species is." The specimens exhibited were found by Mr. O'Meara in a mudhole, near Arklow, in the county Wicklow, in fresh water, remote from marine influences—a fact which decides the question as to habitat.

Mr. O'Meara likewise exhibited a slide of Diatoms from Jutland, which he owed to Rev. T. G. Stokes.

Mr. Archer desired to draw attention to a seemingly remarkable circumstance in a specimen of *Amœba villosa* (Wallich), in which, from the ordinary villous patch, were given off a number, probably about a dozen, of long, very fine, linear pseudopodial (?) processes. These hair-like developments were of different lengths, some nearly as long as the body-mass; the specimen, however, was below an average size, and these were much finer and more delicate than the seemingly somewhat similar, though coarser, processes recorded by Mr. Archer in this form on a previous occasion (see Club Minutes of 15th February, 1866). These gave the example a singular appearance; watched for some time they did not disappear, but the Amœba ere long unfortunately got lost. The observation might, however, be just worth this brief record.

24th March, 1870.

Professor E. Perceval Wright exhibited minute portions of the strange organism *Myriosteon Higginsii* (J. E. Gray); for these he was indebted to the great kindness of his friend Dr. Gray. A careful examination, both chemical and microscopical, showed that the Myriosteon was portion of the cartilaginous skeleton of a fish. On the outer surface of the portion examined there was imbedded a minute foraminiferous shell, and from an examination of this portion Dr. Wright thought for a moment that the whole structure might be foraminiferous, but this idea was soon dissipated on a section being made. It might, therefore, after all turn out to be a portion of some ray, though from a comparison of the tails of several species of rays from Mathè which were at hand for comparison, it did not appear that it could be any portion of a tail of a ray. As Dr. J. E. Gray has sent portions of the strueture to Mr. Carter and Dr. Günther, the solution of this problem cannot be far off. Dr. Wright's great object was to determine whether it was or was not any portion of an echinoderm, and, thanks to Dr. Gray, he had been able to say that it is certainly piscine and not any portion of an echinoderm.

Mr. Archer once more referred to the large green Actinophryan to which he had on a previous occasion drawn the Club's attention (see Minutes of 15th April, 1869), although he had not any examples now to exhibit, for the purpose of recording two sufficiently remarkable circumstances. The first was (what he had for some time supposed) that this form is characterised by the possession of a sharply-defined, spherical, clear, and colourless body immersed in the very centre of the body-mass, which seemed to be a veritable "central capsule." This was the then undecided point alluded to by him on the previous occasion (loc. cit.), and he trusted to be able on some future opportunity to demonstrate the existence of this to the Club, although it would be a matter of some nicety and difficulty, assisted by a certain amount of good fortune, to manipulate a specimen at a meeting so as successfully to extrude it intact; and a specimen freshly prepared previous to the meeting would not keep in good condition for exhibition sufficiently long. This spherical "capsule" resisted various re-agents. It would serve no good purpose (as here it could be done only inadequately) to enlarge to any greater extent at present on the affinities presented or on the resemblances or the specialities of this fine form, which is widely enough distributed in Ireland, but very local and scanty; though mostly *green* and very handsome, Mr. Archer felt that, still less than *Acanthocystis turfacea* (Carter), it was not at all the same thing as *Actinophrys viridis* (Ehr.), and just as little was it, he thought, the oft-debated *A. sol*, though, perhaps, coming closer to *A. oculata* (Stein), which Mr. Archer would venture to hold was distinct from both.—The second circumstance alluded to by Mr. Archer would be infinitely less easy to demonstrate at pleasure, for, though possibly by no means exceptional, it would seem very rarely to present the opportunity of being witnessed. This was the evolution from the body-mass of minute biciliated greenish "zoospores"—these without "eyespeck." These did not first undergo any encysted condition, but became eliminated from the living body of the Actinophryan by a kind of hernia-like protrusion from the surface becoming gradually more and more constricted off, and finally disconnected and swimming free. Not at first were the cilia apparent, yet they may have been there, though in but feeble action, for the little green bodies by degrees became removed to a somewhat greater distance, sometimes as if guided away by slipping along a pseudopodium; it was not usually till they had reached a distance from the body as great or greater than the average length of the pseudopodia (which in this form are comparatively few, but of considerable length), that they evinced a more considerable amount of activity. These then fidgeted about with a wavering jerky kind of movement, sometimes finding their way to comparatively considerable distance. Mr. Archer had observed as many as perhaps 10—20 evolved from a single individual in several instances, on two evenings, during the space of a few hours; kept on a growing slide he regretted he had not been able to follow out any further development, though in some instances he had noticed several very minute Actinophryans hard by some of the examples as they remained on the slide, leading to the view that these were truly *zoospores* or motile germs destined to reproduce this species, though the biggest of these little Actinophryans which he had seen hardly came up to the dimensions of the "central capsule" of the large typical representative of the form in question. Be then this observation worth what it may, it is at least worth recording. Mr. Archer was not, indeed, aware of any similar one having been made, for that by Stein in his work 'Die Infusionsthiere auf ihre Entwickelungsgeschicte untersucht' (p. 164), "Ueber die Swärmsprösslinge der *Actinophrys sol*," &c., does not really seem to be a case in point, nor to apply here; in fact it would almost seem that that observation refers to Podophrya or to an Acineta, and not truly to a Rhizopod at all.—Perhaps some other observer encountering this form would be able to repeat the observation and to throw a further light on the phenomena; and contenting himself meantime with this crude record Mr. Archer would, perhaps, be excused for bringing the matter once again before the Club.

Dr. Moore drew attention to the recurrence, and showed specimens, of the minute protococcaceous form which periodically makes its re-appearance on the water-troughs in the warm houses at the Botanic Gardens, almost with the punctuality of a deliberately sown annual. Some remarks on this form, one not readily to be identified, and one combining several puzzling and to some extent seemingly contradictory characteristics, are recorded in the Club minutes of July, 1866.

Rev. E. O'Meara exhibited a slide containing many interesting diatomaceous forms taken from the stomachs of Ascidians, dredged in Kilkerron and Roundstone Bays by Alex. G. More, Esq. Although the material was scanty, the variety of forms contained was considerable. Some of the most remarkable were *Cocconeis punctatissima* (Grev. 'Micr. Journ.,' vol. v, p. 3, Trinidad), *Cocconeis coronata* (Brightwell, 'Mic. Journ.,' vol. vii, p. 9, fig. 2, Shell-gleanings, West Indies), *Cocconeis fimbriata* (Brightwell, loc. cit. fig. 3, Corsican Algæ), *Coscinodiscus marginatus* (Ehr., America, Cuxhaven). Special attention was invited by Mr. O'Meara to a very pretty Navicula, which he could not find described, and which he proposed to name *Navicula Morei*, after the gentleman to whom he was indebted for the material.

Mr. Archer showed a new Cosmarium, first taken in County Westmeath, and again only a few days ago in County Tipperary. It would be useless to give here any imperfect "description" of it for the present, as such things require exactitude and comparison with other forms to fasten their specialities on the recollection, if not figures. He would only here note that though it must be regarded as truly a Cosmarium (perhaps coming nearest *C. commissurale*), it was amongst other characters marked by the possession of a very minute but evident mucro at each opposite lateral extremity of the segments, thus somewhat *arthrodesmoid* in its external characters. The specimens taken for these wide-apart sources were absolutely identical in the most minute detail, and he did not think that in all the species there was one more readily recognised than this, once one had got a grasp of its specialities in all aspects, nor were there many prettier *wee* things often under view than that now claiming the Club's attention.

Literary and Philosophical Society of Manchester.

Ordinary Meeting, December 28th, 1869.

J. P. Joule, LL.D., F.R.S., &c., President, in the Chair.

"On Pollen; considered as an Aid in the Differentiation of Species," by Charles Bailey, Esq.

Having recently examined the pollen of several thousand species of plants, I am led to think that the characters presented by these grains might prove useful as a means of differentiation in allied species; my researches, however, have not been sufficiently extensive to form any positive conclusions, but as leisure permits I hope to prosecute the subject further. In the meanwhile the following notes are thrown out as indications of some of the more noticeable distinctions to be drawn from a careful comparison of these organs, and they may serve to draw the attention of others to the matter.

There are four points in one or other of which pollen grains of plants belonging to the same genus may be found to differ from each other, viz., form, markings, dimensions, and colour.

1. Form. It has long been noticed that certain types of pollen are characteristic of the natural order to which the plants which produce them belong, as for instance, the peculiar pitted polyhedral pollen of the *Caryophyllaceæ*, the spherical spiny pollen of the *Malvaceæ*, the large triangular pollen of the *Onagraceæ*, the peculiar pollen of the *Coniferæ*, or the elliptical pollen of the *Liliaceæ* and other monocotyledonous orders; in fact, most orders possess a type sufficiently marked to be characteristic of each. This statement, however, must be accepted with limitations; the *Compositæ*, for instance, have three or more well-marked types, represented by the beautifully sculptured pollen of the Chicory, the minute oval spiny pollen of the Asters, Calendulas, Cacalias, &c., and another form wholly destitute of spines as in the *Centaurea Scabiosa*. There are, besides, other natural orders where similar variety occurs.

But differences of form are met with in plants of the same genus, by which the one species or the other is readily marked off by its pollen; thus the pollen grain of *Anemone sulphurea* is roundish, but that of *Anemone montana* is elliptic; the pollen of *Aronicum Doronicum* is much more elongate than that of *A. scorpioides;* and while the grains of *Ranunculus philonotis* are round and yellow, those of *R. platanifolius* are elliptic, white and smaller.

2. Markings. Here again there is endless diversity, and a boundless field lies open for the researches of tired-out dot and line hunters of diatom-valves. A few instances only of the more striking differences can be given here.

The pollen of the *Geraniaceæ* and *Campanulaceæ* is for the most part globular, but while some of the grains are quite smooth others are covered with spines; thus the pollen of *Campanula Media* has a number of short spines sparsely scattered over the surface of the grain, but *C. rapunculoides* is wholly destitute of them. In other plants these spines are replaced by tubercles, and both spines and tubercles vary greatly in length and number; for example, in *Valeriana tuberosa* the spines are only half the length of those on the pollen of *V. montana*, the grains being also slightly smaller. The pollen of the *Liliaceæ* is often covered with a more or less prominent reticulation, which is subject to much variation; compare, for example, the coarse network which invests the pollen of *Lilium croceum* with the finer reticulation of *L. canadense*, the grains of the latter species being much more globose and smaller.

3. Dimensions. Some instances of the differences observable in the size of pollen grains have already been published by Professor Gulliver, whose measurements of the pollen of various species of *Ranunculus* show the help that may be derived from this character; *R. arvensis* is nearly twice the size of *R. hirsutus*, their dimensions being respectively $\frac{1}{470}$th and $\frac{1}{888}$th of an inch.

I have not had the time to make similar careful measurements with the micrometer, but I have seen sufficient to be satisfied that while there is considerable variation in dimensions between the pollen of one species and that of another, they are tolerably constant in size in the same species.

For some noticeable differences compare the smaller pollen of *Epilobium brachycarpum* with the larger pollen of *E. Fleischeri* or that of *Senecio gallicus* with *S. incanus*, the spines on the latter species being also much coarser. Again, the pollen of *Silene acaulis* is but half the size of that of *S. alpina*, the latter having some beautiful markings in addition; the pollen grains of this genus differ from the usual caryophyllaceous type in not having the pits or depressions common in the order, so that the grains become spherical rather than polyhedral.

4. Colour. This is not so reliable a character for differentiation as the others noticed, since species differ amongst each other according to the soil, &c., of the place where they have grown. I remember gathering some years ago, near Ashbourne, Derbyshire, a variety of *Stellaria Holostea* having a dark purple pollen instead of the ordinary pale yellow. An example or two under this head will suffice.

The pollen of *Ajuga genevensis* is yellow, but that of *A. pyramidalis* is usually white; again, while the grains of *Ornithogalum umbellatum* are large and yellow, those of *O. nutans* are small and white.

Some objection may be raised to any reliance being placed upon the dry shrivelled-up grains of herbaria specimens—such specimens being in most cases the only ones obtainable for purposes of investigation; but the structure of pollen is such as to bring into greater prominence the pores, folds, valves, and other markings which are met with on their surface after the grains have collapsed by the discharge of their contents.

In regard to the mounting of these objects for the microscope, they show to the best advantage when put up perfectly dry; the cells should be sufficiently shallow to admit of no more than a single layer, and at the same time deep enough to permit the grains to move about. If pollen is mounted soon after it has been discharged from the fresh anthers the fovilla is apt to condense on the covering glass, and the slide soon becomes useless. The stamens taken from an unopened flower-bud furnish the best and cleanest pollen, and these should be selected in preference to those taken from the fully developed flower.

Canada balsam, glycerine, and other media are occasionally helpful in making out structure; thus the pores of *Campanula rotundifolia*, *Phyteuma Halleri*, and other allied species are made much more distinct when mounted in balsam.

A large series of slides illustrative of the above remarks was exhibited at the meeting.

MICROSCOPICAL AND NATURAL HISTORY SECTION.

December 6th, 1869.

JOHN WATSON, Esq., President of the Section, in the Chair.

Mr. W. Boyd Dawkins, M.A., F.R.S., was elected a member of the Section.

Mr. Charles Bailey read a paper "On Pollen, considered as an Aid in the Differentiation of Species." [This paper was afterwards read at the Ordinary Meeting of the Society, held December 28th, 1869. See page 309.]

Mr. J. B. Dancer, F.R.A.S., read a short paper on some of the new Hydro-Carbon compounds from which he had obtained very beautiful polarising objects for the microscope. These were exhibited to the members, and a more detailed account promised when the experiments are complete.

Brighton and Sussex Natural History Society.

April 10.—Mr. Glaisher, Vice-President, in the chair.

The President, Mr. T. H. Hennah, F.R.M.S., read a report on soundings made by Sir E. Parry in 1818 in Arctic seas.

The history of these soundings was this :—Mr. J. Cordy Burrows, some years since, purchased from the widow of Sir E. Parry his geological collections, among which were certain soundings. The geological specimens were placed in the Brighton Museum, but the soundings were given to Mr. Peto, who, in January of this year, placed them in Mr. Hennah's hands for examination. They were made by Sir E. Parry in his Arctic Expedition of 1818, in Davis Strait and Lancaster Sound, between lat. 68° N and 76° 15′ W, and long. 73° W and 78° 34′ W, in depths of 22 fathoms to 1058 fathoms. Those from shallow water consisted of fragments of stone and coral water, worn evidently by a strong current, zoophites, a microscopic madrepore, and the tube of an annelid were found in them. From deeper localities the soundings were rich in organic *débris*, much of the sand being in the form of testa of arenaceous Foraminifera of different kinds; diatoms, especially large Coscinodisci, were also abundant. Sponge spicules also abounded, but shelly Foraminiferæ and Polycystina were very scarce. Of the Foraminifera in many cases casts only of the inside of the shells were found. In sand and weed from Lancaster Sound, lat. 73° N., and 670 fathoms, borings of annelids still containing the skins of their inhabitants were found, affording conclusive evidence of the existence of life at great depths in Arctic seas. It was much to be regretted that these soundings, which might years since have taught so valuable a lesson, should have been allowed to remain unexamined until their historical interest and the prominence of deep-sea soundings had caused them to be brought to light. The recent discoveries of Carpenter, Thompson, and Jeffreys were next alluded to, and the fact pointed out that the soundings made by Sir E. Parry corresponded well with what had recently been found in temperate seas having a low bottom temperature.

May 12.—Mr. T. H. Hennah, F.R.M.S., President, in the chair.

An evening for the exhibition of specimens, at which Dr. Badcock exhibited fossil wood recently obtained from Portland Island. Mr. Parley laid on the table specimens of oak-stained green by fungus *Helotium Æruginosum* picked up at Tunbridge Wells. This supplies the green seen in Tunbridge Wells' ware. Dr. Halifax, commenting on this fungus, stated he had raised it from spores obtained from specimens of green-coloured oak. There was no doubt of the true nature of the fungus, for in thin slices of the infected wood the *mycelious* threads, and even spores at times, could be made out.

Mr. Wonfor remarked on the fact that this fungus was only found in woods in England and France, where the Hastings sand cropped out. This would lead to the inference that there was something peculiar in the chemical conditions of the soil, &c., favorable to its development.

Mr. Dennant exhibited a bottle of ooze obtained in the Porcupine expedition, in lat. 47° 35′ N. long. 16° 15′ W., at 2435 fathoms.

Mr. Hennah exhibited two live sea mice, *Aphrodita hystrix*, recently dredged up off Brighton.

Dr. Halifax exhibited very beautiful micro-photographs, taken by himself, the most striking being stomach of fowl, teeth of medicinal leech, sections of proboscis of blow-fly to show rasping teeth within central disk, poison-bag, with poison exuding, of spiders, &c.

Mr. Wonfor exhibited cluster cups in dog violet, nettle, eggs and coccoons of Emperor moth, *Saturnia Carpini*, and forty-three males, attracted by one female, in two days at Polegate and Tilgate, and read a paper on the power possessed by some females of drawing from long distance, and in great numbers, the males of the same species, in which he detailed experiments tried by himself and others with various insects, to find out, if possible, by what sense they were attracted. Sight was out of the question, for they had come up when the female was in a box in the pocket, or shut up in a leather bag. It would seem that scent was more likely, for they always came up against the wind, while it was noticed that if they got far to windward they did not return, but flew away.

Mr. Hennah announced that Mr. Peake had found a *Pygidium* in the lace wing fly, *Chrysopa peola*. This he believed to be an original discovery.

June 9.—Mr. Sewell, Vice-President, in the chair.

A paper, "On Diptera and their Wings," by Mr. Peake, was read, in the absence of that gentleman, by the Hon. Sec., Mr. Wonfor.

While wings are common to the whole order of *Insecta*, the *Diptera* consists entirely of two-winged flies, having instead of a second or hinder pair little thread-like bodies, terminated by knobs, and called *Halteres*, originally thought to be balancers, now considered by some *olfactory* organs and by others organs of *hearing*. From many points of resemblance he deemed them analogous to the hind wings of other insects, and that at present their special use had not been ascertained. Besides the Halteres, they had winglets (*alulæ*), which were thought to be only appendages to the fore wings. Among the Diptera three classes of flies were found distinguished by the form of their bodies and the shape of their wings, first, the slender flies, such as the gnats, having long bodies, narrow wings and long legs, but without winglets; secondly, those whose bodies though slender, were more weighty, as the Asilidæ, having larger bodies, shorter legs, and very minute winglets; lastly, those like the house fly, with short, thick, and often very heavy bodies, furnished with proportionate wings, shorter legs, and conspicuous winglets. From these circumstances it might be inferred that the long legs of the light-bodied flies acted as rudders, while the winglets helped the wings in flying. The wings consisted of two membranous laminæ, united by veins or nerves, and upon their arrangements, and the form of the antennæ, the distinguishing characters of the Diptera were founded.

The several parts of the wings, and nerves, and their appendages, as seen in the great groups Nemocera and Brachycera were next pointed out, and the papera, illustrated by very beautiful camera lucida drawings, made by Mr. Peake, and by microscopic preparations of the wings, &c.

May 26.—The President, Mr. T. H. Hennah, F.R.M.S., in the chair.

As this was the first meeting of the section, Mr. Wonfor, Hon. Sec., announced the objects sought in forming a section.

Mr. Hennah then read a paper on "Systematic *Recent* Examination, with Moderate Powers."

As we glance through the history of the microscope, we cannot avoid noticing how little has depended upon the instruments, and how much upon the method and perseverance of the men who have accumulated so vast an amount of information; and although the wonderful perfection of modern high powers,—and, indeed, of the microscope generally,—has undoubtedly increased both our means of research and the number of observers, the conclusion is forced upon us that—accordingly as we use it—the microscope is either a new sense, or a mere toy. I hope the growing tendency to the latter result may find no place amongst us. Pride in the possession of a fine instrument, and a consequent desire to exhibit its powers, often leads to the exclusive study of conventional test objects, which, while it gives command of the microscope in a special way, and stimulates opticians to improvements, too frequently arrests original investigation.

Whether we use high or low powers we should—in original investigations—be on our guard against the unconscious tendency of the mind to make "the wish father to the thought;" and, although we cannot be altogether free from preconceived ideas, their influence should be limited to the *suggestion of inquiry*. It is necessary for a just appreciation of our own work (and that of others) that we should be well acquainted with the literature of the microscope. It is, however, already so much scattered that it is difficult to ascertain the actual amount of knowledge on any given subject, and much time is wasted in investigations which should be but past steps in our progress. At the same time, we should not too readily accept authority on matters difficult of proof, as a false idea of the state of microscopical knowledge is frequently given and doubt arises as to our powers of observation or the instruments we are using. In the frequent intercourse of men engaged in a common pursuit—such as it is the intention of this section to promote—lies the best substitute for individual experience; the knowledge acquired by any should be available for all, and errors of solitary observation are soon corrected when brought to the test of criticism and comparison. Most conducive to a true knowledge of objects is their examination in a recent state, and an acquaintance with the appearance of ordinary things will be found much more valuable than the settlement of a *Diatom* or *Podura* question. In the food we eat, the clothes we wear, the parasites that plague us, and the very dust about us, there is a large field for investigation as a necessary preparation for other studies, which has also an interest of its own at a time when the President of the Board of Trade tells us that adulteration is but a form of trade competition. The poor Welsh imposter gave evidence of her fatal deception only a short time since, in the starch which the microscope discovered in her stomach; and many other instances might be adduced to recommend the study of common things. Our principal object, however, should be to inquire into the natural history of our own locality, the minute fauna of which has been but imperfectly examined. Our shore offers every inducement to extend research. The smaller Crustaceans are scarcely known amongst us, although two of the most curious and interesting—the *Caprella* and *Ammothea*—abound on the weed at Kemp Town and give promise of allied species of greater rarity as a reward for search. For full appreciation of minute structure comparison with permanent specimens is necessary. Mounted specimens cannot, however, be seen under sufficiently varied conditions, and we may as well take an ancient Egyptian as a specimen man as trust exclusively to the mummies in balsam which fill our cabinets; we must, instead—as students of nature—follow her home and watch her ways patiently, as far as we can. Nothing can be known of the Protozoa, or Rotatoria, unless we examine them in life. Cyclosis in vegetable cells must in like manner be seen in life to be seen at all. The generation of the cryptogams would be actually hidden if the germination of their spores had not been a subject of unwearied attention. The structure of the Foraminifera was not demonstrated by Carpenter without systematic work. The discovery of the alternation of generations was due to careful study, and the knowledge of the fact of the Polyps of our shores having other existences as free swimming Medusæ considerably modified our previous ideas respecting them. "We can all," concluded Mr. Hennah, "make some advance in knowledge by the *systematic* study of recent things, even though it be by tedious repetition; and I trust that by our work we may justify the formation of this section of our Society."

Mr. Wonfor then announced the receipt of 18 slides from Mr. Hennah, 6 from Dr. Halifax, and 38 from himself, for the Society's cabinet, and urged on all the members to contribute.

The meeting then resolved itself into a conversazione, at which the following gentlemen exhibited microscopic objects, under a goodly array of microscopes by the best London makers:—Mr. J.

Dennant exhibited sections of fossil teeth from the coal measures; Dr. Halifax showed some of his sections of insects, in which the internal parts were displayed *in situ*, one of the most striking being the lady-bird. Mr. Cooper exhibited deep sea soundings from different localities, and Foraminifera from Hastings, the Mediterranean, and Australia. Mr. Sewell exhibited injected preparations of Dr. Thudicum's rabbit, an animal possessing a world-wide reputation from being fed at times on muscle containing *Trichina spiralis*; the presence of Entozoa was traced in all parts of the voluntary muscle, but nowhere else. Mr. Aylen showed a number of entomological preparations. Dr. Addison exhibited blood, as acted upon directly by various agents, such as diluted sherry, &c., the effect produced was an alteration in the appearance of the blood discs, and, as some expressed it, the formation of tails; this was considered by no means the least interesting part of the evening's display. Mr. Peake exhibited the *pygidium* of the lace-wing fly, *Chrysopa perla,* discovered by himself. The existence of this peculiar structure has long been known in the flea, but has not been pointed out in any other insect. Dr. Kebbell exhibited with a Nachet's prism the rasping teeth situated on the disc of the proboscis of the Blow-fly. Mr. Smith showed fruit of Hepaticæ and epicarpal Stomata of moss, *Funaria.* These are only found on the fruit of mosses, and never on the leaves. Mr. Hennah showed plant circulation in the hairs of the *Tradescantia,* Spider-wort, and remarked that every microscopist should possess a root of this plant in his garden; pollen showing the production of the pollen-tubes; and *Caprellæ* from Black Rock, Kemp Town, &c. Mr. Wonfor exhibited a slide of Diatom, mounted by Müller, of Holstein, on which, in the space of a quarter of an inch, 408 separate siliceous skeletons of plants were arranged in symmetrical rows; very gorgeous crystals of *Hematoxylin*, the colouring matter obtained from log-wood; artificial alizarine, prepared from an oil of coal gas.

In addition to these microscopic objects, Dr. Halifax exhibited a number of beautiful microphotographs of his own taking; and Mr. J. Howell, pebbles picked up on the beach, showing encrinites, pentacrinites, bryozoa, cidaris spines, &c.

MEMOIRS.

OBSERVATIONS *on the* HISTOLOGY *of the* EYE.

By J. W. HULKE, Esq., F.R.S.

(With Plate XIX.)

THE APPARATUS OF ACCOMMODATION.

The term Accommodation.—Let me, in the first place, endeavour to explain what the term accommodation technically means.

That we cannot see perfectly distinctly at the same instant two objects placed at different distances before the eye, is a fact of the truth of which a moment's attention suffices to convince the most unobservant person. The fact is most easily realised when the objects are near, for when they are at a great distance the minor distinctness of one of them is less appreciable, but when they are relatively close to the spectator it is impossible for him not to become aware of the phenomena. Thus when I look at the nearer of two trees, placed several yards apart, nearly in the same line, in a distant field, the minor distinctness of the further tree is so slight that I may fail to notice it; but when I look at a book through a veil, both being near me and only a few inches apart I find that when my eye is fixed on the print, I see it quite distinctly, while I am scarcely conscious of the presence of the intervening veil; and again, when I look intently at the veil, and perceive its texture distinctly, at that same moment the print becomes confused and unrecognizable. What is the explanation of this? In order to see an object distinctly an exact image of it must be formed on the bacillary layer of the spectator's retina. Every luminous point on the surface of the object turned towards the spectator must be represented by a corresponding image upon his retina. This is effected by the refractive power of the eye—that power its transparent parts possess in common with inanimate transparent bodies, of changing the original direction of the luminous rays which enter it, and of giving those rays a new direction, of a kind dependent on their relative densities and their curves. Luminous pencils coming from remote objects consist of parallel rays, and having regard to the small opening of the pupil, the pencils which enter the eye from an object twenty or more feet distant, may be considered to be composed of parallel rays. Now the refractive power of the eye is such, that parallel rays entering it are collected in exact foci upon its retina, without the exercise of any vital effort, the eye itself being quite passive. It would occur as well in a dead eye, so long as its media remained transparent, and while they retained their proper curves. The luminous pencils which a near object sends to a spectator's eye consist of divergent rays, and the unaided refractive power of the eye which sufficed to unite the parallel rays from a distant object in the bacillary plane of the retina is insufficient to collect divergent rays in exact foci on this plane. The foci of those rays lie behind the retina, which the pencil strike as spots, the sections of cones, called circles of dispersion, and not as points. The result of this is a blurred confused image, and not a clear one, the production of which requires the rays to be brought to exact foci in the bacillary layer.

We are, however, conscious that we possess the power of seeing distinctly near objects as well as distant ones, which proves that the eye has the power to unite divergent as well as parallel rays in exact retinal foci, and this implies the possession of a power of altering its refractive state so as to suit it to the distance of the object we desire to see distinctly, or in other words to adapt it to the degree of divergence of the luminous rays entering the eye from the object. This adaptation of the refractive state of the eye is technically called its *accommodation*. It has been at different times attributed to a change of figure of the eyeball, to an alteration of the curve of the cornea, and to a shifting of the position of the lens; but more delicate methods of observation than were formerly at the command of the physiologist, have shown all these views to be erroneous, and by the direct inspection of the human eye with instruments specially contrived for this investigation, it has been demonstrated that its accommodation for a nearer object is effected by increased convexity of the lens, and this chiefly of its anterior surface —the curve of its posterior surface being altered in a scarcely appreciable degree. With the increased convexity of the lens, its axis is proportionately lengthened, the pupillary region of the iris approaches the cornea, and the peripheral portion of the iris recedes from it. There are not any grounds for supposing that this change of figure is wrought by any power inherent in the lens itself, which is plastic, but not endowed with contractile irritability, and is not dominated by the will, being devoid of nerves.

The active factor of accommodation must therefore be external to the lens. Now in close relation to the lens there are two muscular organs, the iris and the ciliary muscle, the existence of accommodation in persons from whom the iris is congenitally absent, and its persistence where the iris has been in part or entirely removed, demonstrate its independence in man of this diaphragm. There remains, therefore, only the ciliary muscle as the active factor of accommodation in the human eye. To the ciliary muscle and lens I would now invite attention. I shall take the lens first.

The structure of the Lens.—The lens of the human adult has a flattened convex figure. The anterior surface is less convex than the posterior, the radius of curvature of the former being nearly twice as great as the radius of the latter surface. The infant's lens is more nearly spherical, which makes the distance between its summit and the cornea smaller than in the adult's eye. This circumstance is not without influence in the causation of the minute white speck on the front of the lens (the central subcapsular cataract) not infrequent in persons who have suffered from infantile purulent ophthalmia, even when this has not been complicated with perforation of the cornea. The compression of the cornea by the swollen eyelids and œdematous conjunctiva and a slight amount of deep congestion pushing forwards the lens, may bring the lens and cornea together, and thus disturb the nutrition of the growing lens at the point of contact, and induce a perverted growth and retrogressive changes in the lens tissue.

In other mammalia and in the lower vertebrata the figure of the lens is less flattened than in adult man, resembling more nearly the shape of the human fœtal lens.

The lens of all vertebrate animals is formed almost entirely of a peculiar fibrous tissue with a very scanty formless interstitial substance. It is enclosed in a short capsule, the integrity of which is of the highest importance to its transparence. This capsule is a perfectly transparent, very elastic, yet brittle membrane. It is little prone to degenerative metamorphoses, and it never undergoes absorption. I have found it transparent and unchanged sixteen years after the extraction of the lens for cataract. Its chemical constitution and reactions resemble those of the other hyaloid membranes; it is unaffected by weak acids and alkalies, and it resists putrefaction.

With high magnifying powers no indications of structure are discernible in it, except faint marks of lamination in the stout capsules of the largest mammalia. The front half of the capsule, or, more exactly, that part of it which lies in front of the attachment of the suspensory ligament, is thicker than the posterior half; at a rough estimate, its thickness may be taken to be three times as great.

This difference, and the unavoidable implication of the front half of the capsule in operations for the removal of cataract, have led to the adoption of the terms anterior and posterior capsule. As these terms are convenient, their use is not objectionable, if it be borne in mind that they refer to two halves of one and the same size, and not to two distinct ones. Besides being stouter, the front half of the capsule differs from the posterior in being lined with an epithelium. This consists, in the central region, of a single layer of large, flat, polyhedral cells, each enclosing a circular nucleus. These nuclei are remarkably uniform in size and shape. At the edge of the lens the epithelial cells are much smaller, and so closely crowded that their nuclei are separated by very small interspaces. In mature lenses the marginal epithelium is composed only of a single layer of cells; but in young and growing lenses it is formed of several layers of cells with an imbricated arrangement, which constitute the matrix out of which the fibrous tissue of the lens is evolved. The capsular epithelium plays an important *rôle* in the production of the so-called capsular opacities accompanying various forms of cataract, congenital as well as acquired, for out of it are evolved, by what may be called a perverted development, nucleated fibrous webs, often of great toughness and density, underlying the inner surface of the capsule. It must not be forgotten that the capsule itself never becomes opaque, what are called opacities of the capsule being always deposits of opaque substances, or adventitious growths upon its surfaces. These may exceptionally be overlaid by a transparent colloid substance, and their own opacity may seem to be seated in the capsule; but the colloid mass is not a part of the capsule —it is something superadded to it. I am aware that some good observers deny this origin of these intra-capsular fibrous webs, which are not optically distinguishable from a connective substance. They ask—How can a connective substance be the derivative of an epithelium? And they maintain that in all these cases the capsule has not been entire, but it has had some rent through which these fibrous webs have intruded themselves from without. I cannot yield to this opinion, because I have found such webs on the inner surface of capsules which I could not doubt were entire, and where their intrusion was impossible; and because I have, as I believe, been able to trace the convolution of the fibrous tissue from the epithelial cells through intermediate phases. A normal lens-fibre, which all allow is the final phase of a capsular epithelial cell, differs hardly less from its initial phase than do some of the elements of these fibre-nucleated webs. The posterior half of the capsule has not any epithelial lining, but its inner surface frequently exhibits polyhedral marks, which have been mistaken for an epithelium. They really express the hexahedral oblique cross-sections of the swollen posterior ends of the lens-fibres.

The tissue composing the lens itself consists in greatest part of long, flat, ribbon-like fibres. These have two wide surfaces, and four narrow ones meeting on two thin bevelled edges, which give to their cross-sections a hexahedral figure. The fibres are really long tubes filled with the protein substance to which chemists have given the name globulin. When the fibres are broken across, it escapes in the form of globules from their ends. It often accumulates in large masses between the layers of fibres in lenses which have been artificially hardened with chromic acid.

A nucleus is present in all the superficial fibres, near the edge of the lens; but the deeper fibres are more sparingly nucleated. The fibres cohere very closely by their flat surfaces, and still more intimately by their bevelled edges. These latter in some vertebrates are serrated, which renders their union still more secure. This serration is very coarse in fish and in some chelonia; much finer in snakes; so fine in frogs that the edges of the fibres appear only as slightly frayed; it is absent from the human lens.

The other constituent of the lens is the interstitial tissue, a formless substance present in very minute quantity in the axis of the lens, and in those extensions of the axis called the central or axial planes. In young persons the refractive index of this substance agrees with that of the fibrous tissue; hence their lenses are free from the internal reflexions which the lenses of elderly persons exhibit, and which give these an opalescence which an incautious observer may readily mistake for a cataractous opacity. The axis in the simplest form of lens, as that of some fish and of amphibia, is a streak or line of this interstitial substance traversing the centre of the lens. The lens-fibres are grouped around it in the manner of the meridian lines on a globe, the inner fibres progressively shortening towards the centre of the lens. In other fish, and in the porpoise and rabbit amongst mammals, the axial streak is flattened. It extends in two opposite directions, and forms a central plane, the ends of which make a linear stigma on the front and back of the lens. The directions of these stigmata do not coincide; they intersect at a right angle as they would if the plane of which they are the ends had been twisted through 90° in passing through the lens. From the edges of this, which may be distinguished as the primary plane, others, secondary planes, run out in a complicated manner towards the circumference of the lens. The primary central planes in most mammalia diverge at equal distances of 120° from the axis, and their ends form on the front and back of the lens a trifid stigma. The rays of one stigma intersect the angles included between those of the other stigma, as they would do if the tripartite axis had been twisted through 60°. In these lenses the arrangement of the fibres with respect to the central planes is much more complicated than in the simple amphibian lens. All its details are probably not yet known to us; but, so far as they have been ascertained, it appears that the fibres pass between the front and back of the lens, winding round its equatorial edge in such a manner that a fibre starting from the interval between two of the front planes falls behind on the edge of a posterior plane, while a fibre starting from the edge of one of the front planes would fall behind in the angle made by two of the posterior planes. The intervening fibres between these extremes take intermediate positions on the planes.

The tripartite division of the axis persistent in many mammals is present also in the human fœtal lens, which, as has been already mentioned, is nearly spherical. As the lens enlarges, the three primary planes detach secondary and tertiary ones; the multiplication continues during the whole period of growth, until in the human adult the minor planes form an excessively complex frame.

With this excessive complexity of the planes the fibres maintain a general direction between the front and the back of the lens; and, since the surfaces of the fibres cohere less strongly than their edges, most lenses, when artificially hardened, can be split by a coarse dissection into concentric laminæ.

In youth the lens is soft, but with advancing years it acquires greater consistence, becoming in aged persons really hard. It is to this change, in consequence of which the lens becomes less plastic as we advance in life, that presbyopia is mainly due. In birds and in lizards the nucleus only is concentrically laminated; and the outer fibres pass vertically and obliquely between the capsule and nucleus.

Connections and Relations of the Lens.—The back of the lens lies in a hollow in the front of the vitreous humour, which is bounded by an extension of the hyaloid capsule of this humour, which is so distinct from the posterior lens-capsule that I should hardly have mentioned them separately had this not been recently denied. It bends inwards near the edge of the lens, and forms the posterior wall of the space known as Petit's canal. The lens may be removed in its entire capsule without destroying the integrity of this partition between the vitreous humour and the lens-bed. But its chief support is the suspensory ligament which slings the lens to the ciliary processes. This arises from the whole inner surface of the ciliary body, in front of the ora retinæ, from columnar epithelial-like bodies resting on the pigmental epithelium, and it is attached in a plaited manner to the capsule at the edge of the lens, advancing slightly upon the front, and to a less extent on the back of the lens. It is composed of fibres which chemically resemble yellow elastic tissue. They are remarkable for their hard, sharp outlines, and their clear fracture. They break up near the lens into brushes of very fine fibrillæ, the interspaces between which are occupied with delicate membraniform expansions. Kölliker regards the elongated bodies from which these fibres arise as special modifications of the connective-tissue radial fibres of the retina, which are continued in front of the ora, the nervous retinal elements ceasing at this boundary line. The place of the attachment of the suspensory ligament to the lens varies within rather large limits in different animals; in birds and reptiles it is nearer the front of the lens than in man.

The Ciliary Muscle.—Let us now turn to the muscular apparatus of accommodation. In the human eye, the ciliary muscle is the active factor of accommodation. When the cornea and sclerotic are removed, a greyish ring is seen behind the iris on the outward surface of the ciliary body. It was considered ligamentous until Professors Brücke and Bowman, nearly simultaneously, discovered its muscularity. In mammalia the muscular tissue is unstriped. The deepest bundles of muscular fibre, those which are in close relation to the outer surface of the ciliary processes, are very obliquely directed; collectively they form a ring resembling a sphincter, yet not completely separable from the other muscular bundles. Attention was first drawn to these fibres by the late H. Müller, who conceived that, acting through the intervening ciliary processes, they might compress the edge of the lens.

The outermost bundles of muscular fibre run in meridional lines. They stream backwards from the cornea and lose themselves on the outer surface of the ciliary body and choroid. These fibres are connected in front with the middle of the three divisions of the posterior elastic lamina of the cornea (the inner division of this lamina spans the margin of the anterior chamber and forms the pillar of the iris, and the outer division passes backwards and outwards to the sclerotic behind the space known as the circulus venosus, or Schlemm's canal).

The shortening of these meridional bundles of muscular fibre will tend to approximate their corneal and choroidal attachments, and if we regard the corneal one as the more fixed point, a view which best harmonizes with the anatomical facts, the contraction of these bundles tends to draw the choroid forwards and tighten it upon its contents. According to this view, the ciliary muscle (as regards its radial bundles) is a tensor of the choroid, as Brücke named it.

Actions of the Lens and Ciliary Muscle.—Let us next see what light these anatomical data throw on the process of accommodation. You will recollect that in accommodation for a nearer object, the lens as a whole does not shift its place, but its anterior surface becomes notably more convex, and the convexity of its posterior surface is very slightly increased. With this alteration of its figure, the axis is lengthened and the transverse diameter shortened. The pupillary region of the iris approaches the cornea, and the circumference of the iris retreats from it.

The lens with its capsule is elastic but without contractile irritability; its *rôle* is passive. When the suspensory ligament is tight it must exert traction on both surfaces of the lens (chiefly on the front, by reason of the greater stoutness of the fibres and of their attachment to the lens, advancing rather farther from the edge of the lens) tending to compress the lens in the direction of its axis, and to flatten it—the shape of the lens in looking at a distant object, which is a passive act, not requiring accommodation.

When the radial or longitudinal bundles of the ciliary muscle contract, and the distance between their extreme points of attachment is lessened, the previously tense suspensory ligament is relaxed and the lens, no longer compressed by it, becomes more convex by nature of its own elasticity. If, at the same time, the circular bundles of the muscle were to shorten, this would tend to contract the circle of the ciliary processes, by which the suspensory ligament would be still more slackened; I doubt, however, whether they can act as a compressor of the edge of the lens. The ciliary muscle derives its nerves from the lenticular ganglion. They pierce the sclerotic not far from the optic nerve, and gaining the inner surface of this coat they run forwards between it and the choroid till they reach the ciliary muscle, on the outer surface of which they break up and re-combine in the well-known beautiful plexus, a large portion of which is, however, destined for the innervation of the iris. From this coarse plexus bundles of nerve-fibres dip into the muscle, in which they form a finer net, from which single fibres of extreme tenuity are traceable for long distances amongst the muscular bundles, but I have not yet discovered the actual nature of their ultimate connection with the muscular fibre. In my last course of lectures, I adverted to the occurrence of ganglion cells in the plexus. They first became known to me by the beautiful preparations of Schweigger, and are not the coarser gangliform swellings recognisable under slight enlargement described by Dr. R. Lee, jun. The arteries of the ciliary muscle are drawn from the circulus arteriosus major iridis, which distributes many recurrent twigs to it. There are not unfrequently offsets of the arterioles which this arterial circle sends to the ciliary processes. The venous blood escapes in two directions, posteriorly though veinlets which join those of the ciliary process and lead to the vena vorticosa, and in front through veinlets which empty their contents into the circulus venosus in Schlemm's canal.

The Lens and Intra-ocular Muscles of Mammals, Birds, and Reptiles.—Having gained this insight into the apparatus of accommodation as it exists in the human eye, it will not be uninteresting to trace some of its modifications in other members of the vertebrate series.

I should like to take this opportunity of expressing my great obligations to the Zoological Society for the unrivalled facilities they have afforded me by placing at my disposal the eyes of a very large number of the animals which have died in their gardens. Their kindness has placed me under obligations I can never repay.

In all mammalia, monodelphous as well as didelphous, as far as my observations extend,—and through the liberality of the Zoological Society of London, to which I am more indebted than words can express, I have enjoyed unrivalled opportunities of examining the eyes of a very large number of animals dying in their gardens,—the lens and ciliary muscle do not differ in any essential point from those of man. In all other mammalia the lens is more spherical than in man. In most the central planes are three, as in the human fœtus; in a few (cetaceans and some rodents) there is but one plane.

The capsule and its epithelium, and the lens-fibre are essentially like those of man. The ciliary processes are simple, the suspensory ligament and its connections, the arrangement of the ciliary muscle and kind of muscular tissue are such as we find them in the human eye. Birds, however, present very striking differences. In a bird's eye, we are immediately struck with the great extent of the ciliary region compared with that occupied by the retina. It is the stoutest part of the outermost case of the eyeball; its strength is increased by a ring of bony plates, and behind these by a cartilaginous lamina intercalated in the fibrous sclera. The ciliary processes are fringed and papillose, not simple, and a plaited membrane, the pecten, projects like a wedge into the vitreous humour from the entrance of the optic nerve.

All the intraocular muscles are composed of striped fibre, the iris as well as those in the ciliary region.

In the iris the muscular fibres are disposed in two sets, one having a radial direction, the other circularly disposed. The radial fibres pass between the great circumference of the iris and the pupil, coursing along the back of the iris just in front of the uveal epithelium. These are the dilators of the pupil. In front of them there is a stratum of circular fibres forming a continuous sheet from the pupil to the attached border of the iris, stouter here and at the pupil, and thinner intermediately. They are easily demonstrated in the iris of any large bird by dissecting off the thick layer of pigmented connective tissue which forms the front of the iris, and which is, in great part, a derivative of the ligamentous tissue that fixes the border of the iris to the margin of the anterior chamber. It is in this connective tissue that the great blood-vessels and nerves lie. Those circular muscular bundles which bound the pupil are manifestly a constrictor or sphincter pupillæ; but the bundles at the outer border of the iris in contracting not improbably compress the corresponding part of the lens, and so tend to increase the convexity of the uncovered part of the lens in the pupillary area, as H. Müller suggested. This view of their action derives support from the beautiful prints of these bundles which are often found on the lens in dissection.

The primitive muscular fibres of the iris are much finer than those of the voluntary muscles of the limbs, from which they also differ in dividing and combining in nets.

The ciliary region contains two muscles. In the largest raptorial birds these are quite distinct; they are separated by a considerable interval; but in the eyes of smaller birds the muscles are approximated, and in these their distinctness is less obvious, yet, I think, none the less real.

The foremost muscle was described by Sir R. Crampton, and bears his name. Behind, it is always attached to the sclera, and in front to the cornea, either directly or to a tendinous prolongation of the inner corneal lamellæ. Shortening of this muscle would therefore tend to bring them closer together, and as the sclerotic with its strong bony ring is the least mobile of the two, the muscle would tend to retract or depress the cornea, and to counteract any force operating simultaneously in an opposite direction to increase its convexity. The posterior muscle always passes between the sclera and choroid. Separated by a wide interval from Crampton's muscle in eagles, as I have already said, in many smaller birds it appears on a cursory examination to be a continuation of the posterior bundles of this; yet a careful scrutiny always shows that it has distinct attachments. In many birds there is an accessory slip attached anteriorly to a prolongation of the same tendinous band from the cornea into which the inner ends of the bundles of Crampton's muscle are inserted, and in some birds this muscular slip exceeds that which stretches from the sclerotic to the choroid. The contraction of either or both these muscular slips would tend to draw the choroid forwards upon the sclerotic and tighten it on its contents. They are tensors of the choroid and the homologues of the human ciliary muscle. The pecten, which, before the ciliary muscles were known, was ever regarded as the agent of accommodation, is not any longer considered so.

Reptiles have an iris very like that of birds. The primitive muscular fibres, of the striped kind, are extremely fine; they also exhibit divisions and a plexiform arrangement. They are also disposed in two sets, one circular, the other radiating, differing from those of birds mainly in their less development.

Crampton's muscle I have not yet found in reptilian eyes, but all which I have examined, embracing several chelonia and many lizards and snakes, have a striped tensor muscle of the choroid passing from the sclerotic to this coat, occupying the same posterior position as its homologue in the bird, and corresponding functionally to the human ciliary muscle.

I have hitherto been baffled in every endeavour to decipher the details of the muscular part of the accommodative apparatus in batrachia. In the frog's eye, the quantity and the blackness of the pigment offer great difficulties. An unstriped sphincter papillæ is generally demonstrable. It is composed of long spindle-cells enclosing an elongated cylindrical nucleus with some granular pigment. How far it extends outwards, and whether the circular fibres reach the periphery of the iris, as in reptiles, I am unable to say. Radial bundles of spindle-cells, resembling those of the sphincter are also certainly present.

The Coats of the Eyeball.

Conjunctiva.—The conjunctiva is a thin membraniform web of areolar tissue with an external epithelium. The structure of the palpebral part differs slightly from that of the ocular. The former is a tougher, the latter a looser texture. The palpebral part is beset with simple vascular papillæ. The epithelium consists of several layers of cells, the deepest of which are oblong, and stand vertically; while the more superficial are flattened, and obliquely packed. The meshes of the areolar tissue often enclose large numbers of lymph-like corpuscles. The blood-vessels and nerves are numerous. The nerves are remarkable for the specialized terminations some of the tubules exhibit, the terminal clubs (end kolben), named after their discoverer Krause.

The loose fold of conjunctiva which connects the lid and globe and also the palpebral part, but more particularly the former, contains small glandiform bodies—minute spherical capsules, enclosed in a net of minute capillary blood-vessels, and, according to some observers, also surrounded by lymphatics. These, when enlarged, form the transparent bead-like grains which characterize a kind of granular ophthalmia, fortunately for us much less frequent in Great Britain than the pupillar form of this complaint.

Sclerotic.—The sclerotic, in conjunction with the cornea, forms the strong case of the eyeball which supports and protects the delicate inner coats. It is thickest behind, around the optic nerve; becomes thinner from here to the attachment of the tendons of the musculi recti, in front of which its thickness again slightly increases. A funnel-shaped canal pierces it behind for the passage of the optic nerve, and it is perforated by many smaller apertures for the transmission of blood-vessels and of the ciliary nerves. Of these inner openings the only ones which require notice are those by which the venæ vorticosæ choroideæ leave the eyeball. These pierce the sclerotic obliquely, which renders them valvular, a mechanism which lessens the available opening whenever the pressure on the inner surface of the sclerotic rises unduly, and proportionately retards the egress of the venous blood, and raises the pressure still higher.

The sclerotic is principally composed of white fibrous or

common connective tissue in the form of flat fibrillated bundles closely interwoven in planes which cross one another at every possible angle, but which have a general direction parallel to the surface of the coat. Amongst the fibrillated tissue are imbedded simple fusiform and branched corpuscles, which are more numerous in the young than in fully grown persons. The blood-vessels of the sclerotic are not very numerous; the most important are, as before pointed out, the recurrent branches of the posterior ciliary arteries, which unite in a small circle, which communicates with the vessels of the nerve. Whether the sclerotic has any nerves of its own is doubtful. The ciliary nerves all pass through, and do not appear to furnish any branches within their canals.

In birds, lizards, and turtles the fibrous sclerotic is strengthened by the addition of bone and cartilage. The bone is chiefly present in the ciliary region, where it forms the well-known circle of plates. The osseous tissue contains well developed lacunæ, and where the plates are thick they are hollowed by vascular canals and by medullary spaces enclosing fat-cells. It appears to be evolved out of fibrous tissue. The cartilage is always of the hyaline variety. In the back of the sclerotic in many eyes the cartilaginous tissue exceeds the common connective tissue.

The Optic Nerve.

The optic nerve pierces the sclerotic a little below and at the inner side of the posterior pole of the eyeball, the nerve appearing at its inner surface nearly 1‴ to the nasal side of the fovea centralis retinæ, in the form of a disc usually circular, sometimes elliptical, and when so the major axis is generally vertical.

The Physiological Pit.—The common aperture in the sclerotic and choroid through which the nerve passes is a canal narrower anteriorly, where it tightly clasps the nerve, and wider posteriorly, where it loosely encloses it. Around this opening the choroid and sclerotic adhere very intimately, their fibrous tissues intermingling here concentrically round the nerve. Here, too, the minute recurrent branches of the posterior ciliary arteries distributed to the outer part of the sclerotic effect a slight communication with the capillaries in the nerve sheath, and indirectly with those in the nerve itself. Some of these last inosculate with the choroidal blood-vessels in the level of the choroidal opening. Through these collateral channels, where the trunk of the arteria centralis is plugged, a small quantity of blood can enter the retina. The choroidal stroma around the nerve-foramen contains the same stellar pigment-cells which occur in it elsewhere. In some eyes in this situation these are more plentiful, and richer in pigment; and in such eyes the connective tissue corpuscles of the neighbouring sclerotic are also not infrequently pigmented. This excess of pigment expresses itself in the living eye by an incomplete narrow brown or blackish circle around the nerve-disc. In the plane of the choroid and of the inner third of the sclerotic the nerve-opening is crossed by a fibrous web, the *lamina cribrosa,* which peripherally merges in the connective tissues of these two coats. The anterior surface of this perforated lamina is concave, the posterior convex. In the living eye the lamina reveals itself as a white tendinous spot striped with minute grey dots, the bundles of nerve-fibres lying in its meshes. These details of the lamina are recognizable in the healthy nerve-disc in a small central area which corresponds to a depression, which I shall presently describe, known as the physiological pit. A sharply defined image of these details of the lamina overstepping this limit and reaching towards or even to the edge of the nerve-disc is a sign of atrophy.

The nerve-fibres in the trunk of the nerve behind the lamina cribrosa are of the opaque or double bordered kind, while in, and in front of the lamina, they are pale and transparent. Behind the lamina, each nerve-fibre consists of an axis cylinder, a delicate external tubular sheath (the homologue of the sarcolemma of a primitive muscular fibre) and an intermediate cortical substance or medulla—the white substance of Schwann. At the lamina the medulla ceases, the axis cylinder with perhaps a very attenuated prolongation of the sheath passing forward into the nerve disc and retina. The greatly reduced bulk of the nerve-trunk in the lamina, and the transparence of the nerve-bundles in the nerve-disc and retina, are due to this change in the constitution of the primitive fibres.

Exceptionally, as a congenital error, some bundles of opaque nerve-fibres reach the inner surface of the nerve-disc, and are even prolonged for some distance into the retina. It is not a very uncommon defect. The opaque nerve-fibres produce a white patch, which is to be distinguished from other similar white patches due to exudations of its brush-like feathered edge. After emerging from the anterior surface of the lamina cribrosa, the bundles of transparent nerve-fibres bend away on all sides (quoqueversally) from a central point, and curving over the edge of the choroidal foramen spread out on the inner surface of the retina. In doing this they leave a central void, a little hollow—the physiological pit. This pit is usually in the centre of the nerve-disc, but not always so, and, when eccentric, the vasa centralia usually also perforate the disc eccentrically. A normal physiological pit never, however, is so eccentric as to touch the contour of the disc.

The physiological pit is then a small, gentle, funnel-like hollow in the centre of the nerve-disc perforated by the vasa centralia, appearing as a bright hollow spot in which, with an enlargement of 12 or 15 linear, the meshes of the lamina cribrosa and the ends of the bundles of opaque nerve-fibres are plainly discernible.

Blood-vessels.—At a variable distance from the eyeball the trunk of the optic nerve is pierced by a branch of the arteria ophthal., which soon gains the axis of the nerve, and running forwards through the lamina cribrosa perforates the optic nerve-disc, in which it divides into two primary branches which bifurcate, and passing across the boundary of the disc, are distributed to the retina. In the disc, we can distinguish first a short vertical piece of the arterial trunk, and next the branches making a large angle with the trunk and following the surface of the disc. The capillaries of the nerve-trunk and those distributed to its disc-like intraocular end are very numerous; they are sufficiently so to redden the disc when they are distended with blood—a thing which no amount of hyperamia of the retinal capillaries ever does in this membrane.

The very slight diminution which the arteria centralis undergoes from its origin to its final division in the disc, shows that it mainly ministers to the nutrition of the retina. It gives, however, in its course *small* twigs for the nutrition of the nerve outside the eyeball, and these reinforced by others derived from minute nameless arteries distributed to the sheath, ramify in the septa between the nerve-bundles.

The veinlets accompanying the primary branches of the arteria centralis in the retina do not, as a rule, coalesce in a single trunk in front of the lamina cribrosa, but they pierce this separately and first unite in or behind it.

Sheath.—In the arrangement of its sheath, the optic nerve differs from all the other large nerve-trunks. They have but one tightly fitting tube of connective tissue—the external neurilemma; but the optic nerve has a double sheath. It has a thin tightly fitting sheath representing the sheath of other nerves, from the inner surface of which, septa are produced inwards between the nerve-bundles constituting the internal neurilemma or frame which holds the bundles together and carries the nutrient blood-vessels.

In front of the lamina cribrosa in the nerve-disc, the

neurilemma is of the more delicate kind, to which Virchow has given the name neuroglia. Its fibres have two principal directions, one transverse to that of the nerve-bundles and therefore parallel to those of the lamina cribrosa, the other vertical to the front of the lamina and free surface of the nerve-disc. These last fibres correspond to the radial connective-tissue-fibres in the retina.

This sheath and its septal prolongations consist of common connective-tissue fibrillated bundles with corpuscles interspersed.

In front, this division of the sheath blends with the lamina cribrosa and with the inner third of the sclerotic.

The outer sheath is continuous posteriorly with the dura mater, so that coloured fluids injected into the space between the two sheaths soon find their way backwards into the cranial cavity. In front, the outer sheath is continued into the outer two-thirds of the sclerotic. The outer and inner tubes are loosely connected by a very open areolar tissue, composed of curling fibres and coarser bundles, with large fusiform nucleated corpuscles imbedded in them. These give the interstitial or areolar spaces the appearance of having an epithelial lining. They play an important *rôle* in neuritis and in the evolutions of morbid growths.

The space between the sheaths has been lately described as a lymphatic cavity. In severe injuries of the head, blood is sometimes extravasated into it.

The Pecten.—In describing the sclero-choroidal foramen, I mentioned the occasional excess of pigment here giving rise to a dark circle around the nerve-disc, visible in the living eye. The presence of pigment in the disc itself is still more exceptional; I do not allude to pathological formations but to congenital conditions. In some mammalia, however, the neuroglia of the optic disc is always pigmented—granule pigment dotted along the central fibres and in corpuscles. In these eyes the pigment corpuscles are often most numerous along the vessels and in the loose tissue in the sheath close to the globe; but the greatest development of pigment here is in birds, from whose optic nerve entrance a large plaited membrane stands forwards into the vitreous towards the back of the lens, with a forward and downward inclination. The upper edge lies below the level of the fovea centralis. Its general figure is four-sided, but it varies somewhat, as do also its size and the number of plaits.

It is a vascular sheet, consisting of large vessels and a very close capillary net overlaid with a pigmented epithelium, resembling that of the choroid. Its base blends with the lamina cribrosa, and is perforated by the escaping bundles of nerve-fibres. I am unable to offer you any account of its functions. Before the ciliary muscle was known, it was formerly thought to be the factor of accommodation, against which, however, is the absence of muscularity. H. Müller suggested that it may subserve the nutrition of the vitreous humour in the absence of a retinal vascular system. But there are very large eyes with correspondingly bulky vitreous humour and no retinal vascular system, without pectens, and this throws doubt on Müller's suggestion.

Although reaching its maximum development in birds, the pecten is not restricted to them. It is present in lizards. In the gecko, iguano and chameleon it is a little sword-like process, having an intricate structure identical with that of the bird's pecten, but externally unlike this in its surface being smooth and not plaited. Some snakes also have a pecten. I have found it in the boa constrictor and viper, but it is absent from the common snake.

On Spontaneous Generation *and* Evolution.

By W. T. Thiselton Dyer, B.A.
Professor of Botany, Royal College of Science, Dublin.

The value of a theory must be measured not only by the number of known facts which it correlates and explains, but chiefly also by its capacity for adding to actual knowledge in giving new turns to investigation. Whatever may be the different estimates of the modern philosophy of evolution from the first point of view, no one can doubt that it has supplied an immense stimulus to research in often unsuspected fields, while old ones have been reattacked by the help of new ideas and with no less advantage.

Among other inquiries which evolution has more or less directly encouraged, it is not surprising to find the problem of the *de novo* production of living things. As long as spontaneous generation simply served the purpose of concealing ignorance as to the development and reproduction of many of the higher organisms,[1] it was natural that it

[1] "Even as late as 1854 we find Von Siebold stating that he had arrived at the decided conclusion that intestinal worms do not originate by "equivocal generation" from substances of a dissimilar nature—namely, ill-digested

should be more and more discredited, as exact observation gradually drove it to its resting place among those organisms which are minuter and more obscure. The long existence of a belief which was only a shifting supplement to crude and imperfect knowledge, can obviously afford no *à priori* support to the theory in its modern form in which it is directly linked with purely physical views of the nature of life. On the other hand, the very minuteness and obscurity of the organisms among which it is now believed to take place are an argument in its favour. If we agree with Mr. Darwin that "it does not seem incredible that from some low and intermediate form both animals and plants may have been developed,"[1] we are naturally disposed to speculate as to the origin from the inorganic world of such a form itself. Conceiving such a phenomenon once to have taken place, it would be difficult to believe that it has not done so again and again, inasmuch as the collocation of conditions which it requires must necessarily have from time to time recurred. Hence, it might reasonably be supposed to be within the range of possible observation, to demonstrate at least some steps in the actual genesis of life, and this is what the advocates of spontaneous generation believe themselves to have done. As a proof of its purely physical nature, nothing could be more conclusive than the *per saltum* experimental development of life from substances absolutely devoid of it, and under purely physical conditions; yet in connection with a theory of universal evolution, this involves as much difficulty as the supposition of an absolute limit between the organic and inorganic worlds. To affirm that the interval is passed over *per saltum* would not in the least diminish the discontinuity, because it would still imply the existence of an interval; yet that it is only passed *per saltum*, is after all the conclusion that must be arrived at from the statements of believers in spontaneous generation.

A comprehensive theory of evolution would fail most signally of comprehensiveness if it refused to give any account

nutriment and corrupt juices."—('On Tape and Cystic Worms,' translated by Huxley, p. 3.)

Origin from ill-digested nutriment would have been simply spontaneous generation, for which Professor Huxley has recently proposed the term Abiogenesis; the production of new organisms by modification of the living substance of another is Xenogenesis. Mr. Spencer has repudiated Heterogenesis as a synonym of spontaneous generation, and in conformity with his symmetrical terminology has used it as the correlative of Homogenesis for those cases of multiplication which have been described under the name of alternate generation, and in which there is only a cyclical recurrence of the same form.—('Principles of Biology,' i, p. 210.)

[1] 'On the Origin of Species,' 4th ed., p. 571.

of the passage from the lifeless to the living, the inorganic to the organic. But regarding the multitudinous kinds of organisms that now exist and have existed as having "arisen by insensible steps," each step a phase of the moving equilibrium, which is the result of the modification of already modified structures, and which consequently necessitates progression as a general result, it is impossible for evolution to postulate anything like an absolute or discontinuous beginning of life. The strongest claim that evolution has upon our belief is based upon its universality. There can be no breach in its continuity, and accumulated analogies compel us to think that any supposed commencement of organic life must have been as much the result of insensible gradations in something pre-existing as all subsequent developments.

It is remarkable that while on the one hand Mr. Spencer has been criticised for repudiating spontaneous generation, arguments in defence of it based on evolution, and supported by reference to his writings have been used by Dr. Bastian in a recent paper in 'Nature.'[1] The contradiction has been possible, because in neither case has the fundamental principle of evolutional continuity been properly kept in view. Dr. Bastian has no doubt felt that the balance of experimental evidence has so often swayed from one side to the other in relation to this subject, that his facts would have more weight in connection with the *à priori* arguments in their favour, and hence has so published them. No doubt, a mind saturated with the vast series of facts which evolution embraces would be disposed to be more strongly impressed with the probability of new experimental results shown to be conformable with them, and it would be quite justifiable to appeal to such a disposition if the conformity were clearly apparent. While, however, Dr. Bastian's conceptions of the way in which life originates are in reality very different from Mr. Spencer's, it would be entirely erroneous to suppose that Mr. Spencer denies the evolution of living from lifeless matter; though in admitting this, neither he nor any real evolutionist admits the occurrence of what is ordinarily meant by spontaneous generation. Continuity as much forbids us to suppose that living matter has not been evolved from lifeless matter, as to suppose that lifeless matter has ever *per saltum* flashed into life.

Graham with profound acuteness described the colloidal as a dynamical state of matter, a condition of perpetual unstable equilibrium with the environment, and therefore peculiarly sensitive to incidental disturbances. He however went further than this: "the colloid" he observed, "possesses

[1] Vol. ii, pp. 170—177, 193—201, 219—228.

ENERGIA. It may be looked upon as the primary source of the force appearing in the phenomena of vitality." Now, the energy possessed by any chemical aggregate is the force which binds together its component atoms and which is liberated when they are dissevered. The energy of a colloid is therefore nothing more than the amount which is accumulated or set free in its "continual metastasis;" the difference between its energy and that of an explosive compound like potassium picrate is, that the one is actual and the other potential. It is impossible to correlate the energy of a colloid more than of a crystalloid with vitality, because they are one and the same thing. The reason why colloids lend themselves to the exhibition of vital phenomena is not because they possess actual energy, but by reason of their capacity for undergoing small amounts of molecular rearrangements so as to adjust themselves to corresponding small rearrangements of external conditions. In a statically equilibrated crystalloid such a capacity is either impossible or very much restricted. It is easy to represent to the mind at any rate one reason of this; the chemical molecule of the colloid is voluminous compared with the chemical molecule of the crystalloid, the number of actual atoms contained in the one being probably always greatly in excess of that contained in the other. Rearrangements of the atoms as the result of recombinations of the interatomic forces must therefore be possible to a much greater extent in the larger molecule than in the smaller. Yet Graham's principle must not be pushed too far; it must be kept in view that colloidal characters are not the cause, but only a phase of that capacity for responsive molecule readjustment, of which vital properties are the highest exponent, and the exhibition of these last may never be reached at all by substances like hydrated silicic acid, which are nevertheless characteristically colloidal.

According, however, to Dr. Bastian, the life, or in other words, the aggregate set of phenomena displayed by one of the simplest bodies which we call a living thing, is as much the essential and inseparable attribute of the particular molecular collocation which displays it as the properties of the crystal are essential to the kinds and modes of aggregation of the molecules which enter into its composition.[1] But the defect of this view is its one-sidedness, an essential characteristic of life being the duality of its relations—the continued balancing of those which are internal and those which are external. It is not the internal

[1] 'Nature,' vol. ii, p. 174.

relations alone—the molecular collocation—which is never the same from one moment of time to another, but its adaptability to the environment which constitutes life. As Mr. Spencer remarks, "an individual homogeneous throughout, and having its substance everywhere continuously subject to like actions, could undergo none of those changes which life consists of."[1] Colloids which are the nearest approach in inorganic matter to that which is living, tend under uniform conditions to pass from unstable to stable or crystalloid equilibrium, and such a condition of perfect equilibrium is a condition of lifelessness or death. Life does not consist in either the internal or external relations of a body separately, but in their continuous mutual adjustment; molecular constitution is only one of the elements of life.

Dr. Bastian, nevertheless, considers that "monads and bacteria are produced as constantly in solutions of colloidal matter as crystals are produced in solutions of crystallisable matter," and that the difference between the products "may be due simply to the original difference in nature between such kinds of matter."[2] *Primâ facie* the analogy seems a strong one; in either case a substance at first diffused in solution, and therefore amorphous, finally segregates with the assumption of definite form. Further consideration, however, shows that this view is not free from difficulty, and leaves room for doubt whether it really describes the true state of the case. It seems clear that the form of an aggregate so produced must have some relation, whether a crystal or a living thing, to the form of its component units, and must be more or less implicitly determined by it. What explanation otherwise can be given of the constancy with which shapes identical, or generally similar, recur? The form of the aggregate would be purely arbitrary if the units had no influence in determining it. In the case of crystalloids there are many reasons for believing that the moderate number of atoms composing their molecules may, in obedience to their mutual but diverse polarities, arrange themselves in a definite form having a definite resultant polarity. Such a molecular system may be subject to disturbance, and finally rearrangement by the incidence of external forces, such as light or heat, so that a substance chemically the same may possess several allotropic crystalline conditions. But the molecules of colloids are highly complex, consisting of a vastly greater number of atoms, any arrangement of which must tend to be spherical, and therefore with no marked resultant

[1] 'Principles of Biology,' vol. i, p. 286.
[2] 'Nature,' vol. ii, p. 172.

polarity. Hence in an aggregate of colloidal molecules, individual molecules would not tend to any special relative disposition; in other words, could not crystallise, and consequently masses of different colloidal substances exhibit, on the whole, not very diverse external characters.

But colloidal molecules having polarities so variable, what relation can there exist between them and the vast variety of the shapes of living things? In the gelatinous state characteristic of colloids these polarities are hardly sufficient, even to differentiate their external appearances from one another. It is quite evident then that they cannot possess the property of arranging themselves into the special structures of the organisms to which they belong, since the infinite variety which these present would be inexplicable on such a supposition. We must agree with Mr. Spencer in conceiving it possessed by certain intermediate or physiological units composed of chemical units or molecules, but infinitely more complex, and possessing a more or less distinctive character, which ultimately produces a difference in the forms assumed by the aggregate. This helps us to understand the repetition in the offspring of the peculiarities of the parents if sperm cells and germ cells are essentially nothing more than vehicles of small groups of "physiological units in a fit state for obeying their .proclivity towards the structural arrangement of the species they belong to.'[1] If this is true of the higher organisms, it must be equally true of the lower; and as it is impossible to say where a line could be drawn, it probably holds of all living beings, even of those lowest, so destitute of structure as to have no claim to the title of organism at all. Of course it may be objected that physiological units are mere figments of the imagination, but as much might be said against chemical units. In either case the use of symbolic terms is forced upon us by the analysis of our scientific ideas. In both cases they may be purely arbitrary conceptions; but if we use one, it is impossible to object to the use of the other. While, therefore, molecules having definite polarities aggregate into definite crystalline forms, molecules whose polarities are feeble will aggregate into amorphous colloidal masses. The shapes of living bodies are related to organic units more complex than the molecules of colloids. Between a solution of colloidal matter and a bacterium there is a distinct step in integration, which does not exist in the case of the formation of a crystal at all, and this, therefore, is not really analogous to the formation of a bacterium from such a solution, supposing it to take place in the way which is described. A

[1] 'Principles of Biology,' vol. i, p. 254.

particle of protoplasm suspended in a fluid would, just as an oil-drop does, tend to assume a spherical form, on the principle of Plateau's well-known experiments. But this would not account for the definite elongate shape of a bacterium

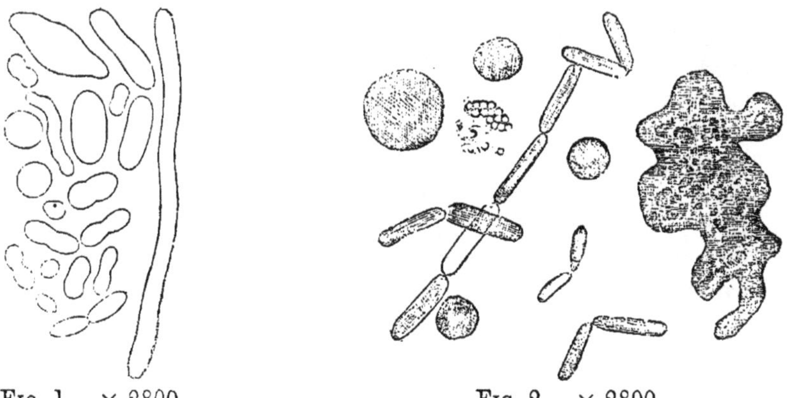

FIG. 1. × 2800. FIG. 2. × 2800.

(Figs. 1 and 2), the difference between which and the rounded outline of a monad or Torula germ must depend on difference in their component organic units.[1] These may be comparatively simple, though not necessarily so, since bacteria may be stages of more complex organisms, capable of attaining complete development in a more favorable environment.

If there is little analogy between the supposed origination of such entities as bacteria and crystallization, still less is there in the case of spores of fungi. The existence of a spore or germ implies the development of something more mature, the form of which is implicitly determined by it. Its component organic units must, therefore, be more complex than those of the simplest living bodies which run through no varied course of development. A germ apparently extremely simple in structure potentially may be rather complex, and in proportion its production by mere segregation from a solution of colloidal matter is *à priori* improbable. There is good reason for believing that even when "under the influence of pre-existing protoplasm an equivalent weight of the matter of life makes its appearance" in the place of carbon dioxide, water, and ammonia, it does so only after a process of evolution by successive integrations. It is alien to the general conception which evolution forms of the mode in which the more complex kinds of matter are derived from the less complex

[1] These figures, which are perhaps the most definite that have been published, are borrowed from Dr. Beale, 'Disease Germs,' Pl. II, fig. 13, and Pl. IV, fig. 26. Fig. 2 represents bacteria with white and red blood-corpuscles from hepatic vein of a cow which died of cattle plague.

to suppose that living matter is capable of moulding, as it were by one operation, inorganic matter, suitable, perhaps, as to ultimate composition, but of the most diverse proximate arrangement into molecular aggregation like its own. It is inconceivable that any given stage of evolution can be reached with equal facility from any inferior stage selected at haphazard.

Compare what takes place in plant nutrition. Under the influence of solar light the inorganic materials appropriated by the tissues of plants undergo successive modifications, which, weakening the stability of their molecular constitution, "give collateral affinities the power to work a rearrangement, which, though less stable under other conditions, is more stable in the presence of these particular undulations."[1] But an essential element in these changes is their occurrence in connection with the colloidal materials of vegetable tissues which bring the carbon dioxide, ammonia, and water, on which the rearrangement is effected in a condensed state into close contiguity. As soon as the bonds which unite their component elements are weakened by the unequal effect of the undulations propagated by solar light on their unequal atomic activities, they are ready to enter into new and interchanged combinations. In this way, by mutual actions and reactions, substances independently more unstable are elaborated from the more stable, and these being more and more like the materials in the presence of which the changes are effected, tend finally to be integrated with them, every stage in the process being a position of equilibrium between the molecular constitution and environing influences. By processes of which this is a rough outline, the conversion of inorganic into living matter is effected; but the presence of pre-existing living matter is a very potent factor in the change. It brings the component materials together, effects their joint exposure to solar influence, shapes the final form of their combinations, and gives them a stability they could not possess apart from it. Probably the merely physical and colloidal properties of the living matter in conjunction with solar action effect the union of the ultimate into proximate constituents, and then the "coercive polar force" of the component molecules of the living matter cause them to aggregate into similar molecules. There is plenty of analogy to induce a belief in such a coercive influence of an aggregate on integrable units; a broken crystal, for example, tends first to restore its shape before receiving further additions.[2] In this

[1] 'Principles of Biology,' vol. i, p. 32.

[2] Another good illustration of the tendency of similar molecules to

case, however, the crystalline units already in existence are coerced merely as to position; living matter being more complex than crystalline, the coercion it exerts is more complex. But if this view be probable, how can we understand inorganic substances, such as carbon dioxide, water, and ammonia or ammonium nitrate, or any such grouping of the necessary materials resisting their normal tendency to crystalline aggregation, and guiding themselves into the complex constitution of living matter.

It cannot be said that synthetical chemistry affords much more support to Dr. Bastian's views. They seem to imply that any collocation of carbon, hydrogen, nitrogen, or oxygen is capable of spontaneous rearrangement into the material substratum of a living thing. Whether the quaternary compound be crystalloid or colloid, stable or unstable, in solution or even a solid, is apparently indifferent. The tendency of all colloids is to settle down into stable crystalline forms; of the reverse passage of crystalloids into colloids Dr. Bastian only assigns a single instance, the metameric change of ammonium cyanate into urea. And urea can hardly be described as a colloid, since it has a comparatively low equivalent, crystallises in slender striated prisms, is not particularly unstable, as it may be boiled without undergoing decomposition, and must, as it is not found in the kidneys, be diffusible to be present in the urine. The general mode by which a chemist attempts to construct substances of high molecular complexity is exactly what evolution would lead us to expect. Modifications are introduced into the simplest types, and these, when so modified, are modified again. Every step must necessarily be a position of equilibrium, since otherwise progress would be impossible, and the structure must necessarily be commenced again from the beginning. If proteinaceous compounds are ever constructed in the laboratory, it will be by such a general process as this. Any substances isomeric with these would be quite as complex, and would be, therefore, as laborious a production. Certainly in no sense can such a substance as ammonium tartrate be an isomer of living protoplasm, and even supposing that that

aggregate is afforded by the pectization of colloid solutions from the gradual withdrawal of the colloid from the crystallized water. A solution of silicic acid divides into 'a clot and serum,' ending in the production of a stony mass which may be anhydrous, or nearly so. According to Graham, 'the intense synæresis of isinglass, dried in a glass dish over sulphuric acid in vacuo, enables the contracting gelatin to tear up the surface of the glass. Glass itself is a colloid, and the adhesion of colloid to colloid appears to be more powerful than that of colloid to crystalloid.'—('Proceedings of Royal Society,' vol. xiii, p. 336.

particular compound were one, ammonium carbonate, and all the other materials of diverse composition from which Dr. Bastian believes life to be evolved, could not be isomeric as well.

Evolution must be consistent with itself. A truly monistic conception of nature, to use Haeckel's word, repudiates the spontaneous appearance of life, as it repudiates everything else spontaneous; but it implies the continuity of what is living with what is lifeless, of what is called vital with what is called physical. The evolution of proteinaceous matter, as Mr. Spencer observes, must have taken place in the early world according to the same laws as those to which chemists have unconsciously conformed, and it is impossible to indicate more clearly than in his words the further changes which, in harmony with general modes of evolution, this proteinaceous matter must have undergone.

"Exposed," he says, "to those innumerable modifications of conditions which the earth's surface afforded, here in amount of light, there in amount of heat, and elsewhere in the mineral quality of its aqueous medium, this extremely changeable substance must have undergone, now one, now another, of its countless metamorphoses. And to the mutual influences of its metamorphic forms, under favouring conditions, we may ascribe the production of the still more composite, still more sensitive, still more variously changeable portions of organic matter, which, in masses more minute and simpler than in existing *Protozoa*, displayed actions verging little by little into those called vital—actions which protein itself exhibits in a certain degree, and which the lowest known living things exhibit only in a greater degree."

It is evident that what is here described requires an amplitude in the range and varieties of the conditions which could not possibly be realised in an experiment. It becomes, in fact, almost as difficult in such conceptions to imagine the evolution of a new plant species in a flower-pot as life in a sealed flask.

Dr. Bastian's first observations deal with the changes of the so-called "proligerous pellicle" of Burdach, and he certainly does not overstate the opposition to "generally received biological notions" involved in the transformation of aggregations of monads and bacteria into larger and higher kinds of living things. His description of the way this takes place is, however, very materially different from that which has been given by other writers, and the difference consists, not

[1] 'Principles of Biology,' vol. i, App., pp. 483, 484.

so much in the forms which he states were produced, as in the general changes preceding the production. The pellicle consists of an aggregation of monads and bacteria, or "granules," in a transparent jelly-like stratum.[1] It is this last which seems really to be of fundamental importance. "Areas of differentiation" are *gradually* formed in it, which are lighter in aspect from an increase of the jelly-like material between the granules (Fig. 3). These areas undergo a kind of segmentation, finally breaking up into "unicellular organisms," one of which exhibited partial amœboid movements. But in all these changes the granules apparently take little if any part, and it would seem far from unreasonable to suppose that they have no more than an accidental connection with the "areas," which it is just as likely are distinct living things, originating, like the bacteria, from germs, and increasing in size by regu-

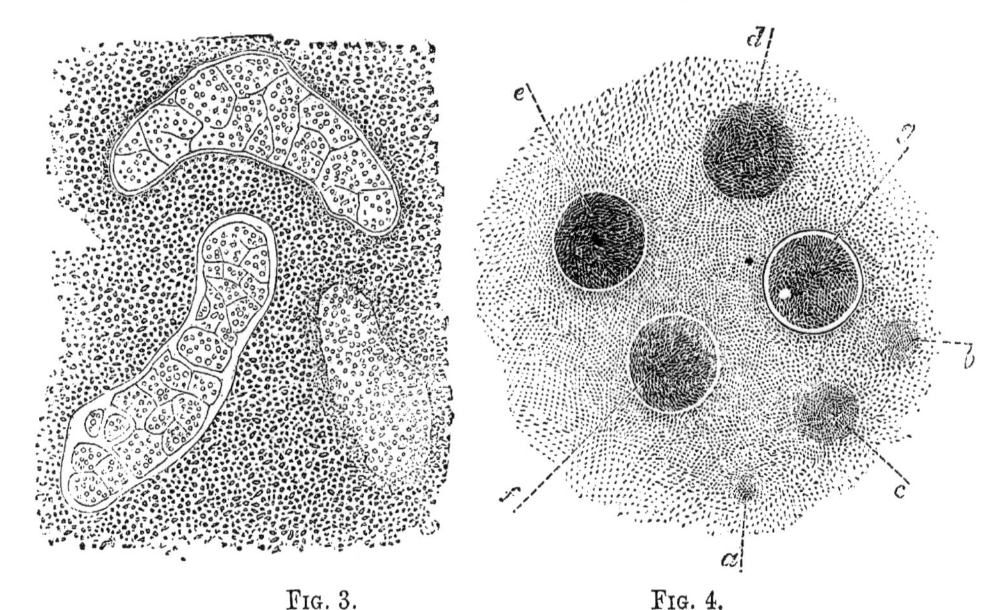

Fig. 3. Fig. 4.

lar nutrition, and finally breaking up into individualized segments. According to Pouchet, who is endorsed by Dr. Hughes Bennett,[2] the development of new forms takes place

[1] Portions of a jelly-like material with granules, are represented in figs. 14*b* and 15*a*, and more unsatisfactory objects it is difficult to conceive. Fig. 11*b* seems to be not a spore case but a similar portion. One of the greatest difficulties in seeing clearly any result connected with spontaneous generation is the apparent inseparableness with it of crude and uninterpretable figures.

[2] "On the Molecular Origin of Infusoria," 'Popular Science Review,' vol. viii, p. 55. This is practically a revival of Buffon's 'Theory of Organic Molecules.'

quite differently in a *second* molecular pellicle produced by the disintegration of the bacteria of the first. The molecules aggregate into these masses, standing out sharply on a lighter ground (Fig. 4). The "spores of fungi" figured by Dr. Bastian in their mode of origin have a very considerable resemblance to the "unicellular organisms," differing only in the segmentation occurring more than once. More definite conclusions might have been arrived at about them if their mentioned production of "ordinary mycelial filaments" had been figured as well.[1]

Results like these, and those of Pouchet, which involve the origination of the most varied forms, in reality prove too much. Granting that the vital phenomena of the simplest living things are purely physical consequences of their material composition, it is still *à priori* incomprehensible that uniform conditions should produce so many diverse and definite specific shapes, since evolution can only conceive of these arising very gradually in response to a varied environment. Conviction in science, as in other matters, results from a balancing of probabilities. To an evolutionist it is much more probable that different specific forms should be genetically related to series of such forms from which they have derived their gradually accumulated distinctive characters, than that they should, within a very short space of time, originate from quite formless matter. Mr. Spencer remarks: "If there can suddenly be imposed on simple protoplasm the organization which constitutes it a *Paramœcium*, I see no reason why animals of greater complexity, or, indeed of any complexity, may not be constituted after the same manner."[2] And a vitalist like Dr. Beale uses language which is little different when he states that he "should as soon think of believing in the *direct* formation from lifeless matter of an oak, a butterfly, a mouse, nay, man himself, as in that of an amœba or a bacterium."[3]

The analogy between the origin of crystals and the *de novo* origin of living things which Dr. Bastian presses so strongly, might have some force if we were quite as much in the dark about the absolute beginning of *all* living as of all crystalline forms. But a belief that specific form is inherited rests on too wide a basis of facts not to afford the very strongest presumption that it is true in the case of any and every living

[1] The bodies represented by fig. 7 *b*, the 'flattened bits of protoplasmic-looking material' of Exp. 4, and the protoplasmic-looking masses of Exps. 9 and 12, are probably to be referred to the 'unicellular organisms.'

[2] 'Principles of Biology,' vol. i, App., p. 480.

[3] 'Disease Germs,' p. 59. The italics are not Dr. Beale's.

thing possessing specific form. Besides, after all, what real resemblance could there be between the earliest stage of a living thing and of a crystal? Whenever a crystalline molecule detaches itself from aqueous adhesion, either by cohering with another molecule, or by adhering to any minute foreign particle, there exists a starting-point for the building up, by successive external accretions, of a predetermined crystalline form. But a germ, even if invisible to the highest powers of existing microscopes, is a fundamentally different thing: no mere aggregation about it of external nutritive matter will be sufficient for its growth; before that can proceed its food must be absorbed, reconstituted, integrated, completely differentiated from that which is outside. A crystal begins with a merely statically disposed foundation; but a germ is from the first potentially as complex as the mature living thing which is finally evolved from it. Nor does mere minuteness abridge the interval between a particle of living and a particle of lifeless matter. The former may possess all its characteristic properties, though practically invisible. Dr. Beale has frequently observed the subdivision of living particles of the higher as well as of the lower forms of life, which could only be seen with difficulty when magnified 5000 diameters. And it is difficult to resist his conclusion that "even if the magnifying power could be increased to 50,000 diameters there would still be seen only more minute living particles growing and dividing and giving rise to particles like themselves."[1] Such particles would convert external lifeless into living matter, and so grow. There is nothing more inconceivable in this than that young plants should be developed from the scales of the stem and leaves of a *Begonia;* the minutest particles of living matter are doubtless equally able with the largest aggregates to exert some degree of that coercive polar force which has already been mentioned. Dr. Hughes Bennett, it is true, argues against the production of the immense numbers of particles of living matter found in solutions by the subdivision of similar particles derived from the atmosphere, in the first place, because "no one has ever seen this remarkable phenomenon," and secondly, because "the idea of their rapid multiplication by division is opposed to that of their power of elongating into bacteria and vibrios."[2] But Dr. Beale has shown, in the case of the yeast cell at any rate, as no doubt in other cases, that the "very minute particles divide and subdivide independently, producing still more minute particles capable of growth and division like

[1] 'Disease Germs,' p. 58.
[2] 'Popular Science Review,' vol. viii, p. 60.

themselves, not one of which, however, may be developed into an ordinary yeast cell,"[1] except it meet with an appropriate medium, when its further development is accomplished, terminating in the formation of a fungus with aërial fructification.[2]

The argument that we may dispense with pre-existent germs in the case of monads and bacteria because we do not suppose them in the formation of crystals, of course derives its force from the assumed identity of the modes in which the matter of either is built up. But even a bacterium is something very different from an aggregation of its lifeless constituent matter. The minutest crystals, on the other hand, differ in no respect from the largest except in size, and give as definite angles and measurements; their shape, in fact, is implicitly determined by that of their constituent molecules, to which it must be simply related. These molecules can be separated by solution and reaggregated again without limit, always with the production of identical crystalline forms. It cannot be imagined, however, that baeterium elements pre-existed in a solution ready to come together into a bacterium; nor is there any reason to suppose that the particles of a bacterium are capable of undergoing separation with the subsequent reconstitution of the bacterium form like a crystal. And even supposing that this were possible it would only show that a bacterium was very different from other living things. Moreover, if a bacterium underwent such a disintegration the fragments would be nothing more than "germs," and would almost certainly reproduce bacteria, not by coalescing, but severally, by growth at the expense of surrounding nutritive matters; and if they did not do this it is difficult to see how they could aggregate into anything more than an amorphous mass.

To Pasteur belongs the credit of having demonstrated the existence of germs and spores in the air, and the method by which he achieved this is well known. It was not, perhaps, perfectly adapted to demonstrate their existence optically, as the minuter particles of living matter could hardly fail to be altered if not destroyed by the treatment with different liquids necessitated by the process, even if they were visible to the low powers which were used. According to Dr. Hughes

[1] 'Disease Germs,' p. 20.

[2] By a curious coincidence, on the next page following the first instalment of Dr. Bastian's memoir, is an abstract of a paper by Dr. Polotebnow, "On the Origin and Development of Bacteria." He finds an unbroken series of forms between the minute round cells which form the mycelium of *Penicillium*, and probably other fungi and fully-developed Bacteria, which he thinks can only occur from these cells.—(*Loc. cit.*, p. 178.)

Bennett, the drawings of dust obtained from air filtered through gun-cotton are only magnified 180 diameters.[1] They demonstrate at any rate the presence in it of fungus spores, a result easily credible to any one who has watched the myriads of them which may be discharged from a single puff ball, darkening the air for a moment or two and then invisibly dispersed, or to any one who remembers the ubiquity of cryptogamic forms. *À fortiori* if the spores of the larger fungi are present in atmospheric dust, the germs of minuter and obscurer organisms cannot be concluded to be absent, especially if we remember that Professor Tyndall has shown the particles of atmospheric dust to be almost wholly destructible by heat, although of course a large proportion are lifeless matter. In his other investigations Pasteur employed a magnifying power of 350 diameters; but Dr. Child pointed out that it is quite possible that living particles might exist in a solution and yet not be detected by this, and a certain degree of uncertainty has been supposed to attach to Pasteur's results on that account.[2] It must, however, be remembered that there is no limit to the extent to which this objection may be urged. Dr. Beale thinks that whatever be the magnifying power we employ we should still be able to see particles more and more minute of living matter, and experience, as far as it has hitherto gone, appears to justify him. With the $\frac{1}{50}$th of an inch object-glass, which with the low eyepiece magnifies nearly 3000 diameters, "particles too transparent to be seen by $\frac{1}{25}$th are distinctly demonstrated."[3] It is difficult to see how, this being the case, we can ever state with absolute certainty from microscopic observations alone, that any given liquid does not contain living matter. But, on the other hand, it may be fairly presumed that in Pasteur's experiments, carried over, in some instances, a year and a half, the minute fragments of living matter would have grown during that time into visible dimensions; so that this objection is not really of so much practical importance. At any rate there was never any difficulty in detecting the existence of life in solutions which had been previously inoculated with germ-containing dust.

Dr. Bastian's results are perhaps the most remarkable of any that have hitherto been published, and they could not fail to have attracted the most serious attention, perhaps even more so, on their own merits alone, apart from any *à priori* discussions. In all the experiments liquids were

[1] 'Popular Science Review,' vol. viii, p. 58.
[2] 'Proceedings of the Royal Society,' vol. xiv, p. 184.
[3] Loc. cit., p. 36.

hermetically sealed in flasks after all air had been expelled by boiling. These conditions would seem to preclude all chance of the subsequent appearance of life in the flasks; yet in only three out of twenty cases did this fail to take place; it is this which is so astonishing. If success is apparently almost inevitable, why have previous observers failed? Why they should do so, in fact, it is hardly possible to explain, dismissing the inadmissible supposition of bad faith, if Dr. Bastian's methods are free from flaw. Dr. Bastian has felt this difficulty, and suggests that in Pasteur's experiments the severity of "the restrictive conditions" produced the negative results; but this explanation, as will be seen afterwards, hardly proves very satisfactory. Professor Huxley has in fact remarked, that "it is probable there must be some error in these experiments, since others similar to them are performed on an enormous scale in the preservation of various kinds of food in tin cases, and with a totally different result."[1] And this suggests what may prove to be a source of error. The tin cases are finally sealed while standing in a bath of calcium chloride. Steam issues from a minute hole in the cover, driving before it all the enclosed air; when this has completely taken place, the whole is closed by first dropping from a sponge a drop of water upon it, which momentarily condensing the steam, is then instantly followed by a plug of molten solder.[2] Dr. Bastian, however, moderated the boiling of his flasks by turning down the lamp-flame at the time of sealing;[3] so that there would seem to be just the possibility of an indraught of air having taken place.

Of the first sixteen experiments, in eight infusions of organic matter were used, and in the other eight saline solutions. Only one of the first set and two of the last gave negative results, in all the rest living things of a rather varied kind were found when the flasks were opened, after the lapse of periods varying from five to sixty-one days, but on the average of twenty. The saline substances were all selected to contain carbon, hydrogen, oxygen, and nitrogen, but were of course on any point of view farther removed than the organic substances employed from the composition of living matter. Notwithstanding this, however, the remarkable result was obtained, that the most evolved organisms were produced by the solutions of the saline, and

[1] "Address to British Association.'
[2] The whole method is described by Dr. Wynter in 'Our Social Bees,' p. 194.
[3] Loc. cit, p. 176.

therefore least evolved substances, certainly from the evolutionist's point of view a most anomalous result. In the concluding part of his paper Dr. Bastian attributes a low evolutional capacity to acid liquids yet in Exp. 13, where no bacteria were found, but a well-marked *Penicillium*, with fructification, and in Exp. 15, where an equally characteristic fungus, with but few bacteria occurred, the solutions were both acid. And in comparing the experiments with organic infusions, the first four, which were alkaline or neutral, show on the whole less "evolutional capacity" than the last four which were acid. So also, all the four cases where spores were observed were in saline solutions, three of which were acid.

Assuming that fungi, which are comparatively high up in the scale, are produced *de novo*, it would be reasonable to expect that they would require the most favorable conditions; and it is therefore remarkable to find them occurring under what on Dr. Bastian's own data we must conclude to be unfavorable.[1] Nor is this the only difficulty about their occurrence. That they should commence with anything resembling a specialised product of ultimate differentiation like a spore, with even, as Mr. Worthington Smith remarks, a neck-like prolongation, which would ordinarily be interpreted to indicate the point of detachment from the parent,[2] is, *à priori*, very improbable. Supposing one of the *Myxogastres* about to be evolved, it would be more reasonable to suppose that the undifferentiated amœboid body, which results from the rupturing of the spore-wall, and which gives rise to the divisionless plasmodium-like mycelium, should be first evolved, rather than the spore, with its distinction of cell-wall and cell contents.

In four cases in solutions of ammonium tartrate and sodium phosphate masses of spiral fibre were met with (fig. 13 *a*). It is almost impossible to resist the conclusion suggested to Dr. Bastian that this is a non-living accidental product, altered by boiling; and it was in all probability introduced with one of the two saline ingredients, as it was only met with when these particular ones were used. It is significant to notice that in two cases (p. 197) foreign bodies were found in the solutions, and that minute shreds of cotton or paper-fibre have often been noticed with surprise, notwithstanding

[1] The reaction of the liquid, as indeed Dr. Bastian admits, has in many instances little influence on the development of living things in it. *Sarcinæ*, for example, are developed indifferently in urine which is alkaline, neutral, or acid ('Neubaer and Vogel,' p. 134).

[2] 'Nature,' vol. ii, p. 276; conf. figs. 11*e*, 13*c*, and 18*b*.

every precaution in cleansing the flasks, and using fresh distilled water (p. 220).

In the four concluding experiments tubes containing one infusion of turnip, the other three saline solutions, were after the exhaustion of the air sealed, and then exposed to a temperature of about 150° C. for four hours. In every case evidence of life was eventually found, the saline solutions as before giving the most conspicuous results. In a solution of ammonium, tartrate, and disodium phosphate, a *Penicillium* appeared, as in Exp. 13, with the same solution, after merely boiling. Spores and *sarcina* were also found in a solution of ammonium carbonate, and disodium phosphate.

What interpretation is to be given to these results? It is evident that one or other of two conclusions must be accepted;—either the fungi origiated *de novo*, or they were produced from germs, which were subjected to a temperature of 150 C. for four hours. It would certainly seem as if these last experiments were absolutely conclusive, especially when, as Dr. Bastian tells us, no fungus spore, or bacterium has been hitherto supposed to be able to withstand mere boiling without disintegration. All that can be said on the other side is that we must keep in mind that fungi, like *Penicillium*, are only ultimate forms, of which yeast globules are a peculiar condition, *Leptorthix*, probably a submerged confervoid form, and from which, if we may believe Dr. Polotebnow, even bacteria are derived. Dr. Beale has stated from observations made with his $\frac{1}{50}$, that minute yeast-cells are capable of throwing off buds or gemmules, much less than the $\frac{1}{100000}$th of an inch in diameter;[1] and these he thinks capable of subdivision practically, *ad infinitum*. All the organisms found in the solutions might have been ultimate stages in the development of such minute atoms of living matter; and as to the influence of temperature upon these nothing can properly be asserted. Disintegration is the result of heterogeneousness of parts; but these particles must be much less liable to alteration in that respect than comparatively large structures, such as spores, the formed material of whose envelope would be acted upon differently from the living matter within. There seems good evidence to show that bodies allied to germs may survive boiling; and perhaps in proportion to their minuteness. Dr. Heisch has described quite lately to the Chemical Society spherical cells found in water contaminated with sewage. They were too minute to be removed by filtration through the finest

[1] 'Disease Germs,' p. 20.

Swedish filtering paper, and boiling for half an hour or more in no way destroyed their vitality.[1]

Not merely does Dr. Bastian believe that from the re-arrangements of the particles of various saline substances in solution living beings may *de novo* originate; but he also states that this may actually take place within crystals of such compounds. He gives the details of his observations in these, as in other cases, with a minuteness which is most conscientious, especially as it is often hardly possible to avoid thinking that they suggest quite a different explanation. Crystals of ammonium tartrate were found after being kept some time to have undergone certain changes, the external portions became more or less opaque, and less soluble, and gradually increasing bubbles are seen in internal cavities, especially in those crystals which are not perfect in shape, and which present a more or less opaque appearance in their interior.[2] It is by no means improbable that a crystal presenting these characters is not homogeneous throughout, and a porosity of the interior, which would explain the opacity, would be likely to be increased by the unequal effects upon a crystal so constituted of changes of temperature. Some of the crystalline layers of agates are pervious to the colouring liquids, which are used to stain them; and these "air bubbles" might be in actual communication with the exterior, although they would be retained during solution by mere adhesion, and would be finally disengaged with all the appearance of being liberated from closed cavities. The existence of these cavities is, however, rather adduced as an evidence of changes which have taken place in the crystals, and which are the result of the development in them of the organic structures, found to be liberated by solution from their very centres. Granting that the development actually takes place, it might have been supposed that the centre of the crystal would have been its least likely seat, as being the part least subject to external influences, and therefore with less determining causes for the necessary re-arrangement of its constituent molecules. Another explanation seems more probable; the presence of foreign bodies, or "nuclei," in a solution at the point of crystallization, when adhesion and cohesion are nearly balanced, often, by making the cohesion preponderate, determines the formation of crystals about them. The character of the bodies which were liberated is not incompatible with this,

1 'Nature,' vol. ii, p. 179.

2 Ammonium tartrate is known to effloresce when kept, from the loss of ammonia.

they consisted principally of fragments of cotton, or paper fibre, mucoid matter, confervoid filaments, and fungus spores. Now, supposing all these present in the original mother liquor, which is allowed to be possible, they would certainly tend to aggregate together on the same principle that a precipitate granulates. Hence a collection of them would be not unlikely to occur at the centre of some crystals, though it would not necessarily be found to do so in every one. Such an explanation is at any rate not improbable, and would seem more easy of acceptance, than the belief that a molecular re-arrangement of the crystals of ammonium tartrate into living, and even organized protoplasm could take place. Dr. Bastian supposes that as ammonium cyanate passes into urea, so ammonium tartrate "may undergo a more or less similar isomeric transformtion."[1] The transformation could hardly be an isomeric one in any case; no proteinaceous body could be properly described as an isomer of ammonium tartrate, at least in the same sense that ammonium cyanate is isomeric, or more properly metameric, with urea. Nor is it easily intelligible how "spontaneously" a crystalline molecule statically stable could pass into the unstable colloidal state. Even supposing that possible, it is difficult to understand its becoming living matter, inasmuch as the molecular structure of living matter must be different from that of similar matter not living, which is all that could arise from any isomeric modification,—otherwise in what does the difference consist? It is still more difficult to understand the "coalescence and re-arrangement" of such matter into definite organisms, such as fungus spores.

All this is something infinitely beyond the scope of evolution, properly so called, which sees in the changes any object undergoes only the correlatives of the changes of the environment, and which banishes the word "spontaneity" from its vocabulary altogether. Here, however, the cycle of changes, with ammonium tartrate at one end, and fungus spores at the other, may for ought that can be seen to the contrary, take place spontaneously in the space of a few months, and with an environment absolutely constant. Indeed, if we may trust the observations of M. Trécul (p. 195), the operation may be still further abridged, as he has seen a tetrahedral crystal within the cells of the bark of the common elder gradually converted into a short fungoid filament.

Dr. Bastian admits that Pasteur's results "seem at first

[1] *Loc. cit.*, p. 222.

sight to be all-convincing," but as soon as the first impressions are got over, he is on the contrary amazed at his "utter one-sidedness." Looking at the origination of life as a purely physical phenomenon, he compares it to evolution, and remarking that the "restrictive influence" of atmospheric pressure at the temperature at which ether boils for example is sufficient to prevent boiling in alcohol or water; he conceives that the "evolutional capacity" of liquids is something similar, and may be possible under restrictive conditions in one case when it is entirely destroyed in others. These restrictive conditions appear to be according to Dr. Bastian pressure, acid reaction, and the influence of a high temperature. He considers, therefore, that it is a quite reasonable supposition to attribute the nonappearance of life in some of Pasteur's experiments, not to the absence of germs in a liquid in which those possibly previously contained had been destroyed by boiling, but to the effect of restrictive conditions. It is to be regretted that Dr. Bastian did not repeat some of Pasteur's experiments with the 'eau de levûre sucrée,'[1] as it would be important to have seen whether they would in his hands have yielded negative results. If so, an explanation of the discrepancy could not have been sought in "restrictive conditions," but must have been looked for in some unseen source of error. As to the presence of air, it must be remarked, that in the experiments of Dr. Child, and in those of other observers quoted in Dr. Bastian's paper, this certainly did not prevent the appearance of life accompanied as it no doubt generally is with considerable gaseous tension arising from putrefactive changes.[2] It is very probable, as Dr. Bastian suggests, that high temperatures might have a more destructive effect on organic matter if contained in acid solutions, but this could only approximate them to those solutions of purely inorganic matter, the evolutional capacity of which seems to have been so high. But an argument is supplied by Dr. Bastian himself which seems suffi-

[1] The composition of this liquid was, water 100, sugar 10, albuminous and mineral matter derived from the yeast of beer 0·2 to 0·7 parts. It was at first slightly acid, becoming more so from gradual oxidation during the experiments.

[2] Dr. Child found that in the presence of carbon dioxide, hydrogen, nitrogen, and possibly of oxygen, no organisms were produced in fluids in sealed flasks, although they were when heated air was introduced. The 'restrictive influence' of pressure in each case would be the same, and it is difficult to see what other repressive influence the gases themselves could have had if life can be produced in vacuo. If the gases were not thoroughly washed the case would of course be different, as chemical substances, inimical to life might have found their way into the solution with them. (See 'Proceedings of the Royal Society,' vols. xiii, pp. 313—314; xiv, pp. 178—186.)

cient to demolish his criticisms of Pasteur's experiments. "The presumption," he remarks, "is a fair one, that solutions which are favorable to the growth and development of certain organisms, would also be favorable to the evolutional changes which more especially tend to the initiation of such living things." Now, the most striking of Pasteur's experiments are those[1] in which after a solution has shown even during eighteen months no trace of life (in consequence Dr. Bastian would say of restrictive conditions) a ball of dusted gun-cotton was introduced without allowing the access of any except calcined air, and in a few hours under precisely the same conditions as before, a development of living things was evident. Certainly, according to the argument above, we ought to infer that the evolutional capacity of the liquid was not really impaired.

In these criticisms of Dr. Bastian's paper, the object has been to point out the difficulties in the way of a reconciliation between spontaneous generation and the principle of evolution. Any one who is thoroughly impressed with the probability of the truth of those principles, finds little difficulty in deductively bridging the interval between a living thing so elemental in its characters as Haeckel's lowest Moner *Protamœba*, and a lifeless proteinaceous substance; but he has little cause for gratitude to observers in the guise of evolutionists who assure him that the matter is quite simple,—the molecules not merely of a proteinaceous substance, but of almost any casual collocation of carbon, hydrogen, oxygen and nitrogen, spontaneously rearrange themselves, and organisms much higher than *Protamœba* appear at once upon the scene. The interval which the evolutionist is modestly content to conceive deductively bridged, is as nothing to the leaping powers of the so-called heterogenist who boldly widens the gap and passes easily from ammonium tartrate to a *Penicillium*, or from a solution of smelling salts to a fungoid mycelium. If evolutionists adopt these results, they will certainly be guilty of inconsistency, "unless," as Dr. Beale remarks, "it be consistent to believe at the same time in the law of continuity and succession, and in a law which involves discontinuity and interruption as applied to the production of living forms at the present time."[2] A believer in spontaneous generation is not indeed really an evolutionist, but is only a vitalist minus the supernatural; the special creation which the one assumes is replaced by the fortuitous concourse of atoms of the other.

[1] 'Annales de Chimie et de Physique,' 1862, p. 42.
[2] 'Disease Germs,' p. 42.

On the RELATIONS *of* PENICILLIUM, TORULA, *and* BACTERIUM. By Professor HUXLEY.

(*Special Report of an Address delivered in the Biological Section of the British Association for the Advancement of Science, Sept. 13th,* 1870.)

THE names Penicillium, Torula, and Bacterium are applied to exceedingly humble things. That which the scientific man terms Penicillium is popularly known as mould; Torula is the name given to yeast; whilst Bacterium is so minute that it has not attracted common attention, and hence has received no popular name. I propose to give a statement of what I imagine to be the relation of these three forms, from work which I have myself lately been carrying on; and I shall have to tell you of some very remarkable facts connected with their growth and development.

It is a fact familiar to every one here that mould makes its appearance on decaying matters. If you examine such a growth of mould, and take some of the finer powder-like matter from its surface, and placing this on a glass slide, apply a high power of the microscope, you observe that the grey powder which you have got on your slide consists of a vast number of small spheres, of various sizes, the biggest as large as the red corpuscles which are floating in my blood—some a great deal less, not more than the $\frac{1}{10000}$th of an inch in diameter, the average diameter being about $\frac{1}{7000}$th of an inch. Such a body is known as a spore or conidium.' On applying pressure to the covering glass which you have placed on the slide, you may burst some of these spores, and you will find that each consists of a transparent coat or bag (fig. 1), which appears to be composed of cellulose. In the

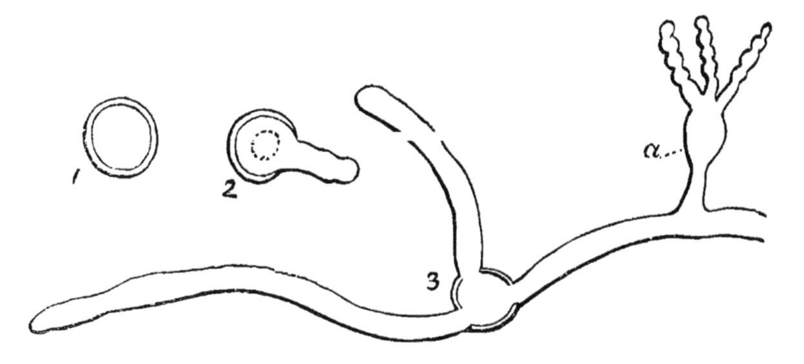

interior of this coat is a delicate soft material, containing nitrogen in addition to carbon, hydrogen, and oxygen, which exist in the cellulose coat, and having the same composition as albumen, fibrin, and such bodies. I am not aware that the analysis has been made in the particular case to which I am alluding, but it has been made in that of yeast very carefully, and yeast is closely analogous to this. The delicate soft contents of the bag are what has been known for many years as Protoplasm. It is in this case homogeneous throughout; a careful examination discloses no granules in it. The middle part only appears to be a little clearer than the rest, that is all, the reason being that the fluid is in this part more watery; such a clear watery part as this is often called a vacuole. The conidium consists, then, of a non-nitrogenous, cellulose outer bag, containing a fluid formed of protoplasm. Now, omitting all allusions to others who have worked at this matter, and who have already determined nine tenths or more of what I have to tell you, I will give you some account of the modifications which this body undergoes. From this spore, under various conditions, proceed a great diversity of forms. Suppose that you make a solution of oxalate or tartrate of ammonium in water, and add some yeast ash consisting merely of phosphates and mineral matters, and then scatter the grey powder from the mould into this—in fact, sow it with conidia—they will find themselves in a suitable condition for development. The spore in from one to twenty-four hours, if kept in a suitable temperature, throws out a process (fig. 2) which becomes longer and longer. Two or three of such processes may be thrown out, which spread at a great rate, and form a ramified star-like mass (fig. 3). This growth in a short time produces thus a body a million-fold bigger than the little spore from which it started. It has taken place under certain conditions of temperature, food (given in the solution), and access of air, though it is possible that this last is not necessary. It is quite independent of light; it will occur equally in the light or in darkness. You have in the solution absolutely dead mineral matters, the oxalate or tartrate of ammonium and phosphates; you place in this solution a minute grain of protoplasm, supply a certain amount of heat, and this little grain puts together the elements in that solution anew; it builds them into a living organism according to a type and pattern of its own. This is the first stage in the growth of this organism, the development of the spore or conidium into the mycelium. The mycelium is made up of these threads, which spring firstly from the spore and branch in various directions, and are called hyphæ. Each hypha has the same composition as the spore, being merely an elongated tube of cellulose with a mass of elongated protoplasm within, and it is continuous throughout. So long as this form of growth goes on the hyphæ are submerged in the fluid, or only floating near the surface; but after a time, in accordance with some change in the conditions of the fluid, one hypha will send up a vertical branch, which grows so as to stick up out of the water; and the aërial processes so formed are most difficult to wet, for they have the power of throwing off the water on account of the adherence of air to their surfaces. This vertical branch grows up and sends out three or more processes, which become constricted and broken up into little balls (fig. 3 *a*). All this takes place simply by prolongation; there are no cells in this matter; there may be partitions in the protoplasm, but there are no nuclei and no subdivision of such bodies. Each of the spherical bodies at the end of the aërial stalk at which we have arrived is a spore or conidium such as we started from, and one would think that this was a complete life-cycle. There is, however, a singular variety of other and secondary developments.

One of the commonest—I found them in all cases where I got the mycelium just described—I will now mention. When a conidium is developing and sending out its first hypha, it will frequently send out from the other side a moniliform process which breaks off and floats about (fig. 4). Sometimes each globule in these detached masses seems to have a nucleus. This (fig. 4 *a*) is the Torula form; it is the same as yeast, differing only in size in this particular case. If the

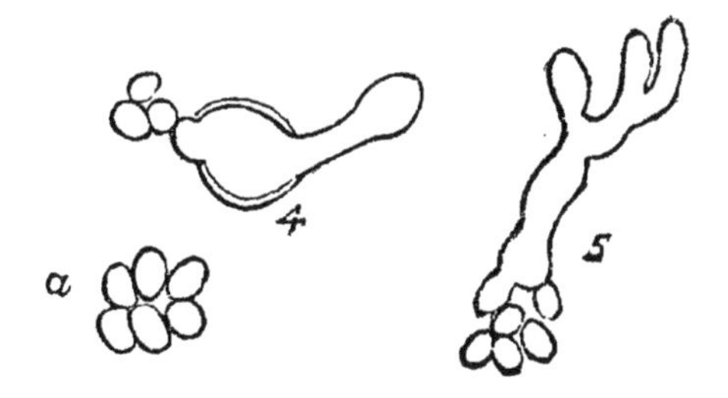

conditions are favorable, absolutely the whole development of the spores or conidia may be Torulæ. The great condition which favours this is the absence of atmospheric air. It is an ascertained fact that this Torula form grows without oxygen and without light: this is a physiological fact of the highest importance, established by Pasteur, with regard to Penicillium-yeast. What is the fate of Torula? what is its further development? Some of these floating masses become very mycelium-like (fig. 5), similar in appearance to what has been called Cladospora. I have never found mycelium without the Torulæ accompanying it, which is an important fact; but I have found Torulæ, and it is a common thing to get them without mycelia. If you break through the felted crust of a mould formed by the interwoven hyphæ of the mycelia, you invariably find Torulæ interspersed amongst these filaments, a condition which suggests a very close relation to the structure of Lichens. In fact, morphologically Penicillium is a Lichen when in this state. This interesting identity appears hitherto to have escaped notice.

Whenever you examine the germinating spores, and especially if you keep away air, and also whenever you examine common yeast, you will find very singular bodies associated with the Torula forms. These bodies are excessively minute. They are only the $\frac{1}{40000}$th of an inch in diameter, the conidia being only $\frac{1}{10000}$th of an inch in diameter—less than a third the size of the minute red corpuscles which are coursing in millions through my smallest blood-vessels. These bodies to which I now allude are but a fourth of the Conidia in diameter, or one twelfth of the diameter of the human blood-corpuscle. You may find one alone, but most commonly two joined together side by side, thus (fig. 6 *a*); or you see them in sets of pairs, bound together by an invisible gelatinous matter, the presence of which is inferred from their adherence (fig. 6 *b*); further, you get little rod-like bodies

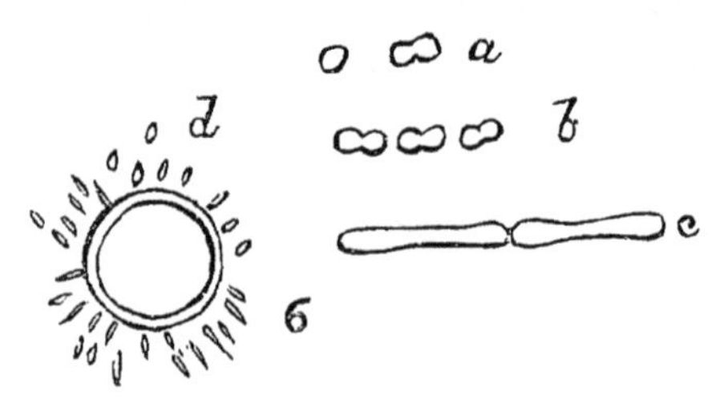

(fig. 6, *c*), also apparently of the same nature. These little bodies we speak of as Bacteria. When you examine these Bacteria with the very highest powers—1200 diameters, or 2000 if you can get it—you see two things with regard to them:—1st, all are in motion; 2nd, they have *two distinct kinds* of movements. The very smallest have merely a trembling movement; those which are elongated oscillate on a central point in their long axis, rotating whilst in an oblique position. That is one kind of movement. The other kind of movement is a darting across the stage of the microscope, sometimes in a straight line, sometimes accompanied by oscillations, which gives a serpentine appearance to the moving Bacterium or chain of Bacteria, whence the name Vibrio. These two kinds of movement are not to be confounded. They must be explained as due to very different causes; and it seems to me that it is a confusion of these two which is at the bottom of the mistakes made in the assertions as to the survival of Bacteria, &c., after the application of very high temperatures. I have made experiments with this matter in view. I boiled a solution containing living Bacteria for two hours. On searching for them after this, I found them unchanged in most respects, but somewhat firmer in texture, like salt beef on board ship after it is boiled. When treated with chromic acid, iodine, or strong alcohol, the Bacteria remain. Their life is undoubtedly destroyed by these reagents. Every one admits that; but there they remain with but a slight change of appearance. Do what you will, however, they retain their *trembling* movement; and this is a very misleading phenomenon. Dr. Bastian was good enough to unseal a flask in my presence, which had been closed at a temperature of 150° Centigrade; and I saw there and then Bacteria exhibiting these active, trembling movements, which, had they come from any other solution, I should have *then* considered as a proof of their being alive. But with regard to the other kind of movements, it is quite otherwise. On raising the liquid in which they are to the boiling-point, it stops at once, or if you add any of the reagents just now mentioned. The first kind of movement is no doubt the Brownian movement, first shown by Robert Brown to be exhibited by minute particles of a variety of substances, when placed in liquid. When you have a rod instead of a granule, vibration must act unequally, and hence come the curious, oscillatory, rotating movements of the elongate Bacteria. This discrimination is of the utmost importance. I cannot be certain about other persons, but I am of opinion that observers who have supposed they have found Bacteria surviving after boiling have made the mistake which I should have done at one time, and, in fact, have confused the Brownian movements with *true living* movements. So, according to my notion, a fluid full of Bacteria, moving like Vibriones, and twisting about in various ways, does not necessarily contain anything alive at all.

How do these Bacteria come about? I speak with caution, in accordance with experiments by other persons, especially that excellent lady, the Frau Johanna Luders.[1] I have never examined yeast without finding Bacteria, and in a sessile state. You may see Bacteria sculling about, and then becoming quiet, and stuck all about in a perfectly motionless

[1] See this Journal, vol. viii.

state, as though fixed in some substance. They are so; they are imbedded in a jelly, like many other low organisms, especially Algæ; they exhibit this phenomenon of a quiescent stage, during which they are encased in gelatinous matter. Pouchet and others have supposed them to be dead, but Cohn showed years ago that they are not dead, but are in a quiescent state, growing and reproducing, but not moving. With Torula, then, we find Bacteria in great numbers in this quiescent state. Usually masses are to be seen adhering very closely and tightly to one Torula cell or another, and such masses are very difficult to separate from the cell to which they are fixed. It seems probable that the Bacteria proceed in this way from the Torula cells as the Torula cells do from Conidia. It is probable that Bacterium is a similar thing to Torula—a simplest stage in the development of a fungus. By sowing Conidia you also get Bacteria in abundance. You get the Bacteria adhering like this to (fig. 6, *d*) the Conidia, and they are, I believe, developed from the protoplasm of the Conidia just as Torulæ are, and we may compare these two forms to the Microgonidia and Macrogonidia of Algæ. They are all terms in the development of Penicillium. This may be set out thus (fig. 7) in a diagrammatic

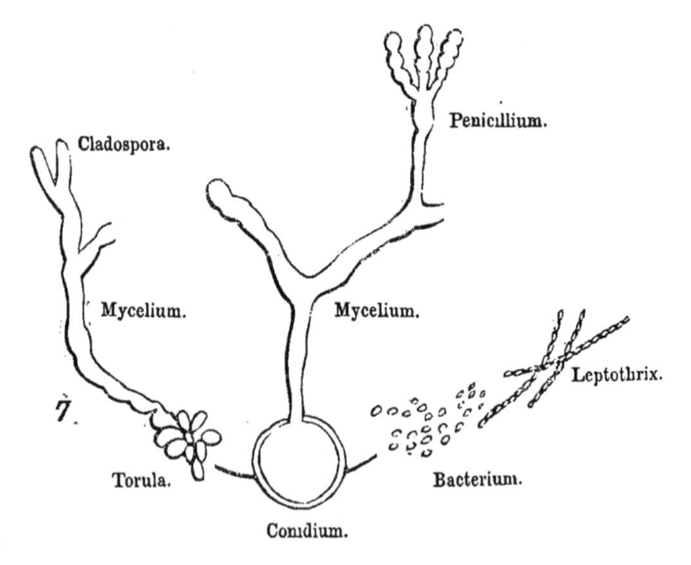

form. But this does not exhaust the whole matter by any means; for it appears that there are many genera which are thus affiliated.

The whole question of spontaneous generation turns upon what we can say *certainly* as to the development of these things, and one naturally turns to works on botany for this information. Finding, however, as I did, the subject in a most unsatisfactory state, and that nothing was definitely known, I set to work for myself. If it is true that Torula forms, Bacteria forms, and such things which are common products of so-called spontaneous generation—if these can be shown to be terms in the development of a *known* form—the probability of the *same* identical form turning up spontaneously becomes by mathematical considerations infinitely minute; and for my part I could as soon believe that the calf I see grazing in a meadow, had been spontaneously generated from the grass and flowers there. In the second place, the discrimination of the modes of movement of bacteria is important. Thirdly, the development of living matter from mineral matter without the influence of light which I have mentioned above is of the highest importance. In relation to the life of the deep sea organisms it is important. We have had all sorts of speculations as to their life in the dark, my own included.[1] The mystery of Bathybius is paralleled by the protoplasmic material of Penicillium which develops in the dark in the same way as it does; just also as Penicillium turns ammoniacal salts in the dark into protoplasm, so may Bathybius do the same at the bottom of the sea, and so we need not trouble ourselves with any special hypothesis to account for the occurrence of a sheet of living matter in this position.

[After some discussion, Professor Huxley added, that he *never* had seen the true vital movements of Bacteria after they had been boiled. He admitted there were flaws in Pasteur's work on account of his having used only low powers of the microscope. The divergence between his (Professor Huxley's) views and that of abiogenists was merely one of time. As a believer in the doctrine of evolution in its most extreme form, he believed that organic matter had at one time developed from mineral matter, but that was no reason why it should do so in the cases cited by abiogenists. He considered that all attempts at what the German's call "rein-cultur" must be abandoned in working at the development of Bacteria, since it was impossible to obtain an optically pure specimen of water. He had never found a drop of water in which he could vouch for the absence of any particles the $\frac{1}{40000}$th of an inch in diameter, and it was particles of this size, the microzymes of some writers, which are the first appearance of Bacterium. If, as he mentioned in his inaugural address, you take an infusion of hay, and examine it *at once*, you find

[1] See this Journal, October, 1868.

swarms of Bacteria, spores, &c.; if you now place it in two flasks A and B, then boil both and stop up A carefully with cotton wool while boiling, but leave B's mouth open, you will find after a few days, that whilst A only contains the original Bacteria—dead, having been killed by boiling, the flask B contains the dead ones and is milky with the development of fresh ones, which exhibit true vital movement. It is such dead Bacteria as exist in A, which by their Brownian movements have led to the assertion that living Bacteria develop after boiling in closed vessels.]

On a NEW METHOD *of* STUDYING *the* CAPILLARY CIRCULATION *in* MAMMALS. By Dr. S. STRICKER, Professor of Experimental Pathology in Vienna; and Dr. BURDON SANDERSON, Professor of Practical Physiology in University College, London.

The science of physiology is based on exact observation of the mechanical and chemical changes which take place in living beings. For this purpose the same methods and instruments must be used as are employed in physics. The observations, however, are much more difficult than those of the physicist, because the subject of observation is a living being, and consequently the conditions are much more complicated than those with which he has to deal. Hence it is of the greatest importance to devise good methods—so much so, that science is as much indebted to those who invent them, as to those who discover new truths.

The purpose of the method we are about to describe is the observation of the way in which the blood streams through the minutest vessels; of the visible changes which take place in the living tissues around these capillaries; and of the relation between these changes and the circulation. The subject is full of interest and importance:—first, in relation to those normal changes in the living tissue which constitute nutrition; and, secondly, in relation to those modified changes which constitute disease.

Hitherto this observation has been made only in the transparent parts of cold-blooded animals. The earlier observers contented themselves with restraining the movements of the animal in an imperfect manner by mechanical means. More recently one of us (Prof. Stricker) introduced the employment of Curare in this investigation, under the action of which voluntary movements cease, while the circulation goes on unimpaired. In the curarized frog we are enabled to explore the phenomena of the capillary circulation for many hours with the greatest precision;—to watch the first commencements of inflammation; to observe the effects of poisons on the blood-vessels; to ascertain the laws of those movements of contraction and expansion, which regulate the whole process of nutrition; in short, to establish a host of facts of fundamental importance in physiology and pathology.

All conclusions, however, which are derived from observations of animals far down in the scale are subject to the objection that their functions are carried on under conditions considerably remote from those which exist in man. It has, therefore, been long desirable to change our field of research to mammalia. Hitherto only one mammalian animal has lent itself to this purpose,—*the bat.* To the use of the bat's wing, however, there are objections which will never be overcome, chiefly because it is covered with an integument so thick, that although the vessels can be seen, the tissues around them are indistinguishable.

It was, therefore, clearly necessary to look in other directions. To this work we set ourselves at the beginning of last month,—both of us with a practical, *i.e.* pathological object in view,—one of us with a view to his researches on inflammation, the other with reference to inquiries about tuberculosis.

The first requirement was a suitable animal, the second an anæsthetic, the third a method of placing the tissue to be explored under the microscope.

We will begin by speaking of the anæsthetic.

It was at once obvious that, the bat being discarded, no other mammal could be procured in which any external part is sufficiently transparent for observation under the higher powers of the microscope. Therefore, the method must necessarily involve the use of the knife, *i.e.* (let us not shudder at the ugly word) vivisection. Hence an anæsthetic was absolutely necessary. Most happily chloral was found to be complely adapted to our purpose. About three grains of chloral under the skin we found to be sufficient to render a full-sized guinea-pig motionless and insensible for many hours.

Next the selection of an animal was to be considered. This was not difficult. One of us (Dr. Sanderson) was familiar with the remarkable structure of the guinea-pig's omentum, and had already described it in connection with another inquiry. The omentum of the guinea-pig is a membrane of extent relatively comparable to that of man, but its structure is entirely different. First it is attached, not to the transverse colon, but to the greater curvature of the stomach. Secondly, it consists, not of four layers of membrane, but two; and lastly, it contains very little fat, but in place of it a great quantity of cells, which are collected in a peculiar way about the blood-vessels and in their neighbourhood, partly in the form of perivascular sheaths, partly in the form of little collections or nodules consisting of cells lying in the meshes of a plexus of capillaries.

Hence, from the simplicity of its anatomical relations, and particularly from its being attached on one side only to the stomach, in which respect no membrane is comparable with it; from its perfect transparency; from its abundant vascularity; and from its containing not only vessels, but living cells, and these cells of two kinds, namely, epithelial and parenchymatous, it is obvious that the omentum of the guinea-pig offers a splendid field for observation.

We have now only to speak of the mechanical arrangements which are necessary in order that the omentum may be placed under the microscope under natural conditions. It is to be borne in mind that the guinea-pig is a warm-blooded animal, that the omentum is part of the peritoneum, that the tissue is extremely delicate, and readily becomes inflamed when touched.

Hence for observation the membrane must be immersed—immersed, however, not in water, for water would at once irritate and kill the tissue, but in solution of common salt of proper strength. Such a solution is what physiologists call an indifferent fluid, because, when it comes in contact with living cells, it does not appreciably interfere with their vital processes. Secondly, it cannot be in a natural condition unless it retain the temperature of the living body. The arrangement for securing this is somewhat complicated. The membrane is laid out in a glass dish, which is supported on the stage of the microscope by a hollow brass plate, through which a stream of water flows at a rate and temperature so regulated that the dish and its contents are maintained at a temperature closely agreeing with that of the body.

For commencing the observation this is all that is necessary. If, however, it is continued, the observer soon encounters two difficulties, both of which must be overcome. The one arises from the clouding of his objective when it is brought near the warm surface of the saline solution; the other from the rapid evaporation of the solution, and the consequent alteration of its density, and eventual desiccation of the membrane. The first difficulty is obviated by warming the objective; the second by providing for the renewal of the water contained in the bath by the constant influx of fresh water at a rate corresponding to that at which it wastes.

The operative procedure is extremely simple. The guinea-pig having been thoroughly chloralised, is laid on a support or block, the upper surface of which is in the same horizontal plane with that of the microscope stage. An incision is made, which extends for two inches at most, from the outer edge of the left rectus muscle, a little below (behind) the end of the ensiform cartilage horizontally cut out so as to divide one or two costal cartilages. The muscles must next be divided and the peritoneum carefully opened. The stomach can then be drawn out of the abdominal cavity without difficulty, especially if the additional precaution be employed of first removing some of its contents. In doing this, very little manipulation is necessary, and special care must be taken to avoid touching the delicate structure attached to its border which is to be subjected to observation. The moment that the organ is fairly out of the abdomen, the membrane must be floated into the warm bath prepared for it, and is then ready for examination. It is, however, found very advantageous to cover those parts of it which do not lie directly under the microscope with sheets of blotting-paper. This arrangement has two advantages; the risk of evaporation is diminished, and the undulatory movements of the water are prevented, so that the object is rendered much steadier than it would otherwise be. The enlargements we have hitherto employed are inconsiderable, the most useful object being the quarter of Ross. We have no doubt, however, that we shall eventually be able to apply both air and immersion objectives of higher magnifying power.

The objects which present themselves to the observer in the omentum of the guinea-pig are manifold. We content ourselves with barely enumerating them. Veins and arteries may be studied of various diameters, some of them surrounded by sheaths containing fat-cells, some by similar sheaths containing the cells of which mention has already been made, others so free that their structure can be perfectly studied. Labyrinthine capillaries of surpassing beauty can be studied both in the sheaths of the vessels and in the little nodules of tissue in their neighbourhood, and finally the epithelial elements with their characteristic spheroidal nuclei by which the wonderful connective-tissue network of the omentum is everywhere covered.

The apparatus used will be best understood from the accompanying sketch, which has been kindly prepared for us by Mr. Lankester.

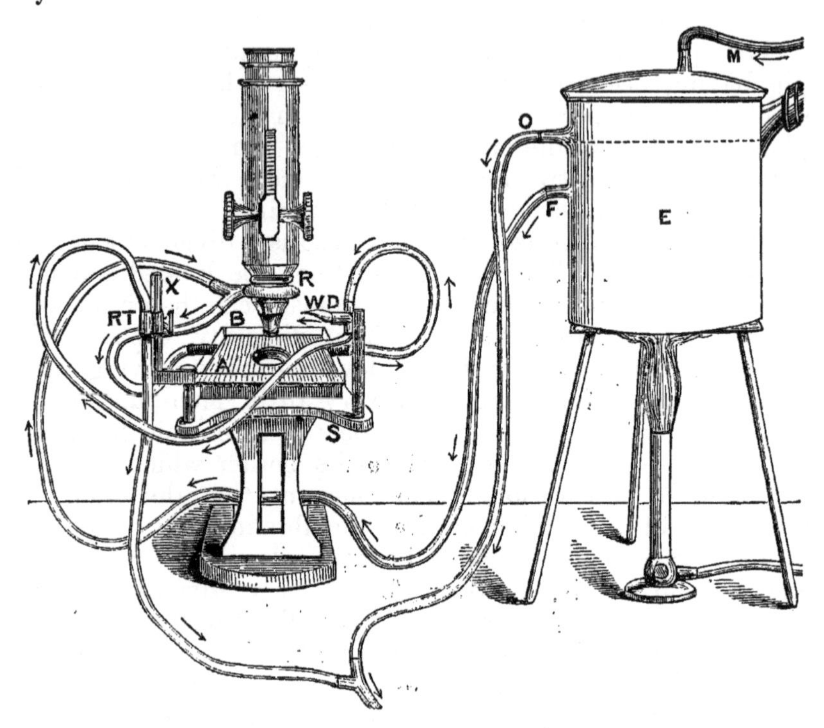

Explanation of the Drawing.

The sketch represents Hartnack's microscope with the apparatus adapted to it.

E is a vessel containing boiling water. *M*, a tube by which water is constantly supplied to the vessel *E*, at such a rate that the supply always exceeds the loss by evaporation, and by the out-flow from the pipe *F*.

O, Outlet tube, leading to waste pipe; consequently the water in *E* always remains at the level indicated by the broken line.

F, Tube by which water is conveyed to a loop-shaped metal tube (*W*) which surrounds the upper part of the objective, for the purpose of keeping it warm, the vulcanite ring *R*, serving to prevent the heating of the microscope tube. From this loop the hot water passes by a second tube into the hot water plate (*A*), a full description of which will be found in Professor Stricker's 'Histology,' just published by the Sydenham Society: on the plate stands the glass dish *B*. From the plate the stream passes, as indicated by the arrows, to *D*, and is thence continued to the regulating tube (*R T*) on the opposite side of the microscope stand, where it terminates in the dropper, the construction of which will be readily understood from the additional sketch (fig. 2). in which it is represented of its actual size. The dropper consists of a glass tube (*T*), having an aperture (*o*) on one side. Into the upper end of this tube a second (*t*), drawn out so as to be capillary at its point, is inserted, from which the water drops constantly. The frequency of the drops, *i.e.*, the velocity of the current through the whole system, is regulated by varying the height of the dropper, for which purpose it is attached to a movable support (*X*), which slides up and down on a vertical stem inserted in the stage of the microscope, and may be fixed at any position by a screw. Hence, the temperature of the water which is supplied to the warm stage (*A*) being constant, and the loss of temperature by radiation and otherwise being also tolerably constant, the arrangement just described affords to the observer a means of so adjusting the flow from *T* that the temperature of *A* is also constant.

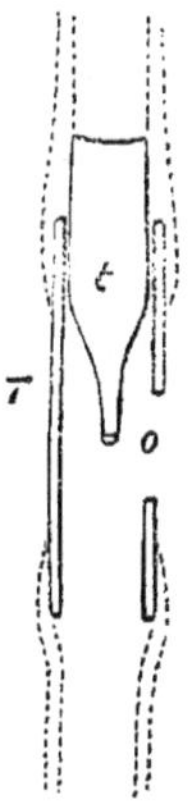

All that remains is to describe the means by which evaporation from the surface of the saline solution, contained in the bath in which the mesentery is laid, is compensated. For this purpose the hot-water tube gives off, at *D*, a branch which immediately terminates in a capillary end. From this end hot water drops constantly, at a rate which must be regulated at the beginning of the observation, once for all, by means of a constricting screw at *D*, which for want of room is not shown in the drawing.

Before beginning observations, the bath, *B*, must be placed on the warm stage charged with saline solution, and allowed to remain there until it has acquired a constant temperature. The rate of flow must then be adjusted by varying the height of the dropper until the reading of the thermometer in the bath is about 37° C.

On the Cellular Elements *of* Tendons *and of* Loose Connective Tissue. By Dr. L. Ranvier.[1]

With Plate XX.

I purpose in this paper to describe the forms and relations of the elementary cells of tendons and of ordinary connective tissue. The results which I have obtained by new means of analysis differ in many respects from those which now-a-days are generally believed and taught on the subject of connective tissue. They considerably modify the manner of investigating this tissue, as much as to its physiological as to its pathological condition, and render the interpretation of certain facts much clearer.

I. It seems to me necessary, first of all, to sketch broadly the history of the fluctuations of histology on the subject of the cells of connective tissue. It is necessary, in fact, that readers should be acquainted with the discussions on this subject, and know how to appreciate the state of our knowledge on connective tissue, which, from a pathologist's point of view, is the most important tissue of the whole organism.

[1] Translated from the 'Archives de Physiologie,' tome 2.

The preparations of connective tissue made after Gerlach's method (dessication of the tissue, section, colouring with carmine, washing, and the action of acetic acid) show star-like figures, in the centre of which is perceived a red irregular corpuscle. Virchow saw in this star-like figure a real cell, in the irregular corpuscle a nucleus; and he gave to the whole the name of plasmatic cell. He had discovered that the bone-corpuscle is a cell, and, struck with the analogy of the form presented by the plasmatic cell, the bone-corpuscle, he was led to consider these two elements as histological equivalents. He also admits that in connective tissue, as in bone, there is a system of channels destined to conduct the plasma, and to put it in connexion with the different parts of the tissue.

Under the influence of the works of Von Recklinghausen, the idea of plasmatic channels in connective tissue received great confirmation; but these same observations tend to show that the star-like figure of connective tissue is not a simple cell.

It was in creating a new method—the impregnation of tissues by nitrate of silver—that Recklinghausen modified the ideas of most histologists on the subject of connective tissue. The impregnation of the cornea of the frog by nitrate of silver renders apparent star-like figures, united one to the other by prolonged ramifications. According to the inventor of this method, these prolongations are the canals in which the cells (the real cells) can make their way in virtue of their amœboid movements. Recklinghausen applied the same method to different parts of connective tissue, and throughout with the same results. Connective tissue seemed to him invariably to contain canals in which circulate the plasma and the cells. These canals (canaux du suc, Saftkanalchen) were held to be the real origin of the lymphatic system.

The researches of Kühne on the intermuscular connective tissue of the frog teach us that in this animal the cells of the connective tissue have no membrane, nor are they comprised in the canals; and that they are free in the meshes between the bundles of connective tissue. The result is that, at least in the frog, we cannot apply to all the connective tissue the general idea of the plasmatic system furnished by studying the cornea with the aid of silver impregnation.

As soon as Virchow's first works on connective tissue appeared, Henle (Canstatt's 'Jahresbericht,' 1851, vol. i, pp. 23, 24) declared himself strongly against his theory. He took, as an example, tendons which, when cut transversely, show star-like figures, which are never distinguished in longitudinal sections. In these last he saw only longitudinal striations or grooves filled with ovoid nuclei, or in the form of rods. In these grooves he says rectangular plates may be met with like scales from the surface of hair, containing no nuclei, and having the appearance of a fine line when they are looked at sideways, and like a plate when seen in front. The star-like forms which are seen upon the transverse section would not (according to him) be produced by cells, but could be marked out only by bundles of fibres cut across. There would be true cells no more in the tendons than there would be in connective tissue, but only nuclei or their derivatives, the fibres of nuclei, annular fibres, and spiral fibres. In a more recent work upon connective tissue (Henle's and Pfeuffer's 'Zeitschrift,' 1858), Henle maintains still the same ideas upon the structure of this tissue.

II. I intend to enter into rather minute details upon the structure of the tendons, to show how far the analysis of an apparently simple tissue requires precautions on the part of the observer. It is necessary, in the first place, to choose tendons so fine that the examination may be conducted without requiring any section to be made.

The tendons which in the tail of the small mammalia terminate the spinal muscles will serve the purpose perfectly well.

The tail of a young rat, of a mouse, or, better still, of a mole, being cut off near to the base, filiform tendons may be extracted of several centimetres in length by tearing off the end of the tail with the fingers. These tendons are placed upon a plate of glass, and fastened at each end with sealing-wax. After colouring with carmine, washing, the addition of acetic acid, and slight pressure with the covering glass, a preparation is obtained upon which may be observed parallel stripes, red, and apparently without intermission, if the examination be made with a magnifying power of 100 diameters. With a magnifier of 250 diameters, the red stripes are observed to be cut by transverse or slightly oblique lines, which divide them into segments of equal length. If a rather strong pressure is then exercised upon the covering glass with a needle or a scalpel, the little segments split lengthwise, the two edges of the crack become separated, and the cylinder is unrolled and reduced to a flat rectangular cell. In the centre of this cell is found a flat nucleus, at first rectangular, which soon, under the influence of acetic acid, recovers itself, and takes a rounded form. The cell is formed by a plate of protoplasm; it is very evident that we have not here a cell, in the sense in which that word was understood some years ago.

The nucleus is deeply coloured with carmine; the protoplasm is slightly coloured; the bundles of fibres become transparent and appear only of a pale pink if the washing after the action of the carmine has been sufficient. In the thickness of the tendon there are no other cells but those of which I speak. In accordance with this description, it is clear that the tendons are traversed throughout their whole length by tubes formed by the flat rectangular cells rolled up and placed end to end. The two edges of a rolled up cell which touch each other are soldered; the two other edges are similarly soldered to the corresponding edges of the cellules placed above and below. This soldering is demonstrated by the impregnation of silver; when one of the little tendons is placed in a solution of nitrate of silver of two parts to the thousand, and is examined as soon as the impregnation is produced, there is seen at the point of union of the tubular cells, a dark line indicating that at this point there is an intermediate substance, as in epithelia (fig. 6, page 480). The aspect of the cellular tubes of the tendons vary according to the age of the animals, the mode of preparation, and according to the tendons submitted to examination. In young mammalia, and during all the period of growth, the tubular cells are very easily demonstrated in the tendons of the tail, by employing the method which I have indicated. But in adult animals it is very difficult to obtain the opening out of the tubular cells and their separation. This arises from the fact that an amorphous, resistant and elastic membrane is formed around the cells and throughout the whole length of the tube.

To render this elastic membrane very evident, it is necessary to submit the tendons to a boiling process for five or six hours. The connective bundles then become very transparent; and a great number of fine elastic fibres are distinguished in their place; the sheaths of the tubes are very clean and filled with granulations, the remains of cells swollen and destroyed by the boiling. After a prolonged and powerful compression made upon the filiform tendons from the tail of an adult dog, the little tendon may be exposed and even removed without the elastic tubes being broken, and upon a similar preparation coloured with carmine one can discern only the red stripes cut by transverse lines. With adult rodents it is much easier to tear the elastic sheath. The tubular cells open, and have the appearance of red plates, in which one cannot at first discover the nucleus (fig. 3). But when the preparations are preserved in a mixture of glycerine 100 and formic acid 1, at the end of some weeks may be perceived upon each plate a rectangular surface, more deeply coloured, which represents the nucleus. Moreover, these cells may be restored to their primitive form by means of experiment. Upon a living animal, it suffices to pass a thread into the sheath of the tendons, or to inject into it a solution of iodine or of weak nitrate of silver, then to wait twenty-four or forty-eight hours before killing the animal and making the examination. We find in this experiment another application of this general law, to wit, that *irritation restores cells to their embryonic form.*

As has been mentioned above, Henle had perceived in the tendons quadrangular plates arranged in longitudinal series. But, not employing carmine, the application of which to histology was discovered later by Gerlach, he could not distinguish the nucleus, and, in consequence, did not consider them as cells. Moreover, using an imperfect method, he could not discover that these rectangular plates proceeded from unrolled tubular cells, and he was forced to remain in absolute doubt upon the subject of the nature and of the histological significance of these remarkable elements. He did not even consider them as having a permanent existence, and met with them amidst other evidently similar elements, which he believed to be completely different. If all these precautions which I have suggested be not taken in the preparation of the tendons, in place of the regular forms, the most extraordinary figures may be seen. It is important to pause here for a moment in order to understand completely the importance of these methods in histological research. The little tendon being fixed by the two extremities, and examined in acetic acid if it be cut in such a manner as to release it from its ligatures, swells up, recedes and loses a third or half of its length. This retraction is due to the fibrous substance; the cellular tubes do not participate in it; and in order to compress themselves into a shorter space they are forced to assume zigzag or undulating forms. At times they even twist themselves into corkscrews. It is not to be doubted that Henle has taken such figures for fibres of nuclei, and for spiral fibres. When, before allowing the tendons to curl up under the influence of acetic acid, they have been compressed in such a way that the tubular cells are unrolled, at the moment of retraction it may be observed that the cells, and especially their nuclei, are thickened lengthwise. These last, formed, as we have seen, by an extremely fine lamina, become plicated transversely, and appear striped (fig. 2).

In order to see that it is not really striped, but simply folded like a cloth, it is necessary to use a powerful object glass of considerable angle of aperture. These different facts show that the elements which appear contorted after the action of acetic acid upon the tissue, for example, the elastic fibres, and the germs of the smooth muscular fibres, owe this form to the action of the reagent upon the intermediate substance. It is this of which we may be assured by direct observation of the tissues, and employing tension before using acetic acid. The cellular tubes exist in all tendons, but they vary a little in their form, according to the tendons submitted to examination. The rods which represent the rolled up nuclei are more or less long, and the interval which separates them one from another is more or less extended; this interval has often a length equal to, even superior to the little rods themselves. Level with the interval, the sheath becomes contracted, in such a way that the tube taken as a whole presents a moniliform appearance. Almost always the little rod which represents the rolled up nucleus is bounded at its extremities by a transverse line and appears cylindrical. The interval which separates the little cylinders is occupied by a granular substance, which is nothing else but a portion of the flat rolled-up cell itself. The tendo Achillis of a frog supplies cellular tubes, such as have just been described; by compressing them after the addition of acetic acid, the opening and the unfolding of the cellular tubes is brought about.

After having placed the tail of a Rodent in a concentrated solution of picric acid, at the end of some days the vertebræ have lost their calcareous salts, and are become soft, whilst the fibrous parts have acquired consistency. It is easy then to effect very fine transverse sections, which comprehend at once the bones, the tendons, the nerves, and the vessels. These sections, placed for several hours in the picrocarminate of ammonia, steeped, and examined in the mixture of glycerine and formic acid, show at the surface of the tendons the most beautiful plasmatic network (fig. 5); but this is merely in appearance. It is necessary to be very careful, as Henle said some time ago, not to mistake these star-like figures and anastomoses for canaliculated networks and for plasmatic cells. Already, with a little attention and an object-glass at a considerable angle of aperture, it may be observed that the star-like figures are limited by partitions extending through the whole depth of the preparation. This is seen still better when the cutting is slightly oblique to the axis of the tendon.

These partitions are simply the superficial layer of the bundles of connective tissue, which in the tendons are all parallel. This layer, upon the nature of which nothing has been decided, is coloured more easily by carmine than the connective fibrils, and it preserves its colour in an acid liquid; whilst the fibrils lose it completely, even immediately after they have been coloured. I shall return to these particulars when I have to speak of the subcutaneous cellular tissue. When in the tendons two bundles of connective tissue adhere, it might be said that they are separated by a single partition. The cellular tubes pass between the bundles, and in the points which they occupy upon the transverse sections it seems as if there were simply a thickening of the partition. These are the points which appear to be the bodies of ramified cells. In this section (fig. 5) may be perceived an irregular corpuscle, which has certainly until now been taken for a nucleus. But on the thicker oblique sections it may be observed, on lowering the object-glass, that this corpuscle corresponds to the surface of a section of a cylinder which intrudes itself into the thickness of the tendon, and is parallel to its axis. In one word, this corpuscle corresponds to the section of a cellular tube. The partitions which diverge from the space containing the cellular tube separate the connective bundles, as has been said above. But from these partitions arise fibrils which run in different directions, of which some (fig. 5) have been cut transversely by the razor. The interpretation of such preparations, in the sense of plasmatic cells, depends, then, upon an illusion. On the surface of the tendons a bed of ordinary connective tissue is invariably met with, containing flat cells, yielding star-like forms by the impregnation of silver. This tissue establishes a communication between the tendon and the ambient cellular tissue, or serves to support an epithelial layer (fig. 6, A) in the case where the tendon slides in a synovial sheath, such as may be observed in the tendons of the tail of mammalia. It is the ordinary connective layer of the surface of the tendons which has probably yielded to Recklinghausen the star-like figures limited by the deposit of silver, and has induced him to admit this form in all the cells of the tendons.[1]

III. Between the structure of the subcutaneous cellular tissue and that of the tendons there are very material differences; yet between the transverse section of a tendon and

[1] On the contrary, Recklinghausen distinguishes clearly enough between the two, as any observer may also do for himself. Experienced microscopists in England accept von Recklinghausen's observations as exact.—Ed. Q. J. M. S.

a section made in the subcutaneous cellular tissue, after colouring with carmine and the action of acetic acid, the resemblance is striking; the star-like figures are still observed, ramified and anastomosed, in the centre of which a red coloured corpuscle is met with. The limit of the figures is very well defined, and appears to be formed by a membrane which is coloured by the carmine; while, if the preparation has been well made, the intermediate substance is colourless. This truly striking cellular appearance is *simply an optical delusion.* In order to be convinced of this, it is necessary to have recourse to other methods of preparation.

Dissociation in water, a process employed since the beginning of histology, had already furnished facts to Henle which are not favorable to the theory of plasmatic cells. By the aid of this process he saw in the subcutaneous cellular tissue only connective bundles surrounded by annular or spiral fibres, simple varieties of that which he calls the fibres of the nucleus, and finally elastic fibres, arising also, according to him, from the fibres of the nucleus.

But we must allow that such an analysis was very inadequate, and that it passed over the most important elements —namely, the *cells.* Even though the cells of connective tissue have not the form nor the affinity which Virchow had assigned to them, there remains no less to that celebrated professor the great merit of having established their existence and signification in pathological transactions.

The dissociation of the cellular tissue in water, as it was practised by Henle, and as it is still done at the present time, gives bad preparations; because, in teazing out the tissue with the needles, the fibres are mixed, and thus lose their relation. At the same time, the cells are swollen, and even destroyed, and their remains are hidden by the intermixed fibrils.

To avoid these inconveniences, I have had recourse to another method. It consists in injecting into the cellular tissues, by the aid of a Pravaz syringe, gelatine maintained at a temperature of 37° Centigrade, a solution of nitrate of silver of 1 to the 1000, or simply serum. It is necessary to make this injection in an adult animal which has just been killed, and before the body becomes cold. Thus is produced an artificial œdema. If the injection has been roughly made, the substance accumulates in a circumscribed spot, in the centre of which there are a few fibres of connective tissue. If, on the contrary, the injection has been carefully made, and if the part be rubbed at the moment of applying the injection, the substance expands, and embraces a greater quantity of elements of the connective tissue. It is difficult, at first sight, to believe that an injected liquid forms a circumscribed mass in the cellular tissue, if it be constituted simply of fibres. But if it be believed that these fibres are very soft, and are easily separated, it will be seen that the liquid, whilst rushing in, forces back these fibres, and that then, being collected together great numbers, they form by their reunion a membrane, more or less complete, which holds back the liquid, and prevents it from spreading further.

In order to obtain the dispersion, it suffices to displace the fibres by rubbing the skin with which they are continuous.

The injections of gelatine have the advantage of becoming solid so soon as the animal is cold. One can then make sections which show the different parts of the connective tissue spread out by this new means of dissociation.

The star-like figures (plasmatic cells) exist no longer, and cannot be discerned by colouring with carmine and the action of acetic acid.

What we do observe in these preparations consists of bundles of connective fibres cut across obliquely, or showing their full length; besides these, cells of a fusiform appearance, or resembling those of lamellar pavement-epithelium; lastly, numerous cells which are round, or of an irregular form. Elastic fibres are distinguished also in these preparations; they are rectangular, or slightly curved.

In order to discern the cells, and fully to understand their relations, it is better to add nitrate of silver to the gelatine, according to the method which Chrzonszczewsky has recommended for the study of capillaries (Paris gelatine softened in cold water, and afterwards melted, one part; solution of nitrate of silver, 2 to the 1000, one part). After colouring with carmine and the action of glycerine and formic acid, the cells make their appearance in the neighbourhood of the bundles; they are flat, very irregular in their contour, destitute of membrane, and they contain a flat and ovoid nucleus. When these cells are seen in profile, they appear fusiform, and their nucleus resembles a rod. Certain of these cells present prolongations, some of which appear to be in relationship with similar prolongations arising from neighbouring cells. But this disposition is very rare.

Good preparations may also be obtained by injecting into the cellular tissue a solution of nitrate of silver (1 to 1000). The portion of tissue in contact with the solution becomes slightly paler, and forms a globular mass filled with liquid, as in œdema, and of which portions may be cut off with sharp scissors, which are easily displayed upon a plate of glass. These preparations, treated by the picrocarminate of ammonia, and preserved in glycerine acidified by formic acid, show elements similar to those which are represented in fig. 7.

Swollen bundles of connective tissue may be seen there surrounded by annular fibres coloured with red. This colouring indicates that these fibres are distinct from the elastic fibres which are not coloured by carmine. Flat cells may also be observed, seen in front or in profile; finally, the rounded, or irregular cells. The flat cells are only slightly united to the bundles of connective fibres. The round and irregular cells appear entirely free in the spaces comprised between the bundles.

It is convenient to study these different elements in a neutral liquid, having already recognised them by the aid of the preceding methods. In order to do that, serum must be injected into the subctaneous cellular tissue. In the œdematous parts fragments are to be raised by means of scissors. These placed upon a plate of glass and covered by a smaller one, present a very irregular appearance, and, with the microscope, granular plates which represent the flat cells which have been mentioned above may then be distinguished along the bundles.

Lastly, irregular corpuscles, very much smaller, appear free in the spaces left between the fibres. These corpuscles resemble the white globules of the blood and the embryonic cells. When picrocarminate is added to the preparation, nuclei are discovered in the interior of the granulated plates and in the corpuscles.

The following are the facts which result from the preceding observations:—

1st. Cellular tissue is essentially formed by the connective bundles, elastic fibres, and cells. Neither laminæ nor pores are observed in it, therefore the words laminar tissue and cribriform tissue are incorrect.

2nd. The bundles of connective fibres are cylindrical; they are very variable in diameter; they are limited by a special layer, a sort of membrane of annular fibres or of spiral fibres. These fibres appear to be a simple thickening of the membrane; like the latter, they are coloured by the carmine, and in this respect they differ from elastic fibres.

3rd. All the cells observed in connective tissue are formed by a mass of granular protoplasm; they contain perfectly formed nuclei. They are not all alike. Some are flat, present an irregular appearance and even prolongations; sometimes they are folded and their edges may be turned back; their nuclei, which are ovoid and very much flattened, enclose one or two well defined neuclei. Other cells, less numerous, are globular or irregular in form, and contain spherical nuclei; certain of these last are in every respect similar to the white blood-globules.

4th. These different cells are placed between the connective bundles, but all do not seem to hold the same relation to them.

While the globular cells appear to circulate easily in the spaces left between the bundles, the flat cells, on the contrary, occupy a position along the bundles, which they abandon with greater difficulty; nevertheless teasing out with needles, or otherwise, suffices generally to make them lose their relationship.

The star-like figures (plasmatic cells) which are observed upon the preparations made by section on pieces dried or hardened, can now be easily explained.

In order to obtain a satisfactory result, I advise that a preparation should be made according to the following method:—

Harden in a solution of picric acid, make a section, colour with picro-carminate of ammonia wash, examine in glycerine 100 and formic acid 1. The connective bundles cut across are become transparent, but remain distinct. At the points where several of these bundles touch one another, may generally be observed a corpuscle formed by a red nucleus, slightly shrivelled by the action of the acid, and around it and enclosing it a mass of protoplasm slightly coloured with yellow by the picric acid, having a semicircular form; nucleus and protoplasma are comprised within a space limited by the edge of the red bundles. This edge is very sharp, and this it is which has induced persons to believe in the existence of a cellular partition. Since upon a transverse section of the bundles, these as they touch each other necessarily form outlines disposed as a net-work, it was believed, and I myself believed, in the existence of a canalicular net-work in which the cells were supposed to be included.

IV. It is not to be said on this account that there is not a plasmatic circulation in connective tissue. This may be carried on around the connective bundles in the very dilatable spaces left between them. The presence of cells similar to the white blood-globules or to the lymph-cells, leads me to think, with Recklinghausen, that the plasmatic circulation is a true lymphatic circulation. Moreover, does not the existence in the subcutaneous cellular tissue of these flat cells disposed upon the surface of the bundles, suggest to us the idea of a vast space enclosed in the connective tissue analogous to the serous cavities? An interpretation founded not only upon the facts which I have just demonstrated, but still more upon the experiments of Von Recklinghausen, in which impalpable bodies introduced into the serous cavities have penetrated directly into the lymphatics and from thence into the blood.

With inferior animals, the frog for instance, the vast sacs which are found under the skin, also inclosed in the abdomen, so similar to the bursæ mucosæ of man, are both serous cavities and lymphatic sacs; this is proved by the most decisive experiments. With man and the superior animals, the experimental demostration is not yet complete, but it will doubtless be forthcoming.

The tubes of the tendons, which at first sight present so peculiar a structure, appear to me to have a very general significance; thus, they may be considered, with some reason, as serous cavities in miniature and at the same time as canals for the circulation of the plasma.

The importance of connective tissue in pathological new formations had not escaped Bichat. The fleshy buds and the tumours of the different organs appear to him to have their point of departure in interstitial cellular tissue. He even perceived perfectly that cirrhosis of the liver is only due to the cellular tissue, and that in this disease, the glandular elements are preserved and continue their function. The passage in his 'General Anatomy,' where he speaks of it, is so striking, that I feel called upon to quote it in his own words: "In many organic affections of the liver, steatomatous tumours are observed, which give to this organ a knotted unequal form, and which occupying only the cellular tissue, leave intact the granular tissue, which separates as usual the bile, which experiences no alteration in its flow."

In his important and numerous works Virchow has developed and extended the idea of Bichat, and further he has shown the importance of the cells of connective tissue in the development of pathological new formations.

In making the fleshy growths of the bones proceed from a cellular tissue that he did not see, but admitted because of this fact alone, that fleshy growths are formed in the bones, Bichat has passed beyond the limits of anatomical observation and has committed an error. Virchow has endeavoured to avoid it, in leaving to Bichat's generalization all its breadth. In order to do so, the celebrated professor of

Berlin, relying upon the systematic conception of Reichert, and upon the observations related at the beginning of this work, has identified bony tissue with connective tissue. But at the present time, since the works of H. Müller upon the development of bone, this identification is no longer admissible. The facts related in this memoir show that there are further some fundamental morphological differences between bony tissue and connective tissue. Between the structure of tendons, for example, and that of the bones which are their substitute in birds, there is no analogy, and the cellular elements are grouped in an entirely different manner in these two kinds of tissue. Histologists who maintain that in a chicken the tendons in becoming bony undergo a simple calcareous infiltration, have been deceived by the delusion, caused by transverse sections. I have undertaken to investigate this subject, and the facts at which I have arrived will form the material for another work.

In conclusion, I must remark that the cells of connective tissue have not the fixity which Virchow believed them to have. They are not enclosed in resisting substance; further, they have no truly specific character; in many respects, they resemble the white blood-globules, or embryonic cells, and are free between the bundles of tissue. Also the rapidity with which, under the influence of an irritant, globules of pus are produced in connective tissue, is truly prodigious. In inflammation, the white globules escape from the vessels and disperse themselves in the spaces left between the fibres, but it is easy to observe the division of the pre-existent cells.[1] It also seems that in a slow movement, such as that which is produced around a simple wound, all the work that is done depends upon the constituent cells of the tissue. Moreover, these cells, once modified by irritation, form masses of embryonic tissue, at the expense of which a new tissue will be formed according to the nature of the pathological movements.

If connective tissue easily yields pathological new formations, it is simply because some of these cellules are embryonic and others closely allied to this state. But the property of engendering these new formations does not belong exclusively to the connective tissue. It suffices that the cellular elements of a tissue are able to return to the embryonic state, for them to take a part in certain cases in the formation of pathological tissues.

[1] Dr. Ranvier is here in accord with Professor Stricker, regarding neither Virchow's nor Cohnheim's view as the whole truth.—Ed. Q. J. M. S.

Virchow's law of connective tissue, the source of all pathological new formations, which appeared so general, fails then especially by its want of generality.

REPORT *on certain points connected with the* HISTOLOGY *of* MINUTE BLOOD-VESSELS. By BREVET LIEUTENANT-COLONEL J. J. WOODWARD, *Assistant Surgeon, U. S. Army.*

Having recently been occupied in the critical examination of certain preparations, in the Microscopical Section of the Museum, illustrative of the minute anatomy of the blood-vessels, I have thought that some of them threw so much light on certain points involved in the recent discussions with regard to the doctrine of inflammation, that a short account of them would be of interest, and might perhaps do good service, in connection with the appreciation of the conflicting statements which have appeared in the Medical Journals since the publication of the paper of Dr. J. Cohnheim,[1] on inflammation and suppuration.

Perhaps the observations of Cohnheim must fairly be regarded as elaborations of the previous experiments of Dr. Augustus Waller, but certainly they produced an impression upon the medical world far beyond that made by the papers in the Philosophical Magazine,[2] and more or less complete accounts of the conclusions arrived at by the distinguished Berlin observer have continued to appear, from time to time, in both foreign and American medical journals, ever since the publication of his paper in 1867.

Recently protests against these conclusions have appeared in various quarters, among which particular reference may be made to the paper of Prof. Koloman Balogh, of Pesth, published in 1869,[3] and that of Dr. V. Feltz of Strasbourg,

[1] "Ueber Entzündung und Eiterung;" 'Virchow's Archiv,' Bd. xl., S. 1.

[2] "Microscopical Examination of some of the principal Tissues of the Animal Frame, as observed in the Tongue of the living Frog, Toad," &c.; 'London, Edinburgh and Dublin Philosophical Magazine;' vol. xxix, p. 271 (1846). "Microscopical Observations on the Perforation of the Capillaries by the Corpuscles of the Blood, and on the Origin of Mucous and Pus-Globules;" Ib., p. 397.

[3] "In welchem Verhältnisse steht das Heraustreten der farblosen Blutzellen durch die unversehrten Gefässwandungen zu der Entzündung und Eiterung?" 'Virchow's Archiv,' Bd. xlv, S. 19. Readers inclined to

in 1870.[1] Both these authors have failed to see the white blood corpuscles pass through the coats of the small vessels in the manner described by Cohnheim, and deny the existence of stomata, between the cells of the vascular epithelium, large enough to permit such a wandering to occur.

After I had perused Cohnheim's paper I procured a number of frogs, and having on hand a small quantity of Wourara, the gift of my friend Dr. S. Weir Mitchell of Philadelphia, I carefully repeated many of the experiments described. I received the impression from what I saw that Cohnheim was a most conscientious observer, who had described as faithfully as possible the impressions made upon him. Certainly the results I obtained, by following his methods of producing inflammation in the cornea and mesentery of frogs, could be described in his very language without drawing upon the imagination. Nevertheless my other duties did not leave me sufficient time for an exhaustive research in this difficult domain, and it is far from my present purpose to enter into a critical discussion of the subject. It is simply my desire to offer a brief description, illustrated by Photo-micrographs, of certain preparations in the Microscopical Section of the Museum, which bear upon some of the points involved, and thus to contribute what is in my power towards the important object of arriving at certainty with regard to the facts on which our future theories of inflammation are to rest.

Most of the preparations here referred to are examples of the results attainable by staining the tissues with a dilute solution of the nitrate of silver. This re-agent has been employed for various histological purposes during the last ten years, and has attracted attention especially in connection with the cornea, the various forms of connective tissue, the ultimate branches of the lymphatics and the boundaries of the cells which constitute epithelial surfaces. General attention was first drawn to its use by Dr. F. von Recklinghausen, of Berlin, in 1860,[2] and further particulars were contributed during 1861, by Prof. His, of Basel,[3] who would appear to have already employed the re-agent for several years. In

attach importance to this paper should read the caustic criticism of Dr. Alexis Schklarewski of Moscow. Ib., Bd. xlvi, S. 116.

[1] "Recherches Expériméntales sur le passage des Leucocytes à travers les parois vasculaires;" 'Journal de l'Anatomie et de la Physiologie,' Jan. & Feb., 1870, p. 33.

[2] "Eine Methode, mikroskopische hohle and solide Gebilde von einander zu unterscheiden;" 'Virchow's Archiv,' Bd. xix, S. 451.

[3] "Ueber das Verhalten des Salpetersauren Silberoxyds zu thierischen Gewebsbestandtheilen;" Ib., Bd. xx, S. 207.

1862 Von Recklinghausen published his work on the lymphatics,[1] which contains a detailed account of many elaborate experiments with regard to the action of silver solutions on the tissues, and in 1863 Dr. Ernst Oedmanson, of Stockholm,[2] gave a description of their behaviour when applied to epithelial surfaces, and described and figured the so-called stomata which play so important a part in the theory of Cohnheim. During 1865 and 1866 the epithelium of the capillary blood-vessels, as shown by silver, was described by several observers, among whom Dr. N. Chrzonszczewsky, of Charkow,[3] may be particularly mentioned.

The perusal of these papers led me to make a number of experiments myself, and to have others made by my assistants at the Museum, the results of which are now to be described.

If a dilute solution of nitrate of silver is brushed over a clean epithelial surface taken from a recently killed animal, and the tissue after washing with distilled water is exposed for a short time to the action of sunlight, it will be found on microscopical examination that a brownish-black precipitate of silver has been produced at the boundaries of the epithelial cells, while the cells themselves are comparatively but little stained, or if the manipulation has been carefully conducted, are not stained at all. For this purpose I have most frequently employed, at the Museum, a solution made by dissolving one part of crystallized nitrate of silver in four hundred parts of distilled water, but considerable variation on either side of this strength does not much modify the result, provided the solution is well washed off before the tissue is exposed to the light.

If the same solution is injected into the blood-vessels, the lining epithelium is handsomely mapped out in all those membranous and superficial parts in which a ready exposure to the action of light is practicable, and although in the parenchymatous organs, such as the liver, the spleen, the

[1] 'Die Lymphgefässe und ihre Beziehung zum Bindegewebe;' Berlin, 1862.

[2] "Beitrag zur Lehre von dem Epithel;" 'Virchow's Archiv,' Bd. xxviii, S. 361.

[3] "Ueber die feinere Structur der Blutcapillaren;" ib., Bd. xxxv, S. 169. C. J. Eberth in his article on the blood vessels in Stricker's Handbook ('Handbuch der Lehre von den Geweben,' &c. Leipsic, 1869. II. Lief: S. 202) enumerates the following microscopists as having described the epithelium of the capillaries prior to Chrzonszczewsky. *Hoyer.* 'Archiv. für Anatomie,' Jan. 18, 1865. *Auerbach.* 'Breslauer Zeitung,' Feb. 17, 1865. *Eberth.* 'Sitzungsberichte der physikal. med. Gesellschaft zu Würzburg,' Feb. 18, 1865. 'Medicinisches Centralblatt,' No. 13, 1865. 'Würzburger Naturwissenschaftliche Zeitschrift;' Bd. vi, 1866. *Aeby.* Medicinisches Centralblatt. No. 14, 1865.

kidneys, &c., the juices of the tissues are apt to interfere with the reaction, yet even here occasional success may be attained. In practice it is often found advantageous to combine the silver solution, intended for injection, with a certain amount of gelatine, by which the blood-vessels are kept handsomely distended, and the beauty of the preparation is much increased. This plan was proposed by Chrzonszczewsky in the paper already quoted. His formula, which I have found to work well, is as follows: Half an ounce of fine gelatine is dissolved in four ounces of distilled water, and to this is added a solution of one scruple of nitrate of silver in two fluid drachms of distilled water. After injecting with this fluid, the tissue is exposed to the light precisely as after the use of the simple silver solution.

There are preserved in the Microscopical Section of the Museum a number of silver stainings in which the epithelium has been thus mapped out on the skin, the peritoneum, the lymphatic-sacs of frogs, and the blood-vessels. These preparations, after the action of the silver, have been mounted in Canada balsam with or without the previous staining of the nuclei with carmine. The detailed steps of the process may prove useful to some readers. The silver staining having been successfully accomplished, the nuclei are tinted preferably by the solution of carmine in borax, described by Thiersch in his work on epithelial cancer.[1] It is prepared as follows: Four parts of borax are dissolved in fifty-six parts of distilled water and one part of carmine added to the solution; one volume of this fluid is mixed with two volumes of absolute alcohol, and after crystals have formed the mixture is filtered. The filtrate may be used for staining, but if the crystals of carmine and borax which remain on the filter are dissolved in a small quantity of distilled water, I find the solution thus obtained answers a still better purpose. The portion of tissue to be studied is soaked in this solution until coloured deep red. It is afterwards treated with a saturated solution of oxalic acid in alcohol, by which all colour is gradually removed except from the nuclei. So soon as this is accomplished the piece is to be carefully washed in alcohol, then soaked in absolute alcohol, and finally mounted in a solution of dried Canada balsam in chloroform or benzole. The treatment by oxalic acid, subsequently to the action of the carmine-borax solution, has the additional effect of altering the purplish-red colour, derived from that fluid, to the brilliant hue obtained by the use of the ammoniacal solution of carmine ordinarily employed in histology. The latter has

[1] 'Der Epithelialkrebs.' Leipzic, 1865, S. 92.

the disadvantage of being apt to dissolve out the previously produced silver staining, an annoyance completely avoided by the carmine-borax solution.

Preparations carefully made by the above process closely resemble the fresh tissues, as they appear after staining if immersed in glycerine or syrup; they are somewhat more transparent, but not inconveniently so, and possess the great advantage of keeping unchanged for an indefinite period of time.

After these preliminary remarks, I proceed at once to the description of the photographs.[1]

1. Photograph representing several venous radicles uniting to form a small vein in the muscular coat of the urinary bladder of the frog. Negative No. 102, New Series. From preparation No. 3378, Microscopical Section. Magnified 400 diameters by Wales's $\frac{1}{8}$th objective, illuminated by the magnesium lamp. The preparation was made by Dr. J. C. W. Kennon.

The principal venous trunk represented in this photograph is 1-400th of an inch in diameter. It is formed by the union of three smaller radicles, of which that on the left hand is much out of focus. Another smaller radicle, also much out of focus, joins the trunk on the left, near the bottom of the picture. The walls of the venous trunk, and of those of its branches which are in focus, are plainly seen to be formed of somewhat irregular epithelial cells, which vary in shape and size, averaging 1-500th of an inch in length, and 1-2200th in breadth. The boundary of each cell is indicated by a zigzag black line. In each of the cells which is accurately in focus, a smooth oval nucleus, 1-2800th of an inch in length, is visible. In examining the original preparation, by changes in the fine adjustment of the microscope, similar nuclei can be seen in each of the epithelial cells. These nuclei, being brilliantly stained with carmine, contrast sharply with the black cell-boundaries resulting from the silver imbibition. By a still further alteration of the fine adjustment, the cells and nuclei of the opposite side of the vein are brought into view.

In the tissue external to the vein, two kinds of nuclei are shown in the photograph. The first are narrow and elongated, averaging about 1-1500th of an inch in length, and 1-9000th in breadth. These are the nuclei of the fibre-cells of the muscular coat of the bladder. The fibre-cells themselves are

[1] Copies of these photographs have been kindly sent to the Editors of this Journal by Colonel Woodward. They will be forwarded for inspection to any microscopical society or club making application.

not shown in carmine stainings, but are readily demonstrated in fresh preparations by the action of solutions of osmic acid,

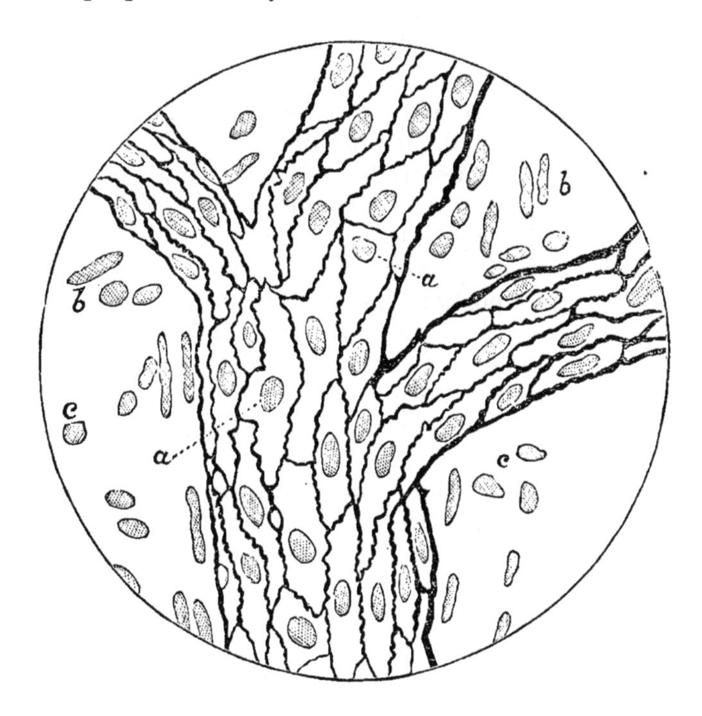

or of chloride of gold. Indications of the muscular bands formed by the union of these fibre-cells are, however, seen in the photograph, particularly on each side of the principal venous trunk. The second variety of nuclei are oval, about 1-3000th of an inch long, and belong to the connective tissue of the bladder. The cells in which these nuclei lie are not seen, the action of the carmine being limited to the nuclei. They can, however, readily be demonstrated in fresh preparations by gold-chloride and some other reagents. The cut represents the outlines of a portion of the photograph; a, a, are the nuclei of the vascular epithelium; b, b, the nuclei of the muscular fibre-cells; c, c, those of the connective tissue.

II. Photograph representing a small vein from another portion of the same preparation. Negative No. 195, New Series. Magnified 1000 diameters by Powell and Lealand's immersion 1-16th objective, illuminated by the Magnesium lamp. One of the epithelial cells near the centre of the vessel is particularly well defined, and shows its nucleus handsomely; the adjacent cells, not lying in the same plain, are many of them partly out of focus, but their boundaries can readily be traced, and the nuclei of several of them are well defined.

Four black spots, seen in the course of the vessel, are blood-corpuscles much out of focus.

III. Photograph representing the stomata between the epithelial cells of a vein 1-50th of an inch in diameter in the mesentery of the frog. Negative No. 40, New Series. From preparation No. 3276, Microscopical Section. Magnified 400 diameters by Wales's $\frac{1}{8}$th objective. The preparation was made by myself.

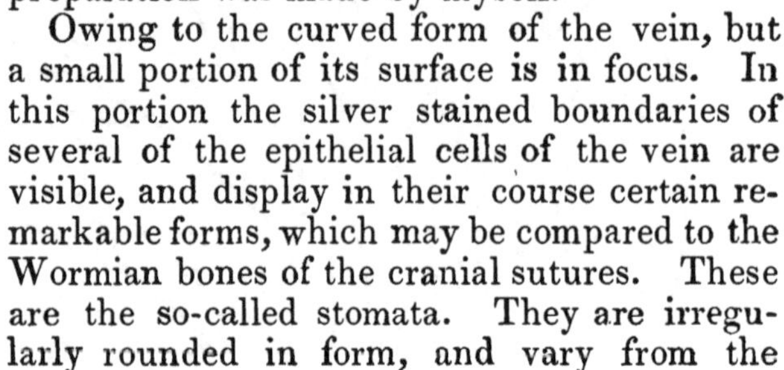

Owing to the curved form of the vein, but a small portion of its surface is in focus. In this portion the silver stained boundaries of several of the epithelial cells of the vein are visible, and display in their course certain remarkable forms, which may be compared to the Wormian bones of the cranial sutures. These are the so-called stomata. They are irregularly rounded in form, and vary from the 1-10000th to the 1-4000th of an inch in long diameter. Those shown in the photograph present a clear central space, bounded by a sharp black outline, which is sometimes even thicker than the boundaries of the cells themselves. The nuclei of the epithelial cells are not shown. The cut exhibits one of these cells, a, with portions of the boundaries of adjacent cells, b, b, b, and the stomata, c, c, c.

IV. Photograph representing the stomata between the epithelial cells of a vein, 1-100th of an inch in diameter, in the mesentery of the frog. Negative No. 224, New Series. From preparation No. 3062, Microscopical Section. Magnified 400 diameters by Wales's $\frac{1}{8}$th objective. The preparation was made by Dr. Kennon.

The vein having collapsed, the epithelial cells of the lower wall come into focus in places, and so somewhat complicate the representation. The stomata are abundantly present, but none of them equal in size the largest shown in the last photograph; in several places, moreover, black spots, similar to the other stomata in shape and size, may be observed in the cell boundaries.

V. Photograph representing the stomata of a vein, 1-1000th of an inch in diameter, in the mesentery of the frog. Negative No. 194, New Series. From preparation No. 3276, Microscopical Section. Magnified 400 diameters by Wales's $\frac{1}{8}$th objective, illuminated by the Magnesium lamp. The small vein represented comes well into view, while the nuclei of the surrounding tissue are seen out of focus on each side of it. Several of the stomata present clear centres, while others are black and opaque throughout; they average 1-10000th of an inch in diameter.

VI. Photograph representing a minute artery, with part of the adjoining network of capillaries, from the muscular

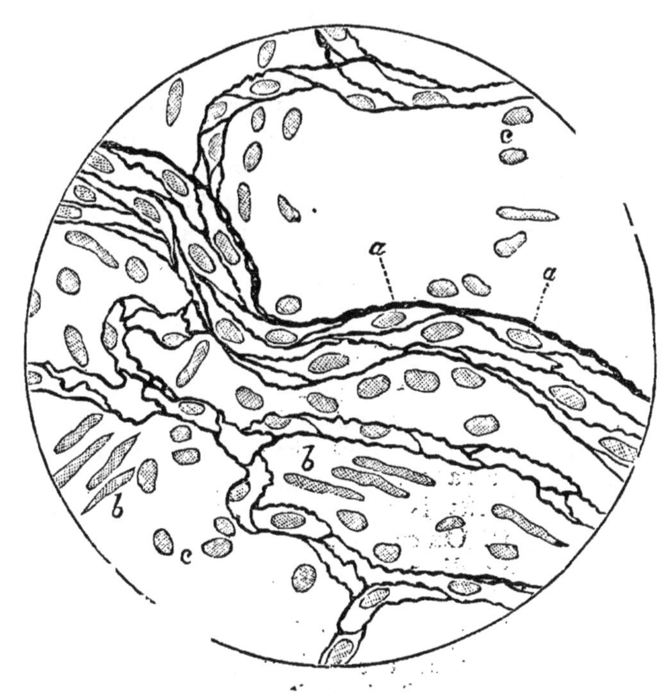

coat of the urinary bladder of the frog. Negative No. 220, New Series. From preparation No. 3378, Microscopic Section. Magnified 400 diameters by Wales's $\frac{1}{8}$th objective. The field is crossed by a small artery, 1-1700th of an inch in diameter. Its epithelial cells are longer in proportion to their width than those of the veins. They average 1-400th of an inch in length, and have nuclei similar to those of the venous epithelium. Wherever the capillaries come into focus the epithelium of their walls is also plainly shown. In the intervascular spaces the nuclei of the muscle and connective tissue appear as in the first photograph. The cut presents an outline of a part of the picture; a, a, are the nuclei of the vascular epithelium; b, b, those of the muscle; c, c, those of the connective tissue.

VII. Photograph representing a portion of the view presented by the last picture. Negative No. 223, New Series. Magnified 1000 diameters by Powell and Leland's immersion 1-16th objective. The spindle-shaped forms of the epithelial cells of the arteries and the characters of the nuclei are plainly shown.

VIII. Photograph representing a small artery in the mesentery of the frog. Negative No. 184, New Series. From preparation No. 3267, Microscopic Section. Magnified 500 diameters by Wales's $\frac{1}{8}$th objective, illuminated by Calcium light. The preparation was made by myself. The artery shown measures 1-280th of an inch in diameter. It is marked by both transverse and longitudinal silver lines. The former are exterior to the latter, as is readily demonstrated in the preparation by the use of the fine adjustment. The longitudinal lines belong to the epithelium, while the transverse markings indicate the boudaries of the circular fibre-cells of the muscular coat, which are usually mapped out in this manner in arteries of moderate size, the silver solution reaching them by imbibition. The epithelial cells of arteries of this size are narrower in proportion to their length than those of smaller twigs, such as that shown in the last two photographs. In the photograph it is somewhat difficult, in many places, to make out their boundaries, as the margin of the cells of the opposite wall come into focus and complicate the appearance. In the study of the original preparation this difficulty is readily overcome by manipulating the fine adjustment. On each side of the artery, the numerous nuclei of the surrounding tissue come more or less distinctly into view.

IX. Photograph representing the epithelium of a capillary, in the muscular coat of the urinary bladder of the frog. Negative No. 216, New Series. From preparation No. 3378, Microscopical Section. Magnified 1000 diameters by Powell and Leland's immersion 1-16th objective, illuminated by the Calcium light. The capillary which crosses the centre of the field measures 1-2300ths of an inch in diameter. The epithelial cells are narrower in proportion to their length than those of the veins. Their nuclei are quite similar to those of the venous epithelium.

The foregoing description of individual photographs will serve to give a correct idea of the epithelium lining the small arteries, veins, and capillaries, as shown in a considerable number of preparations preserved in the Museum, and as observed by me many times in tissues extemporaneously prepared. Both Balogh and Feltz would seem to have been singularly unfortunate in their silver stainings, for they describe the appearances produced as irregular and contradictory. Balogh explains the black lines he occasionally saw in the vessels after silver injections as due to the precipitate of silver occurring preferably on folds in the lining membrane, caused by the irregular shrinkage of the vessel produced by the silver injection, an error readily corrected by combining gelatine with the solution of silver injected, the vessels are thus equaly and smoothly distended, yet the epithelium appears mapped out as usual.

Feltz asserts that if a solution of silver be allowed to dry in the light on a collodium film, irregular black lines are produced, quite like those observed after its action on organic membranes. I myself have examined the irregular figures produced by this experiment, and cannot conceive how any one accustomed to the precise study of organic forms can see any similarity between them and the definite outlines produced by the action of silver solutions on epithelial surfaces.

Besides the preparations exhibiting the vascular epithelium which have been described, the Museum possesses, as I have already mentioned, a number in which the epithelium of the skin, of the lymph-sacs of frogs, of the peritonæum and of other surfaces, are mapped out by silver staining, and the reagent is continually employed by myself and my assistants in the investigation of such surfaces. It is impossible for any one who has had such opportunites for observation, to avoid being struck by the fact that the outlines obtained have a definite form and character for each tissue. It is true that the silver staining does not succeed as frequently as carmine staining does, that its use requires more skill and that failures are more frequent. Sometimes too much action takes place and everything is obscured by the black precipitate produced; sometimes either because the tissues are not fresh, or the light not sufficient, or from some unexplained reason, the solution does not act at all; but the forms above described as characteristic of the arteries are never observed in the veins, never do the outlines produced on the surface of the skin resemble those seen on the peritonæum, in the lymph-sacs, or in the vessels; each membrane permits only the formation of its own characteristic outlines, never of those belonging to another tissue; moreover, in all cases where it is possible to observe the shape of the epithelial cells without the use of reagents, or to isolate them, the forms thus ascertained correspond precisely with those mapped out by the siver solution, and when, after the action of silver, carmine staining is resorted to, the nuclei thus made visible correspond in position to the places they ought to occupy, if in fact the silver had mapped out the cell-boundaries as I certainly believe it does. Whether the discoloration is in the cell wall, or in the cement or matrix by which the adjacent cells are held together, is a more difficult question, and one into which I do not propose to enter at the present time. It is enough for the purposes of this paper that the peripheries of the cells or the substance just external to them, exhibits a much more speedy and intense reaction with the nitrate than the cell contents do, and must therefore differ more or less from these in composition.

Having arrived at this conclusion with regard to the general interpretation of the action of silver on epithelial surfaces, the question of the true meaning of the so-called stomata next demands consideration. They are to be observed most abundantly, as may be inferred from the photographs described, in veins of moderate size. I have found them largest and most numerous in veins 1-50th of an inch in diameter or even larger, and they become smaller and rarer in smaller branches. They are comparatively infrequent in the capillaries and still more so in the small arteries; the Museum, however, posseses preparations showing them in both. I have, moreover, concluded, from my own observations, that in number and size they vary in vessels of the same dimensions in different parts of the body. Thus, for example, in the veins of the mesentery of the frog they are larger and more abundant than in veins of the like dimensions in the urinary bladder of the same animal.

In figure they are rounded, oval or oblong. I have measured them as large as 1-4000th of an inch in diameter, but smaller ones 1-5000th to 1-6000th of an inch are more common, and the smallest and most frequent do not exceed 1-10000th of an inch. Sometimes they present clear centres sharply mapped out by black boundaries, sometimes forms of the same size and character are opaque and black throughout, and this has been interpreted as due to variations in the composition of the fluid by which the opening is occupied, which sometimes precipitates the silver solution while at other times it does not, and the action is limited to the solid margins of the pore. They are almost invariably found in the marginal line between adjacent epithelial cells, and the rare cases in which I have observed them apparently in the cells themselves, are probably to be explained by the adjacent margins having from some cause escaped the influence of the silver salt. From my study of these peculiar inter-cellular forms, I am inclined to regard with favour the opinion that they are actual openings in the epithelial layer. It may aid others in arriving at a conclusion on the subject, if I here present a photograph of the stomata in the external epithelium of the skin of the frog through which, as is well known, a rapid transudation of liquid habitually occurs.

X. Photograph representing a silver staining of the external epithelium of the frog's skin. Negative No. 22, New Series. From preparation 3036, Microscopical Section. Magnified 400 diameters by Wales's $\frac{1}{8}$th. The preparation was made by myself. The epithelium of this surface consists of a number of layers, and the silver has penetrated in different portions of the skin to various depths. In the photograph the epithelial cells of the upper surface are sharply mapped out, while the boundaries of the cells of several of the deeper layers are seen out of focus beyond. The cells are hexagonal in shape, and average 1-1300th of an inch in diameter. Many of the nuclei have been somewhat tinted by the silver, a circumstance which is not unfrequent if the silver action is intense. In the boundaries of the epithelial cells may be seen very many little rings, with black margins and clear centres, averaging 1-5000th of an inch in diameter, and also many similar forms, of the same size and occupying like positions, which are quite black and opaque throughout. In some parts of the preparation, from which the photograph was taken, almost all the rings are black and opaque, while in other portions almost all present clear centres. The view which regards these rings as true pores certainly appears to

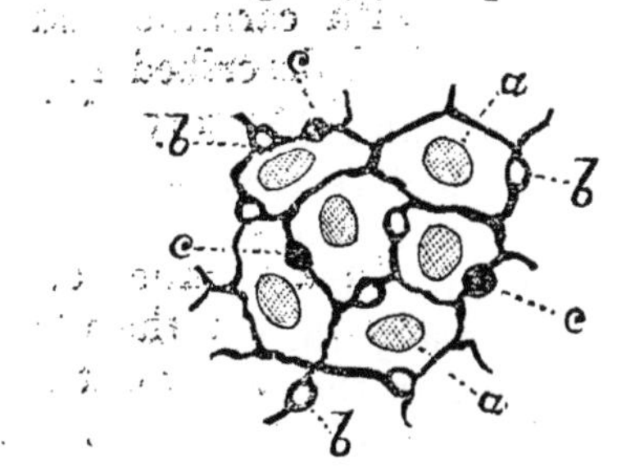

me to require fewer suppositions than any other. The cut represents an outline of a portion of this photograph; a, a, the nuclei of the epithelial cells; b, b, the stomata; c, c, stomata which have become black and opaque throughout. It has been urged, however, by Balogh, that even if the stomata described in the vascular epithelium are admitted as such they are not large enough to permit the passage of the white blood-corpuscles which, as is well known, average about 1-3000th of an inch in diameter. But even if we discard the supposition that the pores may be stretched open and made larger by the distended condition of the vessels of inflamed parts, there appears to me no difficulty in understanding how a white corpuscle might pass through the smallest of the stomata I have described. An opening 1-10000th of an inch in diameter is only a little less than one third the average diameter of the white corpuscles, and any one who has seen the extraordinary modifications of form which these little masses of protoplasm undergo in the course of their so-called "amœboid movements," would readily credit their capability of passing through such apertures. As the amœboid movement does not occur in the white corpuscles while rolled along in the torrent of the circulation, but only when the movement of the blood is arrested more or less completely, the fact that large numbers of white corpuscles do not habitually pass through the vascular walls into the tissues will not militate against the notion of patulous orifices. That a passage of the white blood-corpuscles through the vascular walls does actually occur, is shown by the next picture.

XI. Photograph representing white corpuscles in various phases of the amœboid movement, in the external coat of a small vein of the muscular coat of the stomach of a mare. Negative No. 46, New Series. From preparation No. 3382, Microscopical Section. Magnified 400 diameters by Wales's $\frac{1}{8}$th objective. The preparation was made by Dr. E. M. Schæffer.

This preparation is one of a number of sections made from the stomach of a mare dead of gastroenteritis. In these sections, which are stained with carmine and mounted in Canada balsam after the method described in the early portion of this paper, it was found that many of the small veins of the sub-peritoneal connective tissue and of the muscular coat were surrounded by white corpuscles fixed in all stages of the amœboid movement. In a number of places where the sections pass transversely through the veins, the white corpuscles can be observed in the interior of the vein, and in the vascular walls as in the adjacent tissue. The series of preparations gives a satisfactory demonstration of the wandering of the white corpuscles. I have made efforts to preserve the frog's mesentery permanently, in a number of the cases in which I have observed the same process in that membrane, but hitherto without success.

It will be seen from the forgoing details that, so far as the structure of the vascular walls and the passage of the white corpuscles through them are concerned, the facts appear to be on the side of Cohnheim. How then with regard to the doctrine of inflammation which he builds upon these facts and upon his corneal studies? Does the creeping out of the white corpuscles constitute the essence of the inflammatory process? Do these little moveable masses of living protoplasm furnish the germs for the elements of new formations? Have pus-corpuscles no other origin? Are the processes which go on in the cells of the inflamed tissue purely passive, mere phenomena of retrograde metamorphosis?

I find the evidence insufficient as yet to afford satisfactory answers to such questions. The observations made by Cohnheim on the connective tissue corpuscles of the tongue of the frog are not conclusive in themselves, and Stricker's studies on the same subject[1] show the necessity of further labour in this direction before the possible multiplication of these elements in inflammation can be denied. As to the doctrine that the white corpuscles, after their escape from the blood-vessels, are transformed into the elements of normal or pathological tissues, the facts hitherto brought forward can scarcely be said to do more than raise it to the rank of an ingenious hypothesis. The actual steps of this transformation, if it does occur, have yet to be observed.

In conclusion, I may remark that, as the preparations referred to in this paper form a portion of the Microscopical Collection of the Museum, they can be examined by any professional microscopist who may visit that institution.

The photographs which accompany this paper were prepared by myself in accordance with the methods described on former occasions.

On an APLANATIC SEARCHER, *and its* EFFECTS *in* IMPROVING HIGH-POWER DEFINITION *in the* MICROSCOPE. By G. W. ROYSTON-PIGOTT, M.A., M.D. Cantab., M.R.C.P., F.R.A.S., F.C.P.S., formerly Fellow of St. Peter's College, Cambridge.[2]

THE Aplanatic Searcher is intended to improve the penetration, amplify magnifying power, intensify definition, and raise the objective somewhat further from its dangerous proximity to the delicate covering-glass indispensable to the observation of objects under very high powers.

The inquiry into the practicability of improving the performance of microscopic object-glasses of the very finest known quality was suggested by an accidental resolution in 1862 of the Podura markings into black beads. This led to

[1] 'Studien aus dem Institute für Experimentelle Pathologie in Wien aus dem Jahre, 1869.' Wien, 1870.

[2] Dr. Royston-Pigott has communicated to us a valuable paper and drawings describing this instrument and its performances, which the pressure of matter compels us to postpone until January,—ED. Q. J. M. S.

a search for the cause of defective definition, if any existed. A variety of first-class objectives, from the $\frac{1}{16}$ to the $\frac{1}{4}$, failed to show the beading, although most carefully constructed by Messrs. Powell and Lealand.

Experiments having been instituted on the nature of the errors, it was found that the instrument required a better distribution of power; instead of depending upon the deepest eyepieces and most powerful objectives hitherto constructed, that better effects could be produced by regulating a more gradual bending or refraction of the excentrical rays emanating from a brilliant microscopic origin of light.

It then appeared that delusive images, which the writer has ventured to name *eidola*,[1] exist in close proximity to the best focal point (where the least circle of confusion finds its locus).

I. That these images, possessing extraordinary characters, exist principally above or below the best focal point, according as the objective spherical aberration is positive or negative.

II. That test-images may be formed of a high order of delicacy and accurate portraiture in *miniature*, by employing an objective of twice the focal depth, or, rather, half the focal length of the observing objective.

III. That such test-images (which may be obtained conveniently two thousand times less than a known original) are formed (under precautions) with a remarkable freedon from aberration, which appears to be reduced in the miniature to a *minimum*.

IV. The beauty or indistinctness with which they are displayed (especially on the immersion system) is a marvellous test of the correction of the observing objective, but an indifferent one of the image-forming objective used to produce the testing miniature.

These results enable the observer to compare the known with the unknown. By observing a variety of brilliant images of known objects, as gauze, lace, an ivory thermometer, and sparkles of mercury, all formed in the focus of the objective to be tested with the microscope properly adjusted so that the axes of the two objectives may be coincident, and their corrections suitably manipulated, it is practicable to compare known delusions with suspected phenomena.

It was then observed (by means of such appliances) that the aberration developed by high-power eyepieces and a lengthened tube followed a peculiar law.

[1] From εἴδωλον, a false spectral image.

A. A lengthened tube increased aberration faster than it gained power (roughly the aberration varied as v^2, while the power varied as v).

B. As the image was formed by the objective at points nearer to it than the *standard distance of nine inches*, for which the best English glasses are corrected, the writer found the aberration diminished faster than the power was lost, by shortening the body of the instrument.

C. The aberration became negatively affected, and required a positive compensation.

D. Frequent consideration of the equations for aplanatism suggested the idea of searching the axis of the instrument for aplanatic foci, and that many such foci would probably be found to exist in proportion to the number of terms in the equations (involving curvatures and positions).

E. The law was then ascertained that power could be raised, and definition intensified, by positively correcting the searching lenses in proportion as they approached the objective, at the same time applying a similar correction to the observing objective.

The chief results hitherto obtained may be thus summarised.

The writer measured the distance gained by the aplanatic searcher, whilst observing with a half-inch objective with a power of seven hundred diameters, and found it *two tenths of an inch increase;* so that optical penetration was attainable with this high power through plate-glass nearly one quarter of an inch thick, whilst *visual* focal depth was proportionably increased.

The aplanatic searcher increases the power of the microscope from two and a half to five times the usual power obtained with a third or C eyepiece of one inch focal length. The eighth thus acquires the power of a twenty-fifth, the penetration of a one fourth. And at the same time the lowest possible eyepiece (3-inch focus) is substituted for the deep eyepiece formed of minute lenses, and guarded with a minutely perforated cap. The writer lately exhibited to Messrs. Powell and Lealand a brilliant definition, under a power of four thousand diameters, with their new "eighth immersion" lens, by means of the searcher and low eyepiece.

The traverse of the aplanatic searcher introduces remarkable chromatic corrections displayed in the unexpected colouring developed in microscopic test-objects.[1]

[1] Alluded to by Mr. Reade, F.R.S., in the 'Popular Science Review,' for April, 1870.

The singular properties or, rather, phenomena shown by eidola, enable the practised observer in many cases to distinguish between true and delusive appearances, especially when aided by the aberrameter applied to the objective to display excentrical aberration by cutting off excentrical rays.

Eidola are symmetrically placed on each side of the best focal point, as ascertained by the aberrameter when the compensations have attained a delicate balance of opposite corrections.

If the beading, for instance, of a test-object exists in two contiguous parallel planes, the eidola of one set is commingled with the true image of the other. But the upper or lower set may be separately displayed, either by depressing the false eidola of the lower stratum, or elevating the eidola of the upper; for when the eidola of two contiguous strata are intermingled, correct definition is impossible so long as the aperture of the objective remains considerable.

One other result accrues: when an objective, otherwise excellent, cannot be further corrected, the component glasses being already closely screwed up together, a further correction can be applied by means of the adjustments of the aplanatic searcher itself, all of which are essentially conjugate with the actions of the objective and the variable positions of its component lenses; so that if δx be the traversing movements of the objective lenses, δv that of the searcher, F the focal distance of the image from the objective when δx vanishes, f the focal distance of the virtual image formed by the facet lenses of the objective,

$$\frac{\delta v}{\delta x} = -\left(\frac{F}{\;}\right)^2.$$

The *appendix* refers to plates illustrating the mechanical arrangements for the discrimination of eidol and true images, and for traversing the lenses of the aplanatic searcher.

The plates also show the course of the optical pencils, spurious discs of residuary aberration and imperfect definition, as well as some examples of "high-power resolution" of the Podura and Lepisma beading, as well as the amount of amplification obtained by Camera-Lucida outline drawings of a given scale.—*From the Proceedings of the Royal Society.*

On two New Genera *of* Alcyonoid Corals, *taken in the recent* Expedition *of the* Yacht Norna *off the* Coast *of* Spain *and* Portugal.

By W. Saville Kent, F.G.S., F.R.M.S. of the Geological Department, British Museum.

(Plate XXI.)

Fam. ALCYONIDÆ, *M.-Edw.*

Sub-Fam. Cornularinæ, ib.

Gen. nov. *Gymnosarca.*

Corallum of firm consistence, its surface smooth; attached by its base and partially incrusting; throwing off free cylindrical stolons, which give rise to solitary elevated polyp-cells. Stolons occasionally branching, coalescing with each other, or forming a bond of attachment with any foreign object with which it comes in contact. Polyp-cells cylindrical; animals semi-retractile, forming an ovate head when most retracted.

G. bathybius, Pl. XXI, figs. 1—4.

Diameter of the cylindrical stolons one eighth of an inch; height of polyp-cells one tenth of an inch, of the partially retracted polypes one eighth of an inch. Colour of the general cœnenchyma pale yellow, of the polypes and polyp-cells flesh-pink. Spicula of the cœnenchyma colourless, more or less irregularly fusiform and echinate, occasionally slightly branching; those of the tentacular region of the same type mixed with a few more slender arcuate forms.

Hab. Off the coast of Cezimbra, Portugal, at a depth of 500 fathoms. Parasitically attached to *Lophohelia protifera.*

The firm consistence of the cœnenchyma of this coral is due to the innumerable colourless, irregular, echinate spicula, which occur in profusion throughout its substance, but which lying beneath its surface are not visible exteriorly. The parasites of spicula of the same type mixed with slender arcuate forms, met with in the tentacular region, and where, owing to the tenuity of the investing membrane, they are eminently conspicuous, sufficiently accounts for the inability of the polypes to withdraw for protection within the cavities of the polyp-cells; a fact also serving to demonstrate that the generic or family characteristics based upon the possession or non-possession by the animals of this capability are supported by evidence of histological structure, in addition to that afforded by mere external appearances.

The coral just described possesses a certain external resemblance to *Cornularia crassa* of Milne-Edwards, in which species, however, the animal is completely retractile within the tubular polyp-cells. The specimens on being taken on board the yacht, though immediately immersed in fresh sea-water, refused to expand their tentacles or to evince any signs of life, and I was consequently unable to investigate its structure as thoroughly as that of the species to be next introduced.

Fam. LEMNALIADÆ, *J. E. Gray.*

Gen. nov. *Cereopsis.*

Corallum clavate, attached by a somewhat expanding base. The lower portion of the stem barren, the upper one slightly lobate, bearing scattered semi-retractile polypes. Polyp-cells cylindrical; heads of polypes nearly globular when most contracted.

C. Bocagei, Pl. XXI, figs. 5—13.

Colour of the general surface of corallum cream-yellow, of the individual polypes bright red.

Height of the larger specimen of the two taken one inch; average length of the polyp-cells, independent of the expanded tentacles, one tenth of an inch. Spicula interspersed throughout the corallum of three kinds; those occupying the base of the tentacular region and a portion of the tube closely set, bright transparent red, arcuate or irregular and attenuate, fusiform and echinate; immersed in the substance of the upper portion of the tentacles, minute, scattered, fusiform, irregularly echinate spicula of the same colour; general cœnenchyma crowded with unsymmetrically disposed, colourless, irregularly tuberculate and echinate spicula, the majority of which are stouter but not so long as the coloured forms occupying the tentacular region.

Hab. Mouth of the river Sado, near Setubal, Portugal, taken with the dredge at a depth of fifteen fathoms, attached to dead valves of *Cardium ciliare.*

The comparatively shallow water in which the examples of this last species were taken enabled me to preserve it for some time in the living condition, and to study successfully the structure and appearances then made manifest.

The specimen figured was plaeed in a glass receptacle, with fresh sea-water, immediately upon being discovered among the results of the dredge.[1] At first, the polypes were

[1] I must here acknowledge my indebtedness to Col. Stuart Wortley, whose success in the management of aquaria almost rivals his brilliant

very shy, and refused to fully expand themselves, but after being kept for a few days, the water being continually changed, they appear to have become accustomed to the novel external conditions and unfurled themselves in all directions. Their resemblance then, under the microscope, to some beautiful trumpet-shaped flower of the genus *Cereus*, belonging to the cactus tribe, was most striking. Pl. XXI, fig. 6, though very far from doing justice to their graceful appearance, representing two of these polypes, the one entirely and the other only partially expanded. It was only under such conditions that the relative positions and arrangement of the various forms of spicula could be appreciated; most marked among these were the bright red ones of the basal tentacular region, and which forming dense triangular fascicles, as illustrated in the plate, imparted to the animals the characteristic colour conspicuous even to the unassisted eye. Another circumstance attracting attention were the minor pinnate divisions of the tentacles themselves; these in life are cylindrical and capable of independent motion, and seem, in fact, to fulfil the part of perfect though miniature tentacular organs. The exquisitely transparent tubular body readily permitted the discernment of the enclosed alimentary canal with the dependent filamentous ovaries, as also the circulation of the contained fluids.

This last species, in general outward appearance, would seem to closely approach various representatives of the genus *Alcyonium* proper, and more particularly *A. stellatum* (M.-Edw.). In that genus, however, the polypes are always completely retractile, and their non-retractility, in conjunction with the solitary disposition of the polyp-cells, in my species, renders it requisite to establish a new genus for its reception.

achievements in photography, for the loan of numerous glass receptacles specially adapted for the preservation of living marine organisms, and but for which I should have been deprived of the opportunity of carrying out more than one half of the investigations the expedition afforded me the gratification of prosecuting.

On some Compounds *derived from the* Colouring Matter *of* Blood. By H. C. Sorby, F.R.S., &c.

Several months ago I determined to thoroughly investigate the colouring matter of blood, in order to be able to apply to the detection of blood-stains all the improved methods of experiment which I had devised since the publication of my first paper on that subject.[1] I anticipated being able to finish this inquiry in time to publish an account of my observations in this number of the 'Quarterly Journal of Microscopical Science,' but so many unexpected phenomena presented themselves that I have not yet been able to clear up several important questions. However, not wishing to entirely postpone the subject for another three months, I will now give a short account of some of my conclusions.

In the first place, I have found that one of the chief substances met with in moderately old blood-stains, previously described by me as "brown cruorine," is identical with that formed by adding nitrites to blood, as discovered by Dr. Gamgee.[2]

Hoppe-Seyler's [3] and Preyer's [4] "methæmoglobin" is also in some cases this same compound, but in others was perhaps chiefly a soluble, almost neutral, hæmatin. They considered it to be characterised merely by the position of an absorption-band in the red of its spectrum, not being acquainted with the fact that the position of the chief band in case of acid hæmatin varies to a remarkable extent, according to the strength of the free acid, and not having observed several of the more important peculiarities of the well-defined compound to which I now draw attention. Dr. Gamgee thought it due to a *combination* of hæmoglobin with nitrites as such, but I have been led to conclude that they do not combine with it, but act like many other weak oxidizing reagents. Nitrite of potash is, indeed, a most useful substance in studying the spectra of many colouring matters. Usually it has no effect when added to a solution made alkaline with ammonia, but produces most interesting changes when the solution is neutral or slightly acid. This is just what occurs in the case of blood. It causes no change when the solution con-

1 'Quarterly Journal of Science,' 1865, vol. ii, p. 205.
2 'Philosophical Trans. of Roy. Soc.,' 1868, p. 589.
3 'Zeitschrift für Chimie,' 1865, p. 218.
4 'Archiv für die gesamte Physiologie,' 1868, p. 395.

tains excess of ammonia, but soon acts when it is neutral, and still more quickly when it contains an excess of such a very weak acid as the boric. A somewhat stronger acid, like benzoic, so far increases the oxidizing power that the hæmoglobin is quite decomposed. I have succeeded in obtaining this compound by means of a great variety of oxidizing reagents. It may be obtained very free from other coloured products by keeping for some days a solution of fresh blood with manganese—manganic oxide, prepared by calcining the carbonate in an open crucible. The colour of the solution is gradually changed from red to a sort of orange-brown, and it then shows to great advantage a spectrum with four absorption-bands, the most distinct of which is situated in the red. This spectrum is not nearly so well seen when nitrite of potash is added to a solution of blood, since it is then in rather too alkaline a condition; but if a little boric acid be added, the spectra correspond in all essential particulars.

On keeping this colour produced by the action of manganoso-manganic oxide for a few days in an open tube with iron filings, it is gradually deoxidized, and the colour and spectrum again become exactly the same as those of fresh blood. This and other facts lead me to conclude that it is a sort of *per*oxidized hæmoglobin, containing more oxygen than is taken up by the deoxidized modification discovered by Stokes, when oxidized by exposure to air; but still this extra amount of oxygen combines with the hæmoglobin without its molecular constitution being destroyed, for it may be easily reduced to the protoxidized or deoxidized states by weaker or stronger deoxidizing processes. The alkaline solution gives a very characteristic spectrum, distinguished from the simple oxidized hæmoglobin by a narrow absorption-band in the orange.

Ozone is one of those oxidizing reagents which give rise to the peroxidized hæmoglobin, but at the same time to hæmatin, probably by changing the albuminous constituent of the hæmoglobin, and also to products of the more complete oxidization of hæmatin. This is similar to what occurs when dry blood-stains are exposed to atmospheric air, even in cases where ozone could scarcely be present. The results are very different when the stains are kept damp. The same substances are also met with in scabs formed over wounds of the flesh, and it remains to be ascertained whether the peroxidized compound may not be formed in some diseased conditions of the blood.

In following out these inquiries I have been led to discover a series of new compounds, produced by the more complete oxidization of the colouring matter of blood; but these differ from that just described in not being restored to the original state by deoxidization, as though the whole molecular constitution of the hæmoglobin were changed. So far I have succeeded in preparing at least four such substances. Three of these resemble one another in having oxidized and deoxidized modifications, and when in the latter state in an alkaline solution in being of a dull olive-green colour, with a single absorption-band in their spectra; but they are distinguished from one another by these bands being situated in the red, in the orange, and in the yellow, respectively, as well as by other peculiarities.

There are other products which do not give rise to spectra with definite absorption-bands, but merely cut off more or less of the blue end, and are thus analogous to those substances so often formed by that amount of the oxidization of colouring matters which just precedes complete loss of colour. Such a compound occurs in moderately old blood-stains, and is the chief constituent of some which have been kept many years. My object now is to examine how far these various products of more or less complete oxidization may correspond with any colours formed naturally in normal or abnormal conditions of the blood or secretions; but before doing this I thought it desirable to further investigate such artificially prepared substances, since I have good reason for believing that several others exist, not yet completely understood, which give rise to spectra with special characteristic absorption-bands.

NOTE *on* METHÆMOGLOBIN. By E. RAY LANKESTER.

HOPPE-SEYLER describes Methæmoglobin as an intermediate product of the spontaneous change of Hæmoglobin into Hæmatin. He states that it is soluble in water, is characterised by an absorption-band in red, like that of acid Hæmatin, and is precipitated by acetate of lead. He appears to have seen it only in mixture with Hæmoglobin, and doubts its distinct existence. Preyer, writing since in Pfluger's 'Archiv,' describes Methæmoglobin as produced from Hæmoglobin either by ozone, carbonic acid gas, or spontaneous change in a dry chamber. He says it is *not* precipitated by acetate of lead. He considers it to have three absorption-bands—one like but *not* identical with that of acid Hæmatin, the other two identical with those of Oxyhæmoglobin. He hence regards as Methæmoglobin what Hoppe-Seyler would regard as a mixture of Methæmoglobin and O-Hæmoglobin.

At the meeting of the British Association, at Norwich, Professor Heynsius stated that a pupil of his had succeeded, by passing CO_2 gas through a solution of Hæmoglobin, in converting it *completely* into Hæmatin (as proved spectroscopically); and that then, by addition of a small quantity of ammonia and subsequently of Stokes' reducing fluid, the single band of Stokes' reduced Hæmoglobin could be obtained, and on agitation with air the bands of Oxyhæmoglobin, proving a reformation of Hæmoglobin after it had been broken down to the condition of Hæmatin. (See 'Journal of Anatomy and Physiology,' November, 1868.)

I am led to a different conclusion by repeating this experiment, and find that Hæmatin is *not* formed, but simply Methæmoglobin. When CO_2 gas is passed for an hour through a dilute solution of Hæmoglobin, the two Oxyhæmoglobin bands *entirely* disappear, and give place to two new ones —one in the red near to, but somewhat nearer blue than, the chief band of acid Hæmatin; the second, broad and difficult at first to detect, stretching from near the solar line *b* to F. No precipitation of an albumen occurs in this process. If to part of the solution with these bands ammonia is added, no precipitate occurs. If to another part some strong acetic acid is added, the colour changes to a browner tint, the band between *b* and F fades, and the true first band of acid Hæmatin in the red becomes sharply defined. The other bands of acid Hæmatin are very faint, and it is unnecessary to allude to them here. It is very important to observe that when ammonia in small quantity is added to the solution treated with strong acetic acid *a flocculent precipitate is produced.* Thus we have a radical distinction between the product of the action of carbonic acid and that of strong acetic acid. The first alters the Hæmoglobin molecularly, so as to afford a totally new set of absorption bands, but it does not break it up chemically. The second, on the other hand, distinctly destroys the chemical structure of the Hæmoglobin; it detaches Hæmatin from an albuminoid constituent, and this albumen remains in solution in the excess of acetic acid; it is immediately precipitated on neutralization of the acid by ammonia. The converse experiment of addition of ammonia to Hæmoglobin, and precipitating the detached albumen by neutralizing with acetic acid, is mentioned by Preyer.

A further proof of the merely molecular change of the Hæmoglobin caused by carbonic acid gas is seen in this—that, as Heynsius found, addition of a minute quantity of ammonia, and then of Stokes' fluid, gives reduced Hæmoglobin, as proved by the spectroscope, this on agitation giving the bands of Oxyhæmoglobin in all their original intensity.

I think it is pretty clear that it is the body, with its two perfectly characteristic lines formed by the action of carbonic acid gas on a dilute solution of Hæmoglobin, which Hoppe-Seyler meant to indicate as Methæmoglobin. No one appears to have obtained it free from mixture with Oxyhæmoglobin till now, excepting Heynsius, who mistook it for acid Hæmatin; and no one hitherto, whether regarding it as Methæmoglobin or some distinct body, has recognised the broad band in blue belonging to it, and now described.

I now must mention that it is difficult to get Methæmoglobin free from Oxyhæmoglobin. There seems to be a very strong union between the two, so that after you have formed some Methæmoglobin in a solution, it combines, by a kind of chemical adhesion, with the remaining Oxyhæmoglobin, and further action is stopped. This is the case if a strong solution of Hæmoglobin is used and CO_2 gas passed through it. A four-banded liquid is the result, having the two bands of Methæmoglobin and the two bands of Oxyhæmoglobin also. Such a solution, if left to itself for a day, has a tendency to revert entirely to Hæmoglobin by the development of ammonia and self-reduction. I found, however, that after CO_2 gas had been long passed (eight hours) through such a solution, and the bands of Oxyhæmoglobin were still persisting, so as to give a four-banded liquid, the addition of a very small quantity of acetic acid in a dilute state determined the complete transformation of the Oxyhæmoglobin into Methæmoglobin, and thus a pure two-banded solution having only the Methæmoglobin bands could be obtained. It was clearly shown that the acetic acid had not, when added in this weak form, proceeded to break up the Hæmoglobin as in the formation of Hæmatin, since no precipitate of an albumen was obtained on neutralization with ammonia, but the whole was readily reconvertible to Oxyhæmoglobin, as narrated above.

There is no doubt that Sorby's Brown Cruorin,[1] which was described as three-banded, is a mixture of Methæmoglobin and Oxyhæmoglobin. The product of the action of nitrites on Hæmoglobin, discovered by Dr. Arthur Gamgee, is also

[1] Since making these observations I am glad to find that Mr. Sorby has adopted somewhat similar views, and has confirmed the discovery of a fourth band in what he and others had considered a three-banded body.

apparently of the same nature. Dr. Gamgee described it as three-banded, but I find now, on looking for it, what had escaped both me and him, namely, a fourth band in the blue, which comparison by superposition shows to be identical in position with the second band of Methæmoglobin, as here described.. The band appeared a little fainter relatively than in pure Methæmoglobin in a specimen of the product of nitrite of potassium and Hæmoglobin prepared for the purpose of comparison. It appears that a little Hæmatin also may be formed in the action of nitrites on Hæmoglobin, as I have mentioned in another paper ('Journal of Anatomy,' November, 1869); but it is not to this only, as I was inclined to think, that the optical properties of nitrite-Hæmoglobin are due. Whenever Methæmoglobin is formed there is a tendency of the action to proceed further to the formation of Hæmatin, and hence small quantities of Hæmatin are usually formed simultaneously with it. Gamgee's nitrite-Hæmoglobin, like Heynsius's product of the action of CO_2 on Hæmoglobin, and like Sorby's Brown Cruorin, is, as he demonstrated, reconverted to Oxyhæmoglobin by the addition of ammonia, reducing fluid, and agitation. They are all mixtures (perhaps chemically held) of Oxyhæmoglobin and the pure two-banded Methæmoglobin.[1]

The true relation of Methæmoglobin to Hæmoglobin and Hæmatin is still obscure. I was at one time inclined to regard it as a neutral soluble Hæmatin, but it stands on a higher level than Hæmatin, no separation of an albuminoid occurring in its formation (Preyer's *globin* is possibly due to the formation of *small* quantities of Hæmatin). It is most probably Hæmoglobin in but a slightly changed molecular condition, whence its ready formation and reconversion; whether any loss or addition of *chemical* components occurs in these changes has to be determined

[1] Mr. Sorby, I perceive, in his paper which precedes this, regards the mixtures of Oxyhæmoglobin and Methæmoglobin as Peroxyhæmoglobin. Do not Gamgee's researches on the gas-absorption of this body negative this hypothesis?

REVIEW.

Researches on the Composition and the Significance of the Egg, based on the Study of its Mode of Formation, and of the first Embryonic Phenomena—(Mammifers, Birds, Crustacea, Worms). By EDOUARD VAN BENEDEN, Doctor of Natural Sciences. Presented August 1, 1868; crowned by the Royal Academy of Belgium in public assembly December 16, 1868. Brussels, 1870.

THE very valuable Memoir of Dr. Van Beneden is at length published, consisting of nearly 280 pages quarto, and twelve excellent plates. It is impossible for us here to give an adequate sketch of so extensive and valuable a work. The numerous details and observations which it contains are, however, all directed to establish certain conclusions concerning the signification of the egg, and the various parts which compose it, which we shall state, referring the reader with great confidence to the clear, logical, and interesting details of observation given in the Memoir. "It was not," remarks Dr. Van Beneden, "until after the appearance of the memorable works of Von Baer, Purkinje, R. Wagner, Coste, Prevost, Dumas, and Rusconi on the Vertebrata, of Rathke, Hérold, von Siebold, and P. J. Van Beneden on the lower animals, that the bases of comparative ovology and embryogeny were definitely established. The constitution of the egg of the superior animals, and of a certain number of inferior animals, was known, and it was perceived that throughout the egg consists of the same essential parts: of a membrane, of a vitellus, and of a germinal vesicle, holding in suspension one or several refringent corpuscles. On the other hand, the breaking up or cleavage which Prevost and Dumas had established in the Batrachia came to be discovered in Fishes by Rusconi and Von Baer; Von Siebold pointed it out in certain Nematods; Dumortier, Van Beneden, and Windischman in some Gasteropods.

But what a mystery this segmentation was—manifesting itself everywhere with the same characters! What relation could it have to the formation of the embryo, and what could be its object? This was, indeed, an enigma which seemed to be impenetrable, and one knew no more why the vitellus divided itself up into bits, than one could guess why the egg contained a vesicle destined to disappear. Nothing tended to explain the signification of the egg.

But in 1839 appeared the book of M. Schwann, and the discovery that all the tissues of animals proceed from cells was destined to effect a radical revolution in science; the cell-theory of M. Schwann was destined to throw an entirely new light on anatomy and physiology, as well as on embryogeny, and suffices to immortalise the name of its author. The cell-theory became established, and the profound obscurity which surrounded the question of the signification of the egg and the object of yelk-cleavage disappeared at the same time. Relying on his study of its constitution, M. Schwann was the first to proclaim that the egg is a cell; and since all the tissues are but a mass of cells, it became obvious that the object of the cleavage was to multiply the cell-egg. Bergmann, Reichert, and Remak contributed principally to the demonstration of the part which yelk-cleavage plays in the production of cells.

But although a great number of physiologists—following Schwann—consider the egg as a simple cell, others, with Henle, regard the egg as a combination of cells, and see in the germinal vesicle a complete cell. Among these we may cite Bischoff, Steinlein, Stein, &c.

It is necessary to take the mean between the two opinions, and to consider certain eggs as simple cells, others as complex cells. Can one, for instance, consider as a simple cell the egg of the Trematods, or of the Cestoids, when it is seen to form, by the union in a common shell, of a protoplasmic cell and of other cells, formed by distinct glands, which have wrongly borne the name 'vitellogenous'?"

The study which Dr. Van Beneden has made of the formation of the egg, and of the first embryonic phenomena, clearly gives the solution of the problem.

In every egg, whether of Mammifer or of Bird, of Crustacean or of Trematod, we find a protoplasmic cell, of which the germinal vesicle is the nucleus, the corpuscle of Wagner (germinal spot), the nucleolus. This cell, which Dr. Van Beneden calls the germ or cell-egg (cellule-œuf), and which may be considered as being the first cell of the embryo, arises everywhere throughout the animal series in the same manner; it presents always the same characters; and gives origin by division to the first cells of the embryo.

But the vitellus of the egg is made up of two elements: the one, protoplasmic, represents the mass of the cell-egg; the other, nutritive, forms what Dr. Van Beneden has designated the *deutoplasm* of the egg. This deutoplasm is the accessory part of the vitellus; and so we see it is wanting sometimes, arises in different ways, presents very various relations with the protoplasm, and behaves itself very differently in the course of the first embryonic phenomena, according to the particular case.

Formation.—Sometimes it takes birth in the interior itself of the protoplasm; and is elaborated by the cell-egg itself; at other times it is formed by certain special cells, either in a particular gland (deutoplasmogen of Van Beneden—vitellogenous gland of previous authors), or in the same gland in which the germs are formed, but in a special part of this organ.

Relations to the Protoplasm of the Cell-egg.—Sometimes it is found in the ripe egg, in suspension in the protoplasm; sometimes it does not get mixed up with it. In certain cases it is formed of well-marked cells; more often it is composed of drops or of refringent granules, or even of vesicular elements, which have nothing in common with true cells.

Rôle in Development.—We have now seen that the deutoplasm behaves very variously during the first embryonic phenomena; but its function is always the same. It plays a purely passive part; it serves to nourish the blastoderm and the embryo, and to furnish, by the combustion of the elements of which it consists, the force necessary for the production of all the elements of the embryo, and for the accomplishment of all the phenomena of fœtal life. This deutoplasm, which is sometimes wanting, can be formed of distinct cells, and take origin in a special organ; and although it makes a part of the *egg*, it cannot be regarded as an integral part of the *cell-egg*.

It follows from this that the proposition generally admitted, "*every egg is a cell,*" has not that stamp of exactitude which should characterise any principle of science.

But *in every egg there exists a cell-egg, a germ which is the first cell of the embryo.*

Side by side with this cell there exists in the egg a mass of nutritive matters, possibly mixed with the protoplasm of the cell-egg, and formed in its interior, as is seen in many vertebrate animals. In that case, we may embrace it in the cell, and say, with Schwann, that the egg is a cell as far as Vertebrates are concerned.

But when the deutoplasm is found outside the cell-egg, it cannot be considered as forming an integral part of the germ, and itself may be composed of cells, of which we have examples in many lower animals, which are remarkable for their extreme fecundity. In this case the egg is clearly not a cell, but an aggregate of cells.

In the preceding paragraphs we have the summary of Dr. Van Beneden's view of the structure of the animal ovum—his main point being the recognition of this DEUTOPLASM, which his detailed observations and figures, comprising minutes tudies of the ovogeny of Trematods, Cestoids, Turbellaria, Nematods, Rotifers, Copepods, Isopods, Amphipods, Mysids, Mammals, and Birds, clearly establish.

The mass of important observations contained in this part of the work well deserve careful perusal.

The tubular structure of the Mammalian ovary has been observed by the author, confirming Pflüger; but he differs from that author as to the mode of origin of the *cell-egg* in this case. Dr. Van Beneden has established in a variety of cases, and in that of Mammals (and man himself), that the cell-egg originates in a mass of protoplasm, contained in the ovigerous tubes in the case of Mammals. Several distinct nuclei arising by free development, these enlarge, and each developes a nucleolus; they are, in fact, germinal spot and germinal vesicle respectively; around each germinal vesicle the protoplasm then segregates, and the cell-egg is complete. Its vitelline membrane, bounding the segregated protoplasm, develops later, and in most cases after the addition of the dentoplasm, or vitellogenous matter, as it used to be called.

The very perplexing question of the disappearance of the germinal vesicle at the time of or after fertilization in some eggs—*e.g.*, those of Mammifers, Birds, some Crustacea, and some Mollusca—is discussed by Dr. Van Beneden. It disappears also in marine Annelids, according to Claparède and Mecznikow; and in Batrachia, according to Stricker, who, as well as other workers, has very carefully searched for it in the frog's and toad's ovum at the time of fertilization, but in vain. On the other hand, we have to put against this disappearance in the cases cited the observation of its division into two, causing the first yelk-furrow, as observed by Müller in *Entoconcha,* by Leydig in Rotifers, by Leuckart in *Puppipara,* by Mecznikow in *Cecidomyia* and *Aphides,* by Pagenstecker in *Oxyuris,* by Keferstein in marine *Planariæ,* by Bessels in *Lumbricus* and Leeches, by Van Beneden and Bessels in many Crustacea, by Van Beneden in certain Trematods figured in this work. We particularly refer the English reader to an important paper by Professor Huxley on the ovum of Pyrosoma, to which we are sorry Dr. Van Beneden has not had his attention directed ('Annals of Natural History,' Third Series, vol. 5, 1860, p. 29). Certain authors, building on the first set of facts, have declared that the germinal vesicle and spot are very important for the first stages of egg growth, but that at the period when the egg is ready for fertilisation their function is ended, and they are no more wanted. Others, again, with equal force, pointing to the latter set of facts, declare that the germinal vesicle by its division furnishes the nuclei of the first embryonic cells, and would regard the disappearance as only apparent in the first cases, since a pair of nuclei exactly like the divided germinal vesicle appear immediately after the first cleft is complete in all those cases where the germinal vesicle is found to be wanting at the time of fertilisation.

Dr. Van Beneden lays some stress on this, as well as on the fact that in some observations of his on rabbits, he found a sort of irregularity in the perceptibility or apparent absence of nuclei in the two first cells, due to yelk cleavage. Thus, in one case observed two hours after copulation, the yelk was found cleft into two cells, each with a nucleus, although at the time of fertilization it is not possible to observe a germinal vesicle. In another case, on the contrary, twenty-four hours after copulation, he found the two cells but no nucleus in either. He has also fully established the same sudden disappearance of the nuclei in the cells of the 3rd or 4th generation, reappearing as in the first cleavage, after the process. He also confirms Weismanm's observation of the similar disappearance of the nuclei in the cells of the blastoderm of dipterous insects when they were about to undergo division. He also makes a point, in observing that if the nucleus or germinal vesicle did really break up and perish, there would not suddenly appear in each half of the divided mass, a nucleus equal in size to the bulk of the lost vesicle; if new nuclei were produced by endogenous formation, they would form by degrees, and as small points at first, and would not suddenly jump into existence fully grown. Dr. Van Beneden belongs to that party which would regard the disappearance of the germinal vesicle in the case of Mammifers, Birds, Batrachia, Annelids, &c., as apparent rather than real. We are quite of his opinion here, and in any case would hesitate to accept the assertion of the *death* of the germinal vesicle before fertilisation. The host of cases in which its active life has been clearly traced as the original nucleus of the embryo-cells demands some other explanation than this. On this ground, we wonder that Mr. Hutchinson Stirling in an essay on Protoplasm, which is, if we may say it, overladen with references to the knowledge and wisdom of the Germans—the author accepting all that is presented to him in German gilt, but refusing to admit English work or thought as of the same value at all—should refer to the germinal vesicle as though it were an accepted fact that it dies upon impregnation: "In the egg, on impregnation, it seems to me natural (I say it with a smile) that the old sun that ruled it should go down, and that a new sun, stronger in the combination of the new and the old, should ascend into its place" (p. 33). Had Mr. Stirling never studied the works of Leydig, Mecznikow, and others, he yet might have found in Professor Huxley's paper on Pyrosoma facts entirely discordant with his view of the case. But error is to be expected when bibliographical knowledge only is brought to bear on such a question as the signification of Protoplasm, and the authority of Professor Stricker is quoted as conclusive against that of Professor Huxley.

Dr. Van Beneden does not, however, entirely clear up this question of the disappearance in some eggs and the important cleavage-function in others of the germinal vesicle. We wish Dr. Van Beneden had found it within the scope of his work to look at the *vegetal* ovum, with a view to solving this difficulty. If we may believe the observations of some high authorities (Tulasne) the mass which is fertilized in phænogamous plants at any rate, that which corresponds to the cell-egg in animals, is a simple mass of protoplasm with nothing corresponding to germinal vesicle and spot,—no nucleus and nucleolus. Immediately upon fertilisation, cells are produced in this mass by free-cell-formation, which dividing, give rise to the embryo. But if the observations of Henfrey are to be accepted, the embryo-sac does develop one or more germinal vesicles before ever the pollen-tube touches it, though there is no enclosure by cellulose until after that time. In this case, the embryo-sac with its protoplasm corresponds to the primitive protoplasm of the young ovary of animals, in which germinal vesicles arise, just the same in the plant as in the animal. In the plant, however, the protoplasm never segregates nor is it removed from the seat of its growth. There are some entomostraca (see the recent work of M. Müller on the Scandinavian Cladocera) in which four germinal vesicles are enclosed in one mass of protoplasm; only one of these vesicles develops and forms an embryo, the other being absorbed before fertilisation. So with the embryo-sac of Orchis for instance, three or four germinal vesicles arise in the mass of protoplasm, but only one develops, and the others are absorbed, forming an internal albumen in some cases; the absorption is here, however, *after* fertilisation. Dr. Van Beneden would no doubt speak of the abortive germinal vesicles in Cladocera as contributing to the deutoplasm; but we think a wide distinction ought to be drawn between such deutoplasm and that which is poured round the cell-egg from a distinct gland. From the above remarks, it appears there is much the same obscurity in plants as in animals with regard to the relation of the germinal vesicle to fertilisation, *i. e.*, whether the egg has a nucleus before fertilisation which persists and divides into the nuclei of the embryonal-cells—in all cases alike—and if so, how the apparent absence of such a nucleus at the moment of fertilisation in particular classes is to be explained. One thing may well be remembered in this matter: we have no right to lay any great stress on the mere formal structure of a cell, and the nucleus must be regarded as important only so far as we see it in direct connection with important phenomena. Morphologically it is but the central slightly differentiated part of a lump of viscid matter. Now we have in the non-nucleated red blood-corpuscles of mammalia and the nucleated red corpuscles of all other vertebrata a remarkable instance of the way in which structural units, undoubtedly of the same origin and signification in the two cases, may put on different appearances, and it seems to be just possible that with as little significance as the absence or the differentiation of a 'nucleus' is brought about in these two cases, may the absence or presence of a nucleus in the cell-egg at the moment of fertilisation be produced. This is one way of looking at the matter, but another is suggested to us by Dr. Van Beneden's exposition of the fusion of the deutoplasm with the original protoplasm of the cell-egg, which seems to be worth consideration. In the same way as the protoplasm of the cell-egg is found in some cases to be thoroughly fused with the deutoplasm, and in other cases distinct, may not the nucleus of the same ovum become at a certain epoch of its growth, under certain chemical and physical conditions, *diffused* or mixed up with the surrounding matter *temporarily*, again, contracting, segregating and assuming its nuclear form after a time, that is, after the first contraction of the yelk-cleavage has shown itself. Haeckel, in a recent very interesting 'Essay on the Plastid and Cell Theories,' suggests that we may see in the disappearance of the germinal vesicle of the cell-egg, a return to that elementary ancestral form which must be admitted as preceding the cell, namely, the cytod, the structureless mass of protoplasm with membranous pellicle or without, which he has shown to be the character of several well marked forms, his Monera. The other and more intelligible form of yelk-cleavage, in which the germinal vesicle is seen to persist and divide, has been recently observed by Haeckel in his studies on the development of the Siphonophora. A parallel to the variability of the ovum in the definition or non-definition of a nucleus is to be found in the red corpuscles of some pyrenæmatous Vertebrata. It is undeniable, as pointed out by Mr. Savory, that generally no nucleus can be seen in the circulating red corpuscles of the Frog. It is equally certain that often under the same circumstances they are well *defined*—a condition to be distinguished from that of granulation.

A table giving a summary of the different modes of formation of the blastoderm is given by Dr. Van Beneden, which we extract. The thorough way in which he has gone into his subject, and the importance of the views he advocates, will render this book classical in embryogeny, whilst its clear style, careful historical review, and interesting subject-matter should induce every student to master its contents.

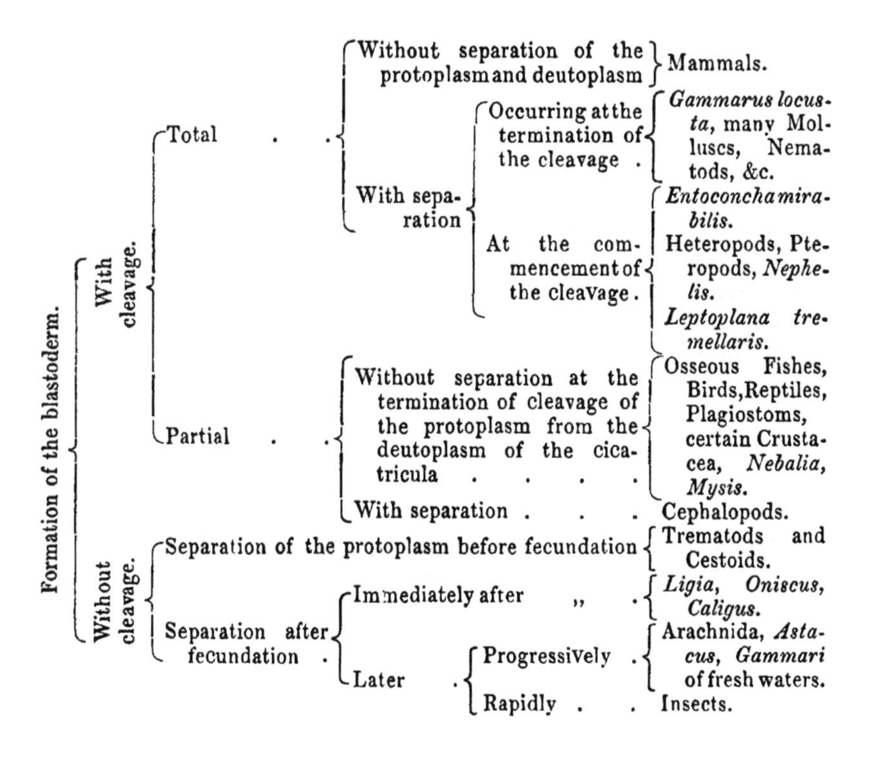

Formation of the blastoderm.	With cleavage.	Total . .	Without separation of the protoplasm and deutoplasm		Mammals.
			With separation	Occurring at the termination of the cleavage .	*Gammarus locusta*, many Molluscs, Nematods, &c.
				At the commencement of the cleavage .	*Entoconcha mirabilis.* Heteropods, Pteropods, *Nephelis. Leptoplana tremellaris.*
		Partial . .	Without separation at the termination of cleavage of the protoplasm from the deutoplasm of the cicatricula		Osseous Fishes, Birds, Reptiles, Plagiostoms, certain Crustacea, *Nebalia, Mysis.*
			With separation . . .		Cephalopods.
	Without cleavage.	Separation of the protoplasm before fecundation			Trematods and Cestoids.
		Separation after fecundation .	Immediately after ,, .		*Ligia, Oniscus, Caligus.*
			Later .	Progressively .	Arachnida, *Astacus, Gammari* of fresh waters.
				Rapidly . .	Insects.

Limits of the Power of the Microscope.—On points of controversy I will not here enter, but I may say that De la Rive ascribes the haze of the Alps in fine weather to floating organic germs. Now the possible existence of germs in such profusion has been held up as an absurdity. It has been affirmed that they would darken the air, and on the assumed impossibility of their existence in the requisite numbers, without invasion of the solar light, a powerful argument has been based by believers in spontaneous generation. Similar arguments have been used by the opponents of the germ theory of epidemic disease, and both parties have triumphantly challenged an appeal to the microscope and the chemist's balance to decide the question. Without committing myself in the least to De la Rive's notion, without offering any objection here to the doctrine of spontaneous generation, without expressing any adherence to the germ theory of disease, I would simply draw attention to the fact that in the atmosphere we have particles which defy both the microscope and the balance, which do not darken the air, and which exist, nevertheless, in multitudes sufficient to reduce to insignificance the Israelitish hyperbole regarding the sands upon the seashore.

The varying judgments of men on these and other questions may perhaps be, to some extent, accounted for by that doctrine of Relativity which plays so important a part in philosophy. This doctrine affirms that the impressions made upon us by any circumstance, or combination of circumstances, depends upon our previous state. Two travellers upon the same peak, the one having ascended to it from the plain, the other having descended to it from a higher elevation, will be differently affected by the scene around them. To the one nature is expanding, to the other it is contracting, and feelings are sure to differ which have two such different antecedent states. In our scientific judgments the law of relativity may also play an important part. To two men, one educated in the school of the senses, who has mainly occupied himself with observation, and the other educated in the school of imagination as well, and exercised in the conceptions of atoms and molecules to which we have so frequently referred, a bit of matter, say $\frac{1}{50000}$th of an inch in diameter, will present itself differently. The one descends to it from his molar heights, the other climbs to it from his molecular lowlands. To the one it appears small, to the other large. So also as regards the appreciation of the most minute forms of life revealed by the microscope. To one of these men they naturally appear conterminous with the ultimate particles of matter, and he readily figures the molecules from which they directly spring; with him there is but a step from the atom to the organism. The other discerns numberless organic gradations between both. Compared with his atoms, the smallest vibrios and bacteria of the microscopic field are as behemoth and leviathan. The law of relativity may to some extent explain the different attitudes of these two men with regard to the question of spontaneous generation. An amount of evidence which satisfies the one entirely fails to satisfy the other; and while to the one the last bold defence and startling expansion of the doctrine will appear perfectly conclusive, to the other it will present itself as imposing a profitless labour of demolition on subsequent investigators. The proper and possible attitude of these two men is that each of them should work as if it were his aim and object to establish the view entertained by the other.

I trust, Mr. President, that you—whom untoward circumstances have made a biologist, but who still keep alive your sympathy with that class of inquiries which nature intended you to pursue and adorn—will excuse me to your brethren if I say that some of them seem to form an inadequate estimate of the distance which separates the microscopic from the molecular limit, and that, as a consequence, they sometimes employ a phraseology which is calculated to mislead. When, for example, the contents of a cell are described as perfectly homogeneous, as absolutely structureless, because the microscope fails to distinguish any structure, then I think the microscope begins to play a mischievous part. A little consideration will make it plain to all of you that the microscope can have no voice in the real question of germ structure. Distilled water is more perfectly homogeneous than the contents of any possible organic germ. What causes the liquid to cease contracting at 39° Fahr., and to grow bigger until it freezes? It is a structural process of which the microscope can take no note, nor is it likely to do so by any conceivable extension of its powers. Place this distilled water in the field of an electro-magnet, and bring a microscope to bear upon it. Will any change be observed when the magnet is excited? Absolutely none; and still profound and complex changes have occurred. First of all, the particles of water are rendered diamagnetically polar; and secondly, in virtue of the structure impressed upon it by the magnetic strain of its molecules, the liquid twists a ray of light in a fashion perfectly determinate both as to quantity and direction. It would be immensely interesting to both you and me if one here present, who has brought his brilliant imagination to bear upon this subject, could make us see as he sees the entangled molecular processes involved in the rotation of the plane of polarisation by magnetic force. While dealing with this question, he lived in a world of matter and of motion to which the microscope has no passport, and in which it can offer no aid. The cases in which similar conditions hold are simply numberless. Have the diamond, the amethyst, and the countless other crystals formed in the laboratories of nature and of man no structure? Assuredly they have; but what can the microscope make of it? Nothing. It cannot be too distinctly borne in mind that between the microscope limit and the true molecular limit there is room for infinite permutations and combinations. It is in this region that the poles of the atoms are arranged, that tendency is given to their powers, so that when these poles and powers have free action and proper stimulus in a suitable environment, they determine first the germ and afterwards the complete organism. This first marshalling of the atoms on which all subsequent action depends baffles a keener power than that of the microscope. Through pure excess of complexity, and long before observation can have any voice in the matter, the most highly trained intellect, the most refined and disciplined imagination, retires in bewilderment from the contemplation of the problem. We are struck dumb by an astonishment which no microscope can relieve, doubting not only the power of our instrument, but even whether we ourselves possess the intellectual elements which will ever enable us to grapple with the ultimate structural energies of nature.

But the speculative faculty, of which imagination forms so large a part, will nevertheless wander into regions where the hope of certainty would seem to be entirely shut out. We think that though the detailed analysis may be, and may ever remain, beyond us, general notions may be attainable. At all events, it is plain that beyond the present outposts of microscopic inquiry lies an immense field for the exercise of the imagination. It is only, however, the privileged spirits who know how to use their liberty without abusing it, who are able to surround imagination by the firm frontiers of reason, that are likely to work with any profit here. But freedom to them is of such paramount importance that, for the sake of securing it, a good deal of wildness on the part of weaker brethren may be overlooked. In more senses than one Mr. Darwin has drawn heavily upon the scientific tolerance of his age. He has drawn heavily upon *time* in his development of species, and he has drawn adventurously upon *matter* in his theory of pangenesis. According to this theory, a germ already microscopic is a world of minor germs. Not only is the organism as a whole wrapped up in the germ, but every organ of the organism has there its special seed. This, I say, is an adventurous draft on the power of matter to divide itself and distribute its forces. But, unless we are perfectly sure that he is overstepping the bounds of reason, that he is unwittingly sinning against observed fact or demonstrated law—for a mind like that of Darwin can never sin wittingly against either fact or law—we ought, I think, to be cautious in limiting his intellectual horizon. If there be the least doubt in the matter, it ought to be given in favour of the freedom of such a mind. To it a vast possibility is in itself a dynamic power, though the possibility may never be drawn upon. It gives me pleasure to think that the facts and reasonings of this discourse tend rather towards the justification of Mr. Darwin than towards his condemnation, that they tend rather to augment than to diminish the cubic space demanded by this soaring speculator; for they seem to show the perfect competence of matter and force, as regards divisibility and distribution, to bear the heaviest strain that he has hitherto imposed upon them.—*Professor Tyndall, 'Address to the British Association,* 1870, *on the Scientific Use of the Imagination.'*

Migration Theory.—A paper was read on this subject by Dr. Caton at the Biological Section of the British Association. The result of a number of experiments on the mesentery of the frog were described, in which the phenomena described by Cohnheim were observed. Inflammation in the fish and tadpole had also been studied; in the former, congestion was found to be absent during inflammation; this peculiarity was referred to the venous heart. Though the formation of pus-cells was observed, migration was never seen. In the tadpole, migration was observed to occur very frequently, produced by the slightest congestion, and even when all local irritation had been carefully avoided. The general conclusions arrived at, were that cell-migration depends on congestion, and that its connection with the suppurative process is very doubtful. Cell-migration in the tadpole was exhibited under the microscope on one of the days of the meeting.

The Application of the Microscope to the Investigation of Meteorites.—The difficulties in the way of the complete investigation of a meteorite resemble those we meet with in terrestrial rocks. In both the ingredient minerals are minute, and are often, especially in the case of the aërolitic rock, very imperfectly crystallized. Moreover the methods for separating them, whether mechanically or chemically, are very incomplete. With a view to obtain some more satisfactory means of dealing with these aggregates of mixed and minute minerals, I sought the aid of the microscope, by having in the first place sections of small fragments cut from the meteorites so as to be transparent.

One may learn, by a study and comparison of such sections, something concerning the changes that a meteorite has passed through; for one soon discovers that it has had a history, of which some of the facts are written in legible characters on the metorite itself; and one finds that it is not difficult roughly to classify meteorities according to the varieties of their structure. In this way one recognises constantly recurring minerals; but the method affords no means of determining what they are. Even the employment of polarized light, so invaluable where a crystal is examined by it of which the crystallographic orientation is at all known, fails, except in rare cases, to be a certain guide to even the system to which such minute crystals belong. It was found that the only satisfactory way of dealing with the problem was by employing the microscope chiefly as a means of selecting and assorting out of the bruised débris of a part of the meteorite the various minerals that compose it, and then investigating each separately by means of the goniometer and by analysis, and finally recurring to the microscopic sections to identify and recognise the minerals so investigated. The present memoir deals with the former part of this inquiry. Obviously the amount of each mineral thus determined, after great care and search, can only be extremely small, as only very small amounts of a meteorite can be spared for the purpose, notwithstanding that as large a surface as possible of its material requires to be searched over for instances of any one of the minerals occurring in a less than usually incomplete form. On this account one has to operate with the greatest caution in performing the analysis of such minerals; and the desirability of determining the silica with more precision than is usually the case in operations on such minute quantities of a silicate suggested to me the process, which was adopted.—*Professor Maskelyne in the 'Proceedings of the Royal Society*, 1870.'

Academy of Natural Sciences—Exhibition of the Biological and Microscopical Section.—The Biological and Microscopical Section of the Academy of Natural Sciences gave an exhibition of microscopes and microscopic specimens, at the Hall of the College of Physicians, on Friday evening, the 10th of June. As this reception was especially intended for professional men, the class of objects displayed included many illustrations in pathological anatomy and histology, of course chiefly interesting to physicians; and about forty-five microscopes in all, each in efficient working condition, were collected from the members and arranged upon the tables. Among the specimens displayed, the Director of the Section, Dr. S. Weir Mitchell, had on exhibition some preparations of teeth, bone, &c., mounted by Professor Christopher Johnson, of Baltimore, and also a collection of blood-corpuscles from different animals. Dr. William Pepper, Vice-Director, further illustrated the subject by a series of blood-crystals. The amœboid movement, lately so famous in connection with both Cohnheim's theory of the origin of pus and Huxley's lecture on "Protoplasm," was well shown by Dr. J. G. Richardson, the Secretary, with a power of 1300 diameters; while the Corresponding Secretary, Professor McQuillen, illustrated various departments of dental anatomy and physiology by sections of teeth of man and animals (showing, among other points, the interglobular spaces), and of bone exhibiting the lacunæ and canaliculi. He likewise displayed specimens of human muscle infested with trichina spiralis from fatal cases of trichiniasis; also the infecting swine's flesh. Professor James Tyson had on exhibition a series of urinary deposits. Dr. Wm. F. Norris contributed some specimens of nerves in the cornea, and capillaries showing their parietal nuclei—respectively gold and silver stainings—by the methods of Cohnheim and Recklinghausen. Some large sections of brain, kidney, &c., were displayed by Dr. W. W. Keen, who also showed sundry illustrations of nerve-structure. Dr. W. B. Corbitt exhibited a valuable series of specimens of various malignant and other tumours, collected in Germany, and many of them classified under the supervision of Professor Rokitansky.

Mr. Walmsley and Mr. T. W. Starr each contributed a number of mounted preparations. Mr. Zentmyer displayed, among other instruments, one of his binocular microscopes. Professor J. A. Meigs exhibited, with others, an injected specimen of the gall-bladder, made by Dr. P. B. Goddard many years ago.

Professor B. H. Rand arranged upon a side table his spectroscopic apparatus, and demonstrated the spectra of different metals, and explained this delicate method of analysis.

QUARTERLY CHRONICLE OF MICROSCOPICAL SCIENCE.

Histology.* — *Cornea.*—Schweigger Seidel ('Berichte der Sächs. Ges. der Wissenschaft,' 1869; Med. Centralblatt, 1870, p. 358) reproduces his often-expressed objections to the use of silver solutions in histology, and the genuineness of the structures displayed by their means, and finds them confirmed by a special examination of the cornea. He believes the stellate corpuscles of the cornea to be purely artificial productions caused by the alterations of form in the part when removed from its connections. Other appearances produced by nitrate of silver, which are regarded as plasmatic canals, he believes to be owing to simple precipitation. These conclusions are certainly very much at variance with the conclusions of most observers.

Lymphatic Spaces in the Eye.—Schwalbe has contributed to M. Schultze's 'Archiv' (vol. vi, p. 261,) a paper of more than 100 pages on the lymphatics of the anterior division of the eyeball, in continuation of that on the lymphatics of the posterior division noticed in our last report.* The most important results arrived at have reference to the communications of the anterior chamber, which Schwalbe regards as a lymphatic space, though it has no connection with the lymphatic vessels. By injection of coloured liquid into the chamber he succeeded in filling a belt of vessels on the surface of the sclerotic, from which the injection passed into a ring of radial vessels in the conjunctiva, and also to certain vessels running in the direction of the musculi recti, all of which were not lymphatics, but veins. That these vessels were really veins was shown, in the first place, by their arrangement, quite different from that of the known lymphatics; and they were also ascertained not to be perivascular lymphatics (such as are described by Lightbody, 'Journal of Anatomy and Physiology,' November, 1866). That the result is not due to extravasation is shown by the low pressure (20mm.) at which the injection is effected. The fact of a lymphatic chamber standing in communication with a venous channel is often met with in the lower vertebrata; thus the caudal sinus of fishes opens into the caudal vein, and the lymphatic hearts of reptiles also into veins; and, according to Schwalbe, it is

* Chronicled by Frank Payne, M.B., Oxon, Pathologist to St. Mary's Hospital.

† See 'Quarterly Journal' for April, p. 195.

here a necessary consequence of the tension (20mm. to 30mm.) of the liquid in the anterior chamber, which would cause it to escape were it in communication with vessels in which the tension is so low as it is in lymphatics, instead of with veins. In seeking for the channel of communication between the chamber and the veins, Schwalbe was led to study the structures situated in the angle between the posterior surface of the cornea and the anterior surface of the iris, namely, the ligamentum pectinatum and the canal of Schlemm (sinus vel circulus venosus Iridis). The ligamentum pectinatum is regarded as corresponding to the so-called canal or space of Fontana in the lower animals. Under these names is understood a trabecular network occupying the groove of the sclerotic which corresponds to the insertion of the ciliary muscle. The laminæ of this structure are a direct continuation of the substance of the membrane of Descemet (membrane of Demours, posterior elastic lamina), and are covered with endothelial sheaths, which are continuous with the endothelium covering that membrane. They are further continuous with the endothelium on the anterior surface of the iris, and thus the endothelial lining of the anterior chamber is complete. The canal or sinus of Schlemm is in most animals only represented by one or two of the interstices of the ligamentum pectinatum which are larger than the rest, but in man it is a more distinct cavity. Its inner wall is formed by a continuation of the membrane of Descemet, which ceases to be a hyaline membrane, and breaks up a fenestrated structure, the openings of which correspond with the laminæ of the trabecular tissue of the ligamentum pectinatum, while its outer wall is formed by dense tissue which is continuous with the sclerotica. It is lined by an endothelium containing few nuclei, but in this respect the endothelium of the membrane of Descemet often resembles it. It is through the fenestrated structure, between the laminæ of the ligamentum pectinatum, and through the canal of Schlemm that an injection passes from the anterior chamber to reach the veins above spoken of. A network of minute veins (the ciliary plexus discovered by Leber) permeates these structures, but the veins are perfectly distinguishable in vertical sections, and the canal of Schlemm itself is, according to Schwalbe, distinctly a lymphatic and not a venous space. Standing, as it does, in communication with veins on the one side, and with the anterior chamber on the other, it may be filled from either of these; but an injection from the anterior chamber enters it under very slight pressure (without filling the ciliary plexus), while an injection from the venous system can only be made to enter it by using very considerable pressure. If an injection be thrown into the long ciliary arteries while the collateral venous channels are tied to increase the pressure, or, what comes to the same thing, the tension in the anterior chamber be diminished by puncture, fluid may be forced through the superficial sclerotic veins into Schlemm's canal, and thence into the anterior chamber. The occurrence of blood in this canal in the eyes of persons who have been hung, and under other circumstances, which have caused it to receive the name of *sinus venosus*, are explained by Schwalbe as consequences of excessive venous tension. It is, therefore, probable that the communication between the anterior chamber and the venous system is regulated by balance of pressure, and not by any system of valves.

The canal of Petit was sometimes, though not always, filled by injection from the anterior chamber. This was found to depend really upon certain small slit-like openings close to the border of the crystalline lens by which the canal communicates with the posterior chamber, and from this chamber it was very easily injected. The success of injections from the anterior chamber depends on altering the convexity of the eyeball and the position of the lens. (By posterior chamber is understood an annular space between the iris and the lens, closed centrally by the contact of the front surface of the lens with the margin of the pupil.)

Lymphatic Spaces in the Brain.—Obersteiner ('Sitzungsberichte der Wiener Akademie,' Math. Naturwiss. Classe 1870, p. 57) describes lymphatic spaces in the brain, not only surrounding the blood-vessels, as described by Robin and His, but also round ganglionic cells. These spaces contained lymph cells, were in communication with the perivascular spaces, and susceptible of injection from these. Similar channels were also seen under the epithelium of the ependyma of the cerebral ventricles of the Frog.

Nerve terminations in the Tongue.—The structure of the tongue is the subject of two papers in M. Schultze's 'Archiv.' Hans v. Wyss (vol. vi, p. 237) has repeatedly studied the tongues of men and mammalia, and describes the goblet-shaped or bud-shaped organs originally discovered simultaneously by Schwalbe and Loven (Schultze's 'Archiv,' IV, 96 and 154). In the human tongue one papilla circumvallata may contain 400 of these structures, arranged in five or six superimposed circular rows. The organs themselves are imbedded in the epithelium, and project slightly beyond its surface. They are composed of two kinds of cells: the covering or protective cells, which are placed outside; and the bacillary cells (sensory cells of F. E. Schulze), which are protected by these. The former are epitheloid structures, arranged concentrically in several rows like a flower-bud, and in such a way that their peripheral extremities form a circular opening, and the central extremities are attached by ramified processes to the fibrous substance of the papilla. The inner, or bacillary cells, occupy all the small space left in the centre of the whole of the protective cells. They are spindle-shaped structures, which are furnished at their outer end with fine hair-like processes projecting slightly beyond the rest, and which, centrally, are believed to be connected with terminal nerve-fibres, though this connection has not been absolutely made out. F. E. Schulze (ibid., p. 407) describes very similar structures from the tongue of an amphibian larva (*Pelobates fuscus*), which were discovered by Stricker in 1857. They are goblet-shaped, or better, bud-shaped structures, composed, like the similar organs in the tongue of Mammalia, of two kinds of cells, to which Schultze gives the name of sensory cells and supporting cells, regarding the former only as connected with the nervous system. They resemble in general the bacillary cells of Wyss, above described. From ten to thirty may be contained in each "bud." Schultze points out that in their fine hair-like processes they resemble the "gustatory cells" of fishes, described by himself in 1862; the "olfactory cells" of the Schneiderian membrane; the "auditory cells" of the ear; and certain cells in the lateral line of fishes, described also by himself. In all these cases the central extremity of the cell is believed, with more or less certainty, to be in connection with nerve-fibres, so that such cells represent the real peripheral terminations of the nerves of special sense.

Dr. Ihlder, of Göttingen, working under the guidance of Professor Krause (Reichert's 'Archiv,' 1870, p. 238), has investigated the terminations of nerve-fibres in the tongues of birds. He traces them into oval concentric clubbed bodies like the terminal clubs seen in other parts by Krause, and names them after Herbst, their discoverer. Krause has previously observed similar structures in various organs of birds.

Epithelium.—Professor Krause (Reichert's 'Archiv,' 1870, p. 232) describes certain peculiar cells from the lowest layer of the corneal epithelium in the sheep, in which the nucleus is replaced by a granulated corpuscle.

Termination of Nerves in Unstriped Muscle.—Krause (Reichert's 'Archiv,' 1870, p. 1) describes the termination of nerves in an unstriped muscle from the rabbit. The examination was best made without the addition of any fluid. The medullated fibres were seen to break up into fine threads with a single contour, which finally ended in peculiar "end platten," composed of three or four nuclei. These are few in comparison with the number of fibres, so that each fibre has not, as in striated muscle, a special terminal apparatus, but hundreds of fibres depend upon a single one of these.

Terminations of Nerves in Salivary Glands.—Krause (Reichert and du Bois Reymond's 'Archiv,' 1870, p. 9) has examined the structure of salivary glands with especial reference to the terminations of nerves and their connection with secreting-cells described by Pflüger, and has not been able to confirm his observations. Krause could in two glands only trace medullated fibres into the proper gland substance, and here they ended in terminal capsules or clubs. He was never able to establish with certainty the connection of medullated fibre with an acinus; and draws attention to the many sources of error which attach to the employment of olmic acid, a reagent which colours many other structures beside nerve-fibres, as well as of chromic acid. With reference to the termination of non-medullated fibres he could not arrive at any certain results. Krause's method of investigation consisted in immersing the perfectly fresh glands in a five per cent. solution of neutral molybdate of ammonia.

Salivary Glands.—Ewald (Inaugural Dissertation, Berlin, 1870, 'Med. Centralblatt,' No. 24, p. 373) publishes some observations on the difference in the histology of salivary glands according as they have or have not been stimulated. He finds the difference (previously observed by Haidenhain) to depend merely on the absence of mucin from the latter.

Placenta.—Langhans ('Med. Centralblatt,' 1870, p. 470; 'Archiv für Gynäkologie,' I, 317) gives a somewhat new view of the structure of the placenta. The fœtal tufts, even some as thick as 1 mm., penetrate the maternal tissue, and losing their epithelial covering, become intimately united with it. The union is not, however, everywhere so close as this. Even when the epithelium is quite wanting, the fœtal and maternal parts can always be clearly distinguished by their structure. This close union does not take place till the later months of pregnancy; and as late as the sixteenth or twentieth week the fœtal tufts showed a clear and continuous covering of epithelium.

Inflammation.—M. Feltz has addressed to the French Academy of Sciences ('Comptes Rendus,' June 6th, 1870) a short account of observations on inflammation, in which he states that he has failed to see the passage of white corpuscles through vascular walls described by Cohnheim. In inflammation of the peritoneum he has convinced himself that the leucocytes are not, at all events, produced by proliferation of the epithelium; but in inflammation of the cornea, the connective tissue corpuscles may give rise to new elements which assume the form of leucocytes.

INDEX TO JOURNAL.

VOL. X, NEW SERIES.

PRINTED BY J. E. ADLARD, BARTHOLOMEW CLOSE.

Ford lith

W. West imp.

Aphrocallistes bocagei sp. nov.

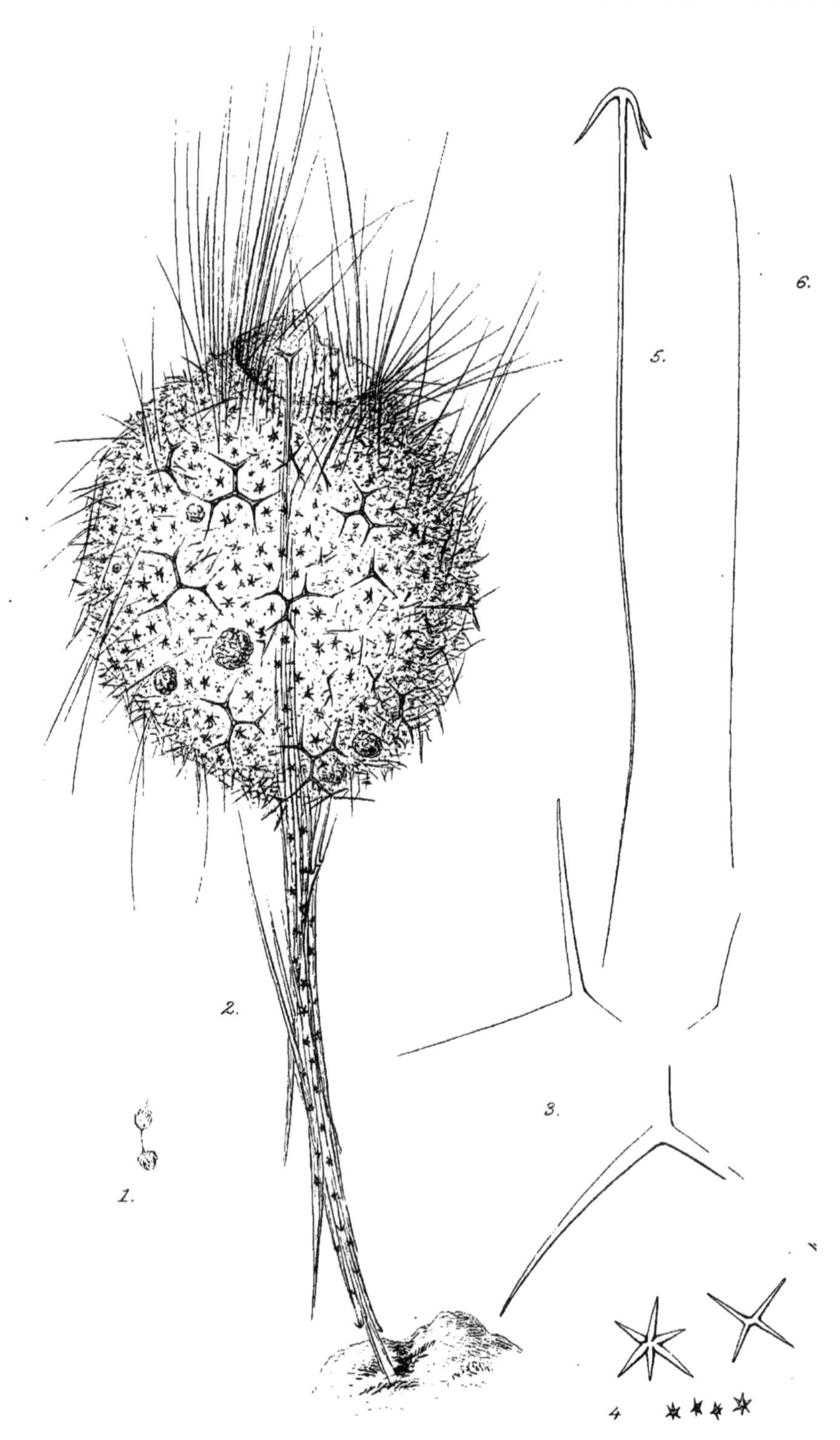

C Stewart, delt. ad nat

W H McFarlane

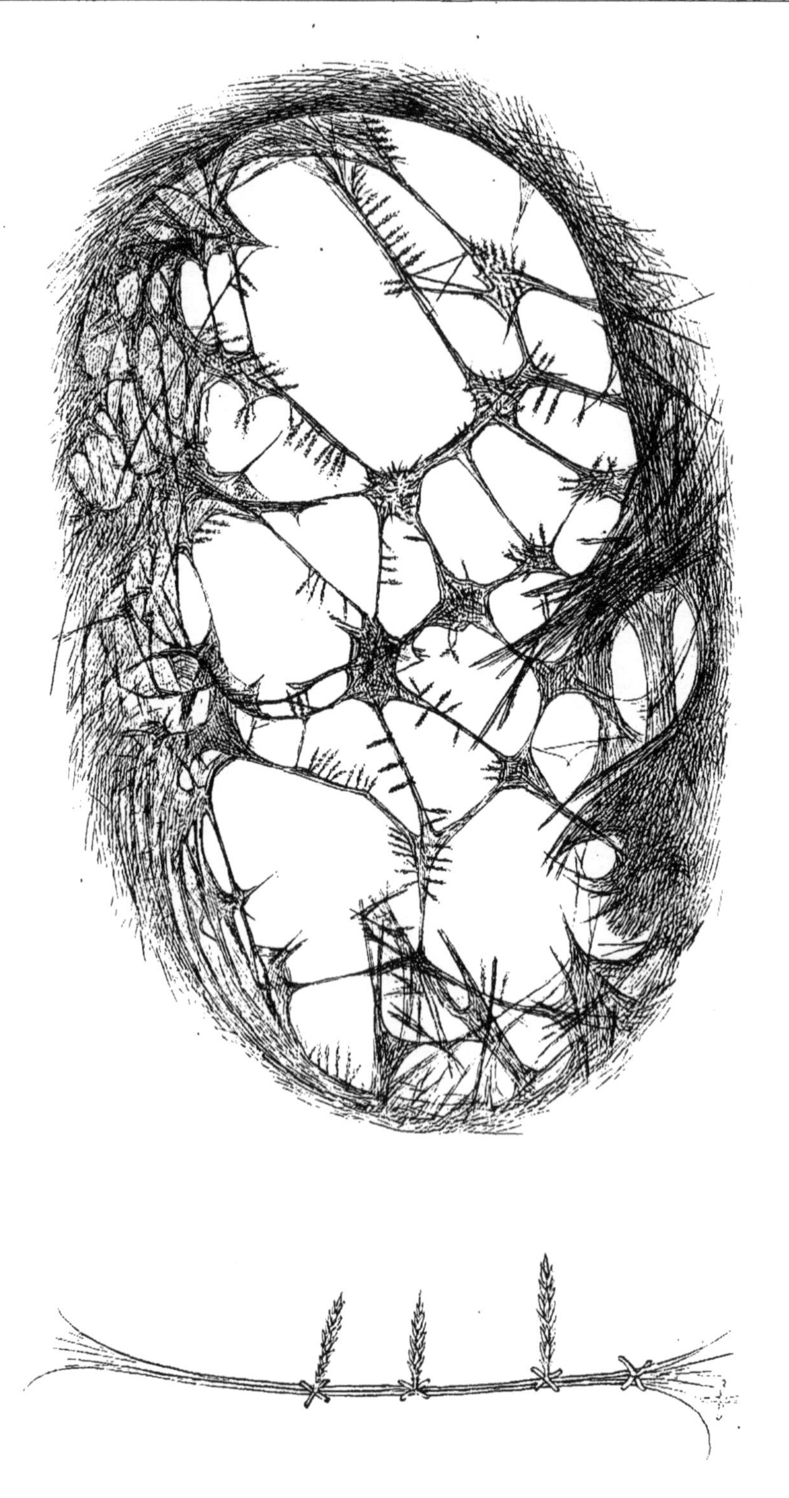

DESCRIPTION OF PLATES II & III,

Illustrating Professor Perceval Wright's Notes on Sponges.

PLATE II.

Wyvillethomsonia Wallichii, sp. nov.

Fig. 1. Specimen in the Museum of the Royal Microscopical Society. Nat. size.

„ 2. Same, enlarged. Some 'gemmule'-like bodies are seen imbedded in the mass. The object is drawn as seen when focussed down nearly to the equatorial margin, hence many of the stellate spicules are not seen.

„ 3. Furcated attenuato-patento-ternate spicule of Bowerbank (50).

„ 4. Stellate spicules of the bark layer.

„ 5. Recurvo-ternate spicule, Bwk. (54).

„ 6. Expando-ternate spicule, Bwk. (128).

PLATE III.

Hyalonema mirabilis, Gray.

Osulum from the interior of the Sponge mass, near the place where the 'glass coil' ends. Upper figure enlarged; lower figure showing three spiculate cruciform spicules, the cruciform portion not spiculate, and the stem more spiculate than usual.

JOURNAL OF MICROSCOPICAL SCIENCE.

DESCRIPTION OF PLATE IV,

Illustrating Mr. Kent's paper on *Victorella pavida.*

Fig.

1.—Small detached fragment of *Victorella pavida,* of the natural size.

2.—The same considerably enlarged.

3.—A piece of *Cordylophora lacustris* magnified much less, and showing the mode in which *Victorella* attaches itself to it.

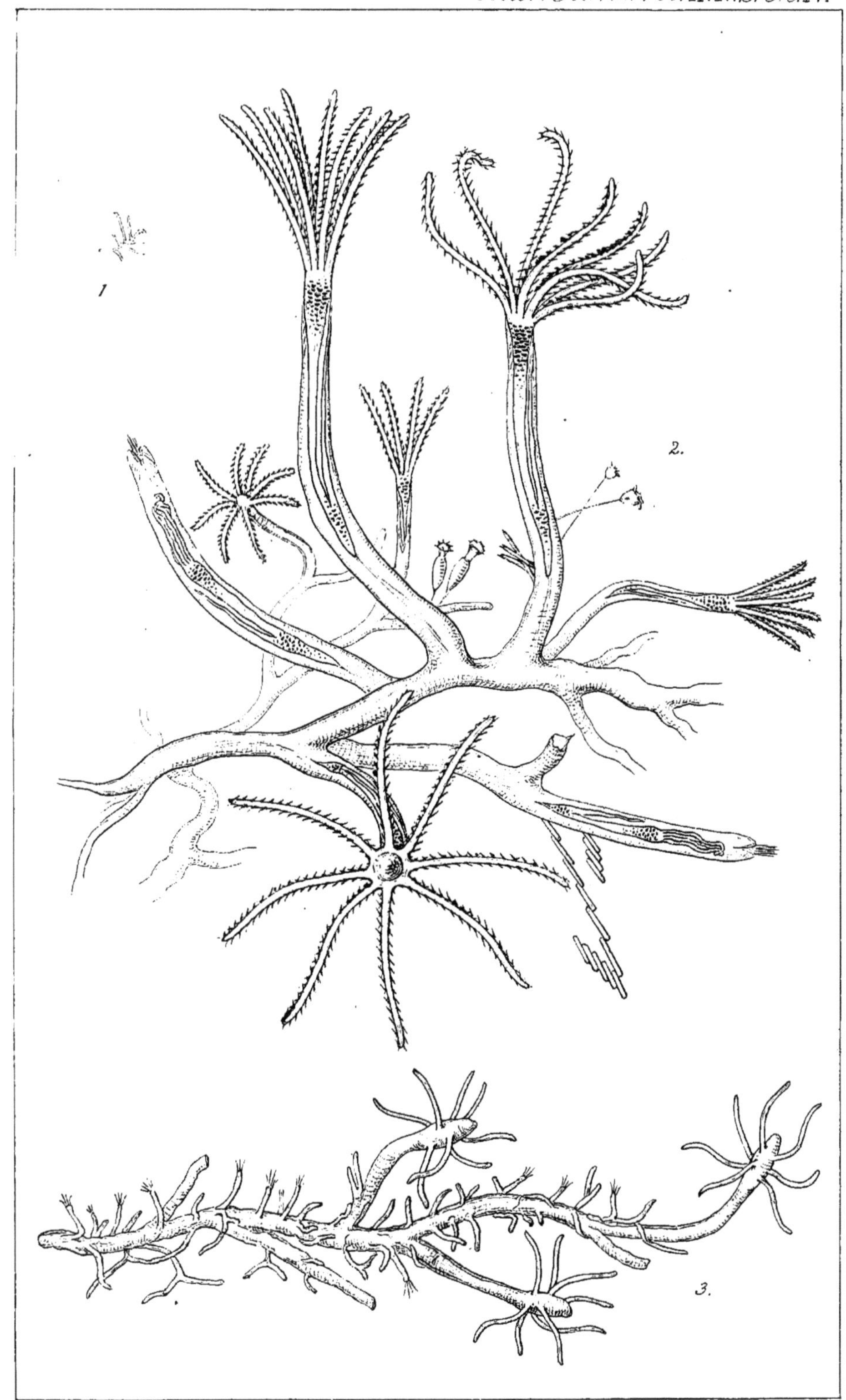

W. S. Kent, del.t W. H. M.c Farlane Lith.r Edin.r

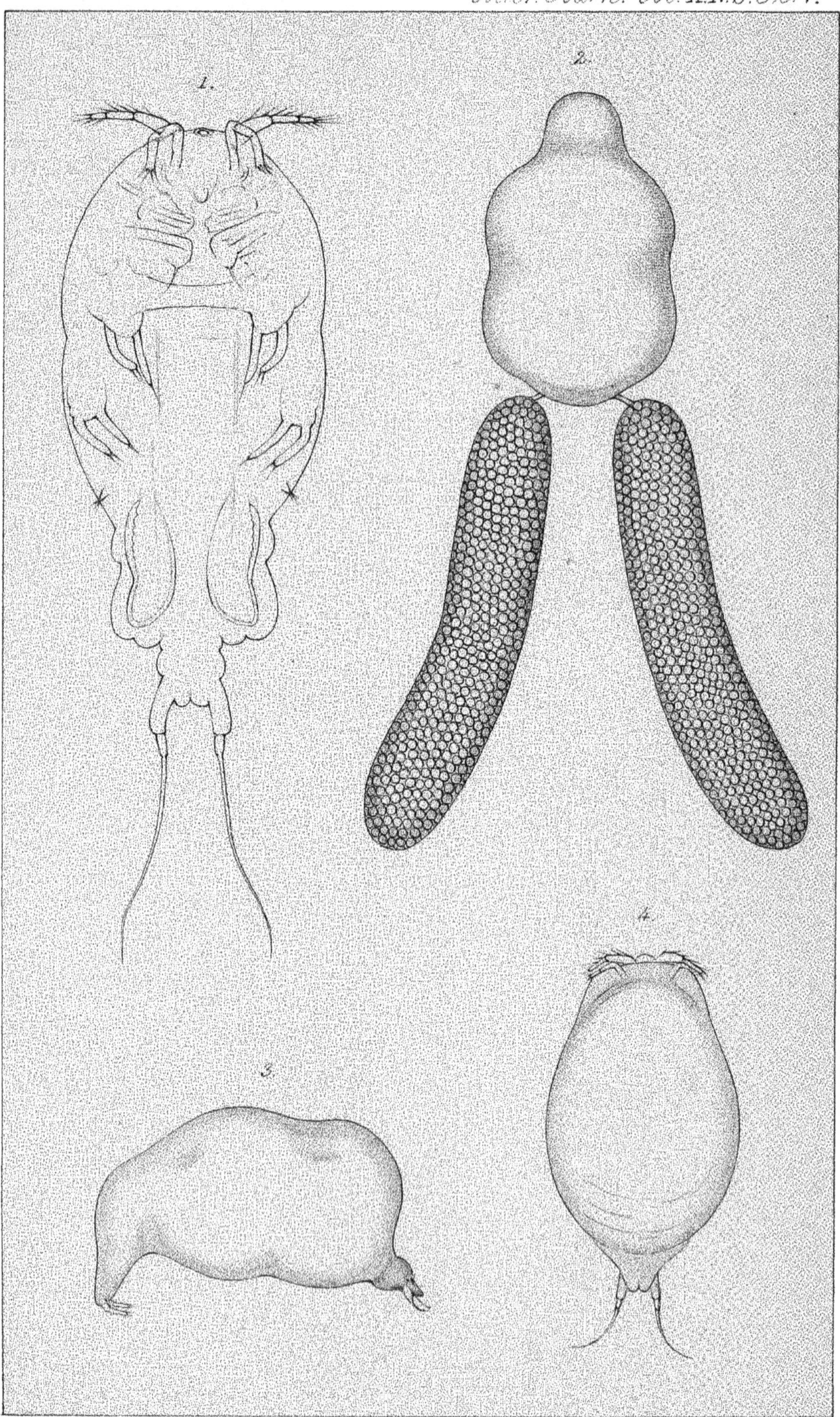

W. S. M. delᵗ

W H. Mᶜ Farlane Lithʳ Edinʳ

JOURNAL OF MICROSCOPICAL SCIENCE.

DESCRIPTION OF PLATE V.

Illustrating Dr. W. C. McIntosh's paper on a Crustacean Parasite of *Nereis cultrifera.*

Fig.

1.—Outline of the supposed male parasite of *Nereis cultrifera,* Grube. × 55 diameters.

2.—Adult female with ovisacs. Enlarged under a lens.

3.—Lateral view of an adult female. Similarly magnified.

4.—Young female. × 44 diameters.

EXPLANATION OF PLATE VI,

Illustrating Dr. Van Beneden's memoir on a new species of *Gregarina.*

Fig.

1 and 2.—Gregarinæ of middle size, as seen under a power of about 250 diameters.

3.—Some individuals of the natural size.

4.—Colossal Gregarina seen with a lens.

5.—Anterior extremity of the body of an individual of average size, magnified 420 diameters.

6.—*a, b, c, d.* Successive forms assumed by the same nucleus of a large individual. The nucleoli are seen to be modified, both as to number and aspect. 350 diameters.

7.—*a, b, c, d, e, f.* Successive stages of the nucleus of a young individual. These modifications were produced successively in the order of the letters in the space of twenty-five minutes. 300 diameters.

8.—A cyst of a Gregarina where no appearance of division has yet been produced. The membrane of the cyst is still very thin. It is closely applied to the granular mass.

9—10.—The same cyst which has undergone, under the microscope, the changes drawn.

11—12.—The same cyst showing the modifications taking place under the microscope, in the form of the spheres.

13.—Two cysts of the second generation, surrounded by the residue of the capsule of the primitive cyst. Each of them is surrounded by a proper membrane, and is itself in course of division.

14.—*Idem.* The spheres of the third generation are completely separated from one another, and well rounded.

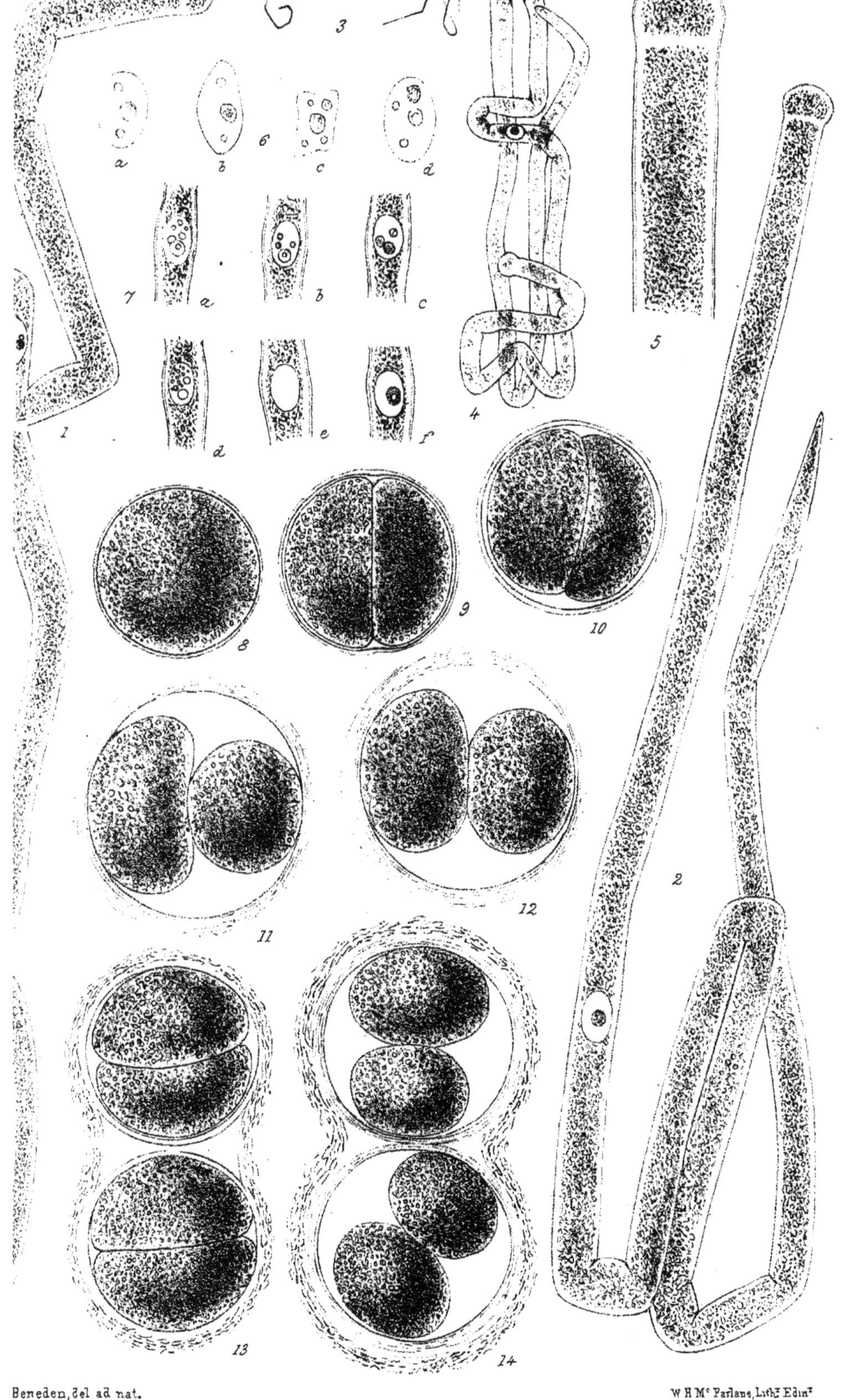

Beneden, del ad nat.

W H Mc Farlane, Lith. Edin.

Gregarina Gigantea, Ed. Van Ben.

1.
2.
3.
4.
5.
6.

JOURNAL OF MICROSCOPICAL SCIENCE.

DESCRIPTION OF PLATE VII,

Illustrating Prof. Cleland's paper on the Grey Matter of the Cerebral Convolutions.

Fig. 1, is a vertical view of the whole depth of the grey matter, not taken from any one specimen, but exhibiting the structure as it may be made out from a number of specimens, only very much shortened.

,, 2, copied from a specimen stained and preserved in glycerine, shows the appearance of some of the smallest nerve-corpuscles, where the external layer of nucleated protoplasm passes into the subjacent layer.

,, 3, taken from a position considerably deeper than that represented in fig. 2, shows various appearances of nerve-corpuscles and nuclei as they lay in the specimen.

,, 4, shows a nerve-corpuscle with three poles, each apparently continued into a medullated fibre.

,, 5, is a similarly shaped corpuscle to that shown in fig. 4, but with the horizontal and deep processes branching.

,, 6, is a large nerve-corpuscle liberated from the surrounding textures, and with small portions of granuliferous matrix, adhering to two of its basal processes. It shows globules within it, produced by the running together of smaller granules in consequence of the action of bile, and concealing the nucleus. It likewise exhibits an appearance of striation like that described by Arndt, which is interesting, as the corpuscle floated free without touching the glass cover.

DESCRIPTION OF PLATE VIII,

Illustrating Dr. Edouard Van Beneden's Memoir on Nematobothrium.

Fig.

1.—Embryo of Nematobothrium, magnified 450 diameters.

2.—An egg enclosing an embryo bent on itself, magnified 450 diameters.

3.—An egg, magnified 450 diameters, to show the mode of dehiscence.

4 & 5.—The embryo of the same, magnified 600 (4) and 650 (5) diameters.

6.—An embryo as it is when folded up in the egg, magnified 600 diameters.

7 & 8.—Muscular disc of the same, under a power of 1200 diameters.

9.—Portion of the muscular disc of another individual; the hooklets are both shorter and stronger than those of the individuals represented in figs. 7 and 8.

10.—The egg of *Distoma tereticolle.* The chorion is surrounded by a thick layer of transparent substance.

11 & 12.—Embryo of the same, highly magnified.

13 & 14.—Embryo of *Distoma filicolle.* The division of the body into two rings, by a circular furrow, is distinguishable. There is no trace of prickles.

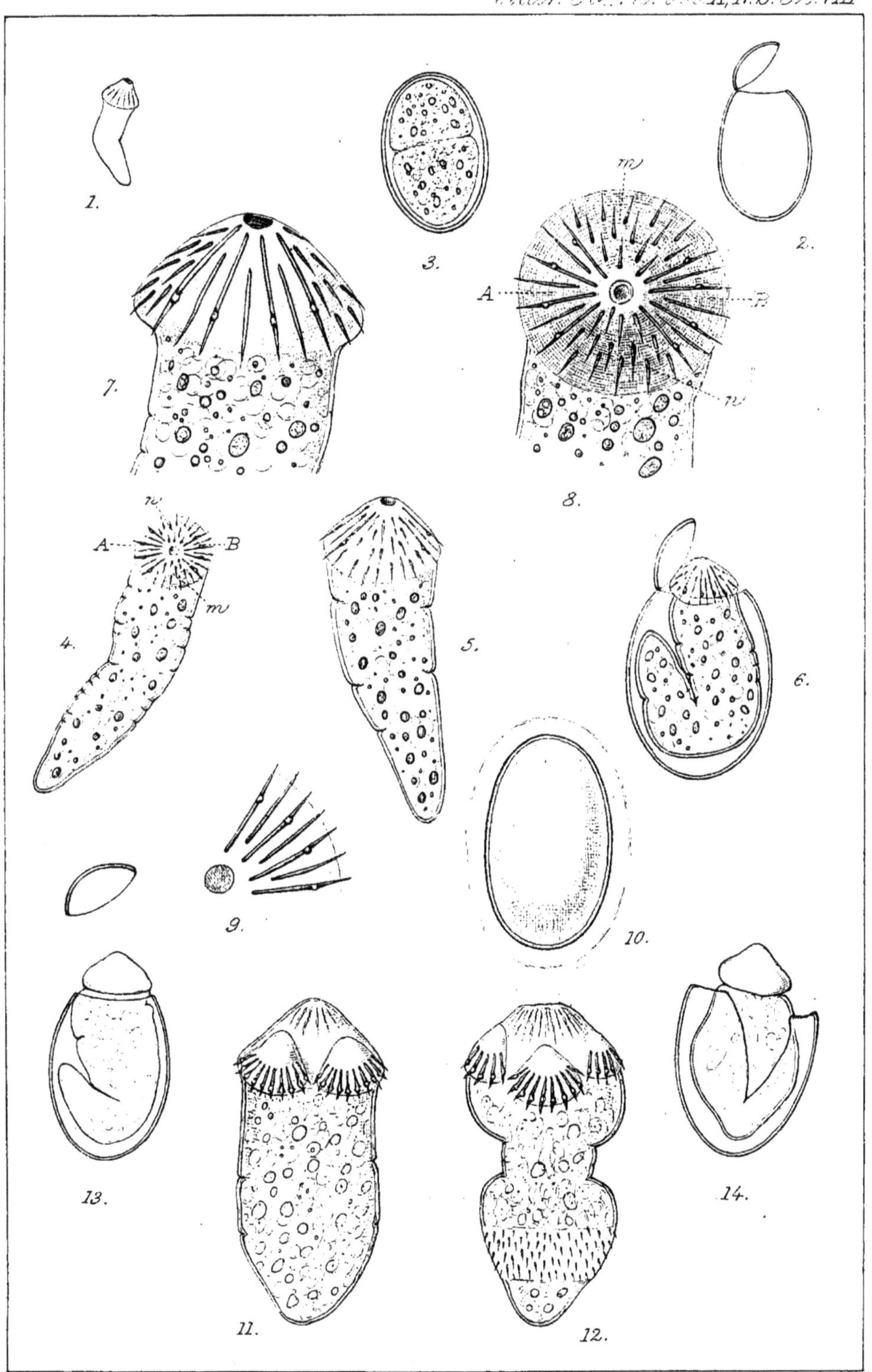

Edouard Van Beneden, ad nat. del^t

W.H.Mc Farlane, Lith^r Edin^r

1-9 Nematobothrium filarina. 10-12. Distoma Tereticolle.
13-14 Distoma filicolle

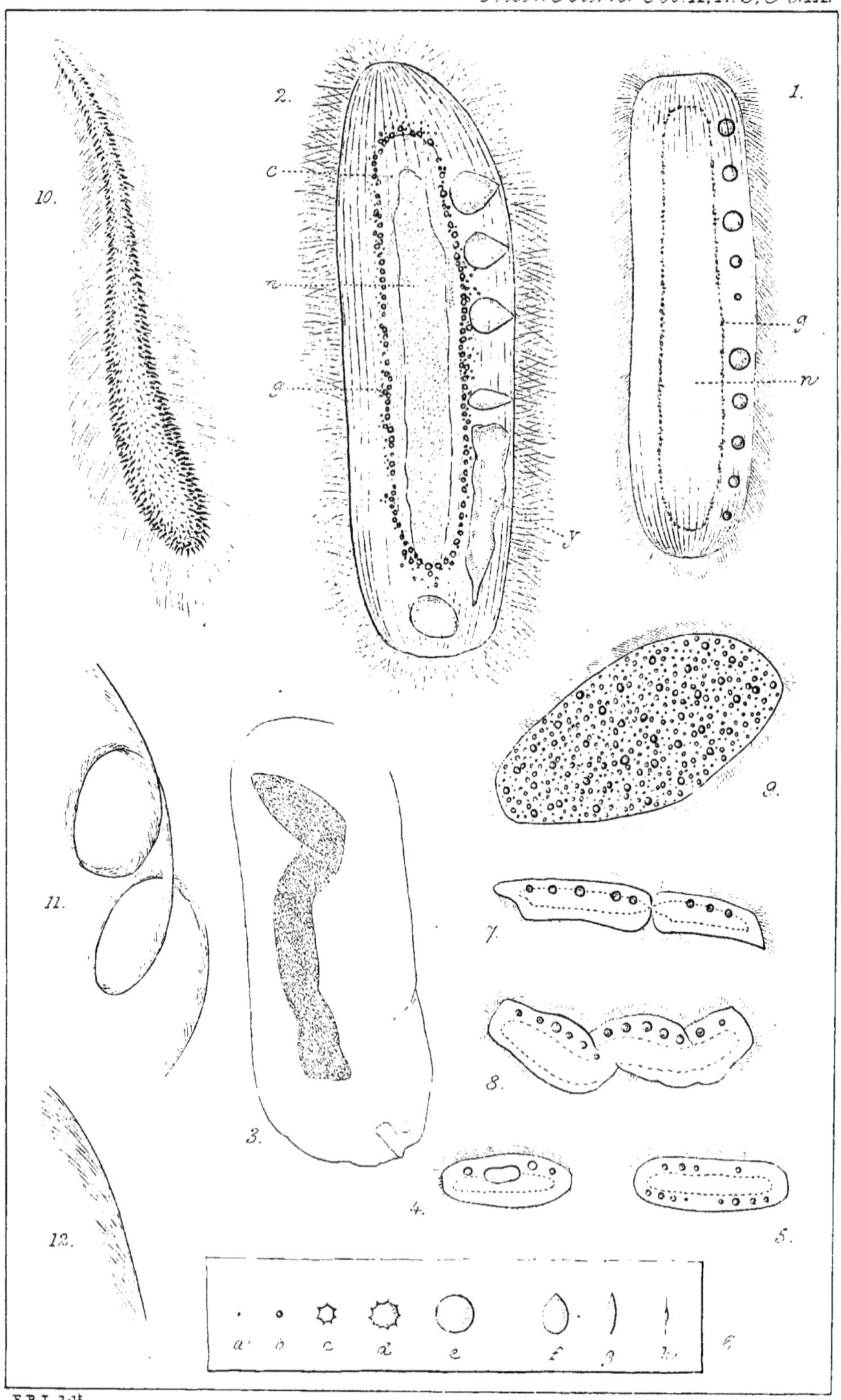

E R L del^t

EXPLANATION OF PLATE IX,

Illustrating Mr. Ray Lankester's remarks on Opalina, and Pachydermon.

The figures are very variously magnified.

Fig.

1.—*Opalina Naidos,* as seen when living.

2.—Another specimen with post-mortem changes. *c,* cavity between nucleus and parenchyma; *n,* nucleus; *g,* granular layer; *y,* distended and elongated vesicle.

3.—Weak acetic acid added so as to bring out the nucleus and its coarsely granular structure.

4.—A specimen with large vesicle; living.

5.—A specimen with vesicles on both sides of the body.

6.—*a, b, c, d, e,* expansion; *e, f, g, h,* contraction, of a vesicle.

7.—Transverse fission of *O. Naidos.*

8.—A living specimen deformed by pressure in its host's body.

9.—*Bursaria Ranarum,* Ehrenberg.

10.—Spermatophor (*Pachydermon,* Claparède) of *Limnodrilus Claparèdii.* This should be at least four diameters larger if drawn to the same scale as fig. 1.

11.—A part of one of the long coiling spermatophors of *Nais serpentina.* This figure is on the same scale as fig. 1, but represents only a small bit of the long spermatophor.

12.—A portion of the same more highly magnified.

DESCRIPTION OF PLATE X,

Illustrating Dr. Royston-Pigott's paper on High-Power Definition and its Difficulties; and the Visibility of Diatomaceous Beading.

Fig.

1 and 2.—Longer and shorter black lines of interference according to the less or greater angle of intersection of ribbing forming *Podura* waviness.

3.—Delusive lines formed by beading imperfectly defined in diatoms and in scales, as in Figs. 6, 7, 8.

4 and 5.—Hexagon shadows developed by ill-defined spherules.

9, 10, 11, 12, and 13.—A variety of appearances shown by diatoms according to their treatment; the nature of the objectives and the direction of the light.

14.—Section of beading of scales arranged in many cases in close contact.

15.—Black shadows of intersecting superimposed beading.

16.—Their accepted though erroneous appearance, as in *Podura curvicollis* test scale.

17, 18, and 19.—Observed diffraction, causing broken lines when fine parallel lines intersect with a small space between the planes in which they lie.

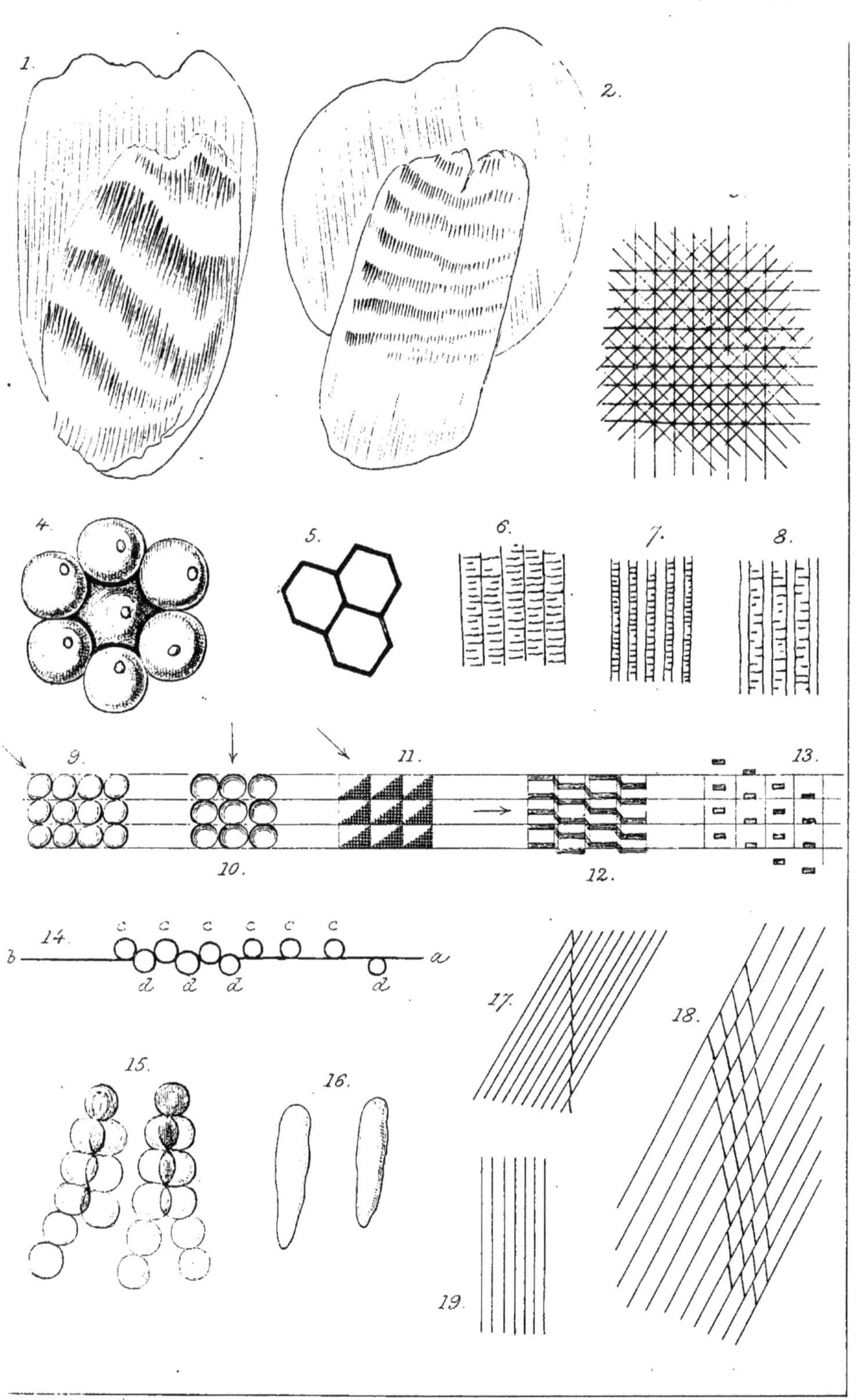

W. H. Mc Farlane, Lithr Edinr

INCREASE OF BIOPLASM.—DEVELOPMENT OF VESSELS AND OTHER TISSUES

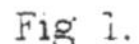

Fig. 1.

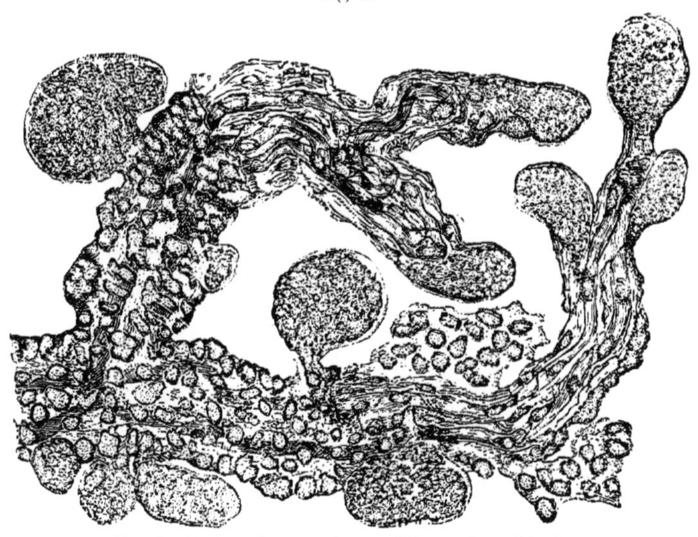

Growing extremities of fœtal tufts from human placenta, 7th month, × 215 At every extremity [illegible] a collection of small masses of bioplasm which are undergoing division These advance [illegible] the vessels and other structures grow in their wake as it were. The masses of bioplasm of the cap[illegible] and other structures entering into the formation of the tuft are also well seen 1861 p 215

Fig. 2.

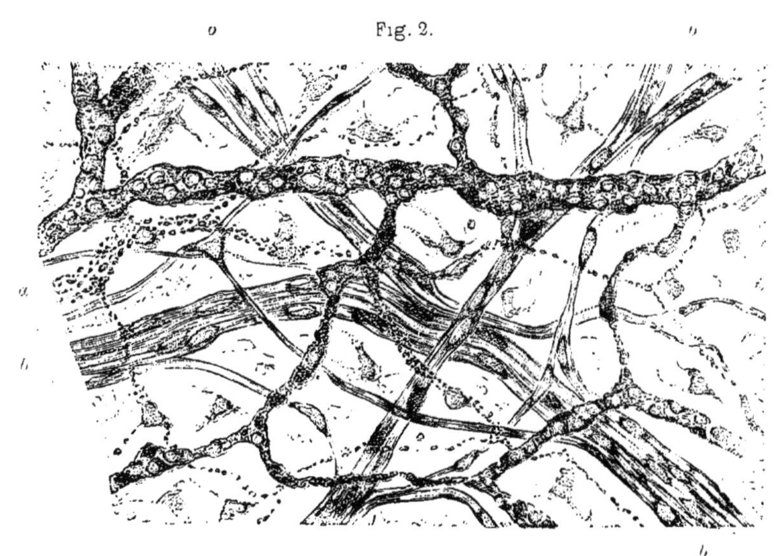

Bladder of a frog which was half starved The capillaries are wasting, and contain bioplasm only No red blood corpuscles could be detected Bundles of unstriped muscle are seen ramifying over the field Some have fibres radiating in three directions, and the bioplasm of these is triangular *b* At *a*, a bundle of very fine nerve fibres is represented Its ramifications may be followed over every part of the specimen The bioplasm of the connective tissue is also represented Thus, *all* the tissues of the thin bladder are demonstrated The drawing was taken from a specimen mounted in 1862 p 217

$\frac{1}{1000}$ of an inch —— × 215 linear.

BLOOD BIOPLASTS, OR, WHITE BLOOD CORPUSCLES IN VESSELS OF EMBRYO.

Fig. 3.

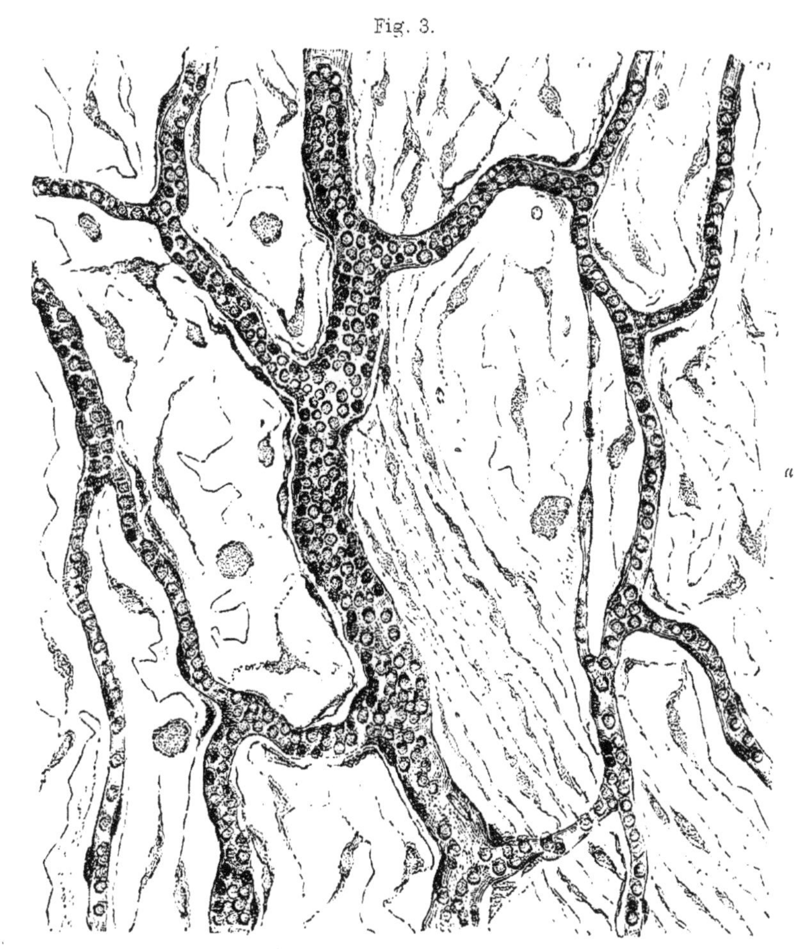

Capillary vessels and small vein from the ovum of the turtle at an early period of development The vessels are entirely filled with white blood corpuscles, and in some places they were completely distended with them Developing connective tissue with connective tissue corpuscles, fat cells and perhaps nerve fibres are also seen To the right of the drawing at *a* will be observed a very young capillary, the tube of which is not yet wide enough to allow a blood corpuscle to pass through it. × 215 1861. p 218

$\frac{1}{1000}$ of an inch —— × 215 linear.

BIOPLASM IN BLOOD VESSELS.

Fig 4.

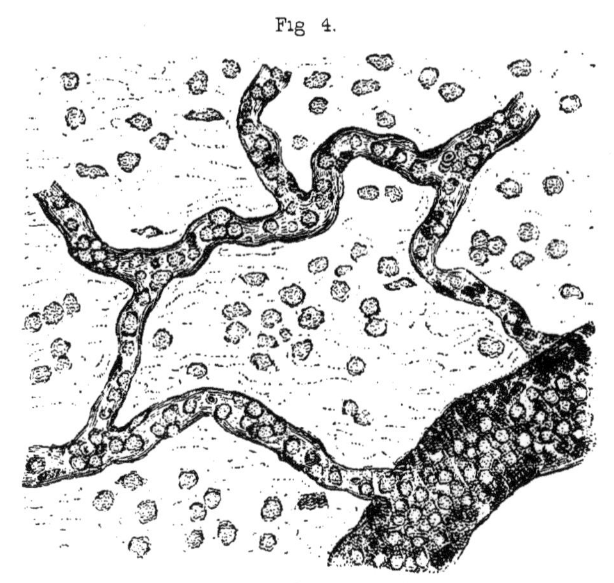

Capillary vessels opening into a small vein Pia mater. Human fœtus, fifth month of intra uterine life The capillaries contain numerous white blood corpuscles (bioplasm) which are coloured with carmine, and the vein is completely filled with them. Very few red blood corpuscles were present × 215 p 216

Fig. 5

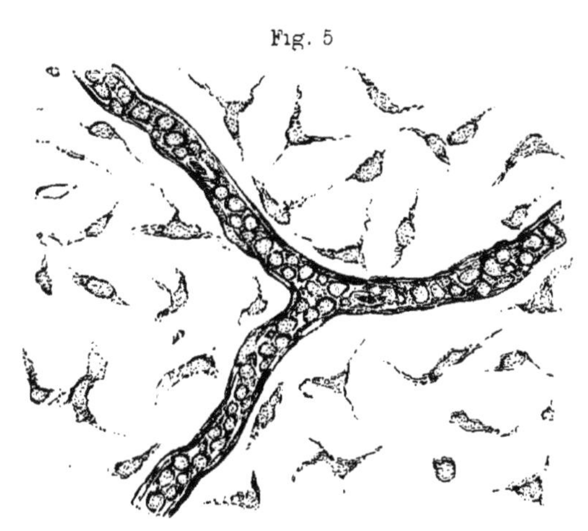

Capillary vessels and connective tissue, and connective tissue corpuscles Mesentery. Frog in winter The capillaries are filled with numerous white blood corpuscles (bioplasm) Only one or two red blood corpuscles were present × 215. p 217.

$\frac{1}{1000}$ of an inch —— × 215 linear.

INFLAMMATION.

Fig. 6

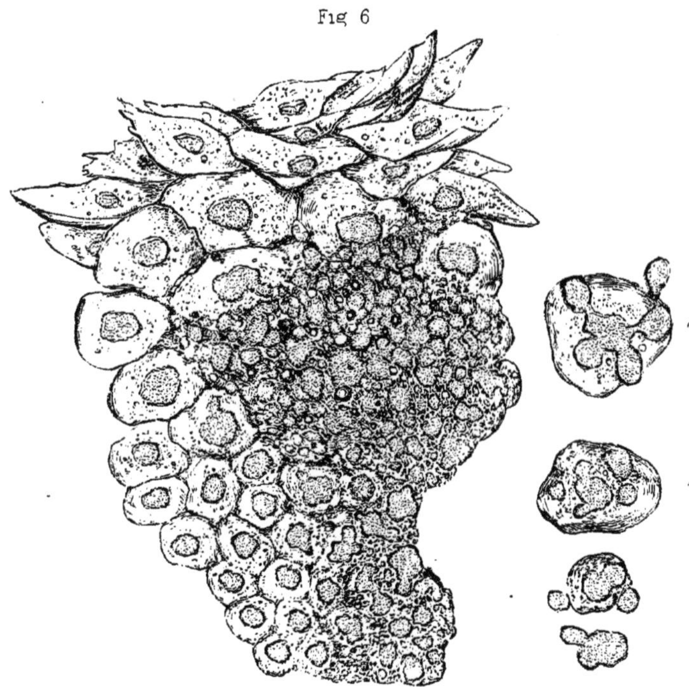

A portion of the epithelial covering of the tongue in a state of inflammation. The bioplasm at the lower part is growing and multiplying very rapidly. The changes taking place in individual cells or elementary parts are represented on the right at *a, b, c, d,* × 700. This drawing has not been copied from a single preparation, but has been completed from the appearances observed in several different specimens, p. 223.

Fig. 7.

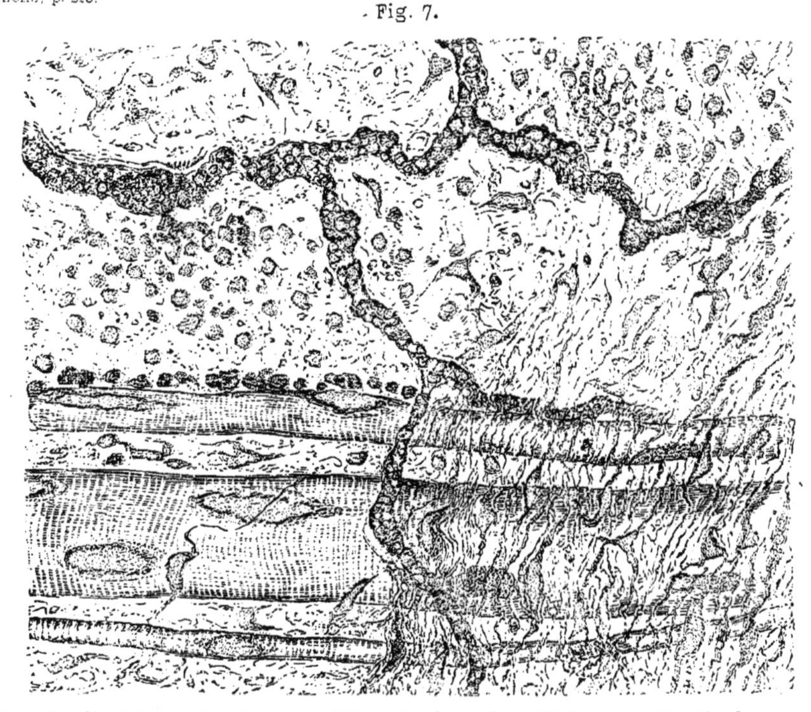

Muscular fibres and connective tissue of the pectoral muscle of the frog in a state of inflammation from the immediate neighbourhood of the seat of injury four days after the muscle had been transfixed by a fine thread. The vessels are seen to be filled with white blood corpuscles and some of the bioplasm particles are escaping from the vessel at two points. The bioplasm of the connective tissue and also that of the muscular fibres is much increased × 215. From a specimen mounted in 1863 p. 225.

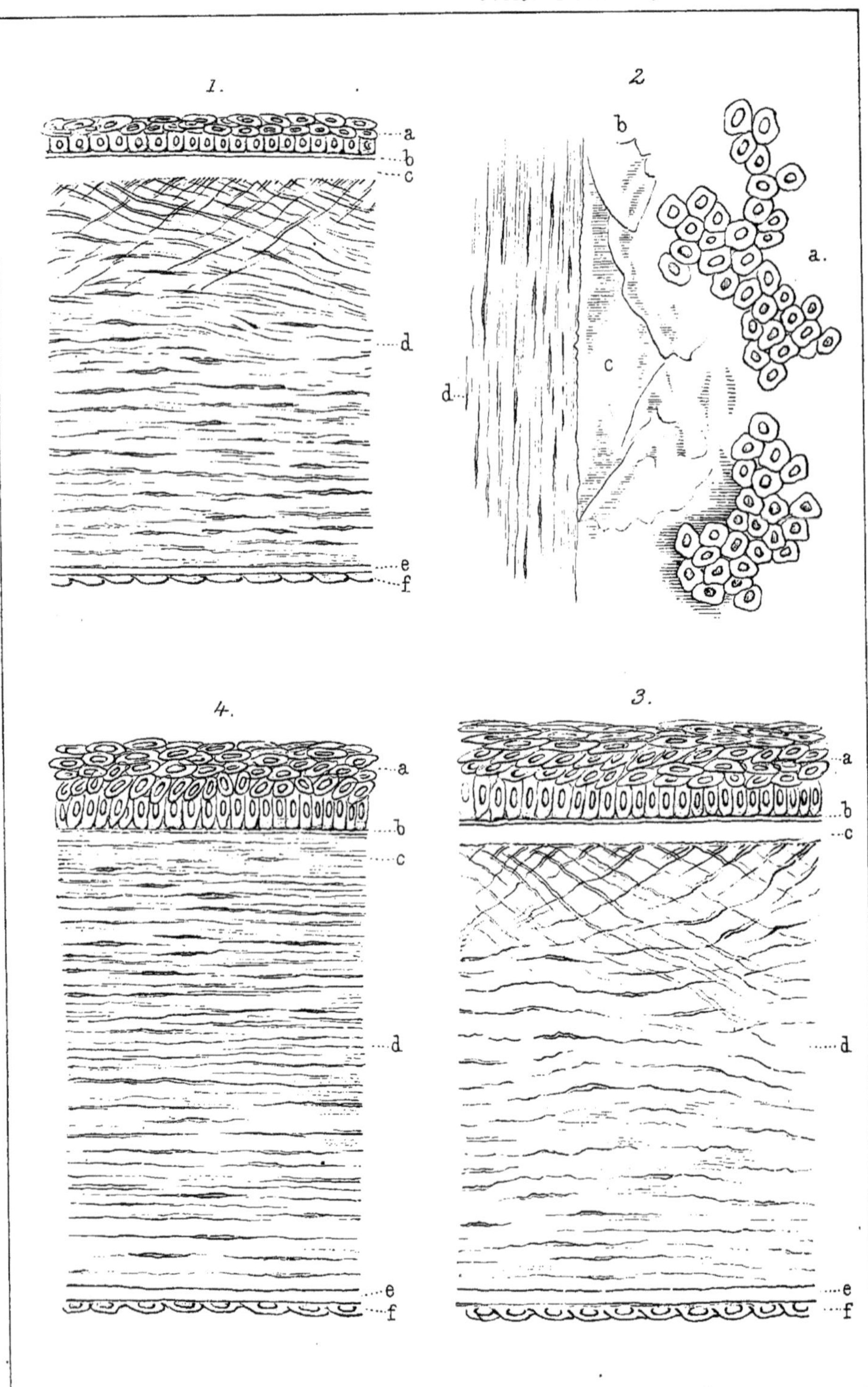

Dr. Macdonald, ad nat. del.t

W. H. McFarlane Lith.r Edin.r

DESCRIPTION OF PLATE XV,

Illustrating Dr. John Denis Macdonald's paper on the Minute Anatomy of some of the Parts concerned in the Function of Accommodation to Distance, with Physiological Notes.

Fig.

1.—Vertical section of human cornea.
- *a.* Epithelium of conjunctiva.
- *b.* Basement membrane of ditto.
- *c.* Anterior elastic lamina.
- *d.* Cornea proper.
- *e.* Posterior elastic lamina.
- *f.* Single pavement of cells.

2.—Portion of human cornea seen in face, near the margin of the preparation.
- *a.* Conjunctival epithelium.
- *b.* Basement membrane ripped up from—
- *c.* Anterior elastic lamina.
- *d.* Cornea proper.

3.—Vertical section of cornea of the pig.
- *a.* Conjunctival epithelium.
- *b.* Basement membrane.
- *c.* Anterior elastic lamina.
- *d.* Cornea proper.
- *e.* Posterior elastic lamina.
- *f.* Single pavement of cells.

4.—Vertical section of cornea of the sheep.
- *a.* Conjunctival epithelium.
- *b.* Basement membrane.
- *c.* More condensed fibrous tissue.
- *d.* Cornea proper.
- *e.* Posterior elastic lamina.
- *f.* Single pavement of cells.

EXPLANATION OF PLATE XVI,

Illustrating Dr. John Denis Macdonald's paper on the Minute Anatomy of some of the Parts concerned in the Function of Accommodation to Distance, with Physiological Notes.

Fig.

1.—Vertical section of the cornea and sclerotic of shark.
- *a.* Conjunctival epithelium.
- *b.* Cornea proper, composed of superimposed laminæ corresponding with the whole area of the cornea.
- *c.* Fibrous part of sclerotic.
- *d.* Cartilaginous portion of ditto.
- *e.* Points of ossification near the surface.

2.—Some few superficial ossific points seen in face, imbedded in the cartilage.

3.—Anterior surface of the iris of the pig in connection with the ciliary muscle.
- *a.* Sphincter fibres.
- *b.* Decussating oblique muscular fibres.
- *c.* Pillars of the iris torn from their origin in the cornea.
- *d.* Annular muscle (*sphincter ciliaris*) occupying the inner wall of circular sinus, and resting upon the ciliary muscle.
- *e.* Fibrous bundles at the posterior part of the circular sinus, torn from their connection with the sclerotic.
- *f.* Ciliary muscle.

4.—Posterior surface of iris, with ciliary processes *in sitû,* also from the pig.
- *a.* Sphincter fibres of the iris.
- *b.* Radiating fibres invested with faint decussating striæ.
- *c.* Ciliary processes and folds.
- *d.* Distribution of vessels with pigmentary deposit in the meshes.

5.—Vertical section of the fore part of the eye of the pig (diagramatic).
- *a.* Conjunctival epithelium.
- *b.* Anterior elastic lamina.
- *c.* Cornea proper.
- *d.* Posterior elastic lamina.
- *e.* Anterior chamber.
- *f.* Iris.
- *g.* Pillars of the iris.
- *h.* The annular muscle (*sphincter ciliaris*) seen in continuity.
- *i.* Ciliary processes in posterior chamber.
- *k.* The pupil.
- *l.* The lens.
- *m.* The canal of Petit.
- *n.* The hyaloid membrane.
- *o.* Ciliary muscle.
- *p.* Choroid.
- *q.* Sclerotic.
- *r.* Fibrous bundles passing into the sclerotic.
- *s.* The circular sinus.

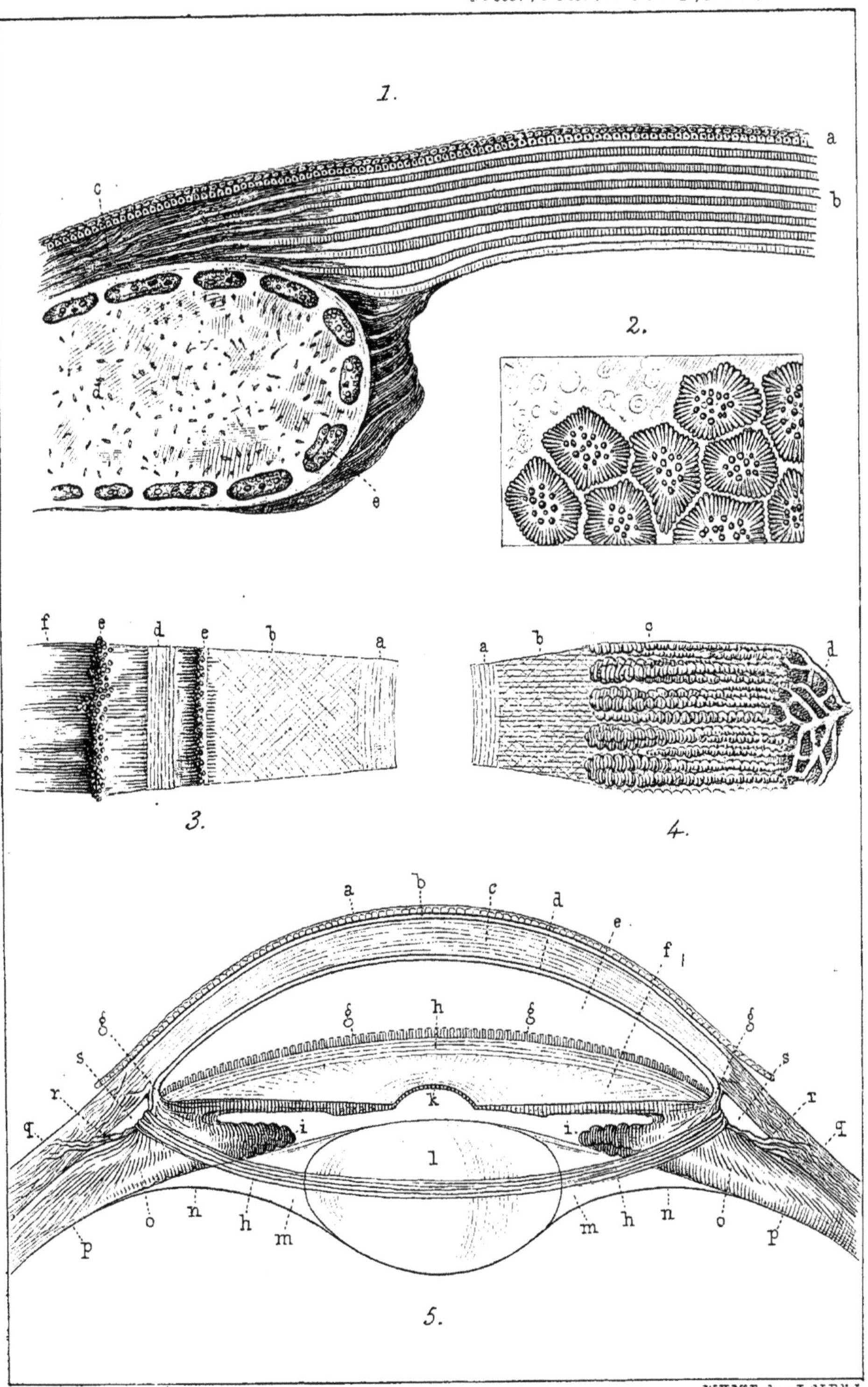

Dr. Macdonald ad nat. delt.

W. H. McFarlane, Lithr. Edinr.

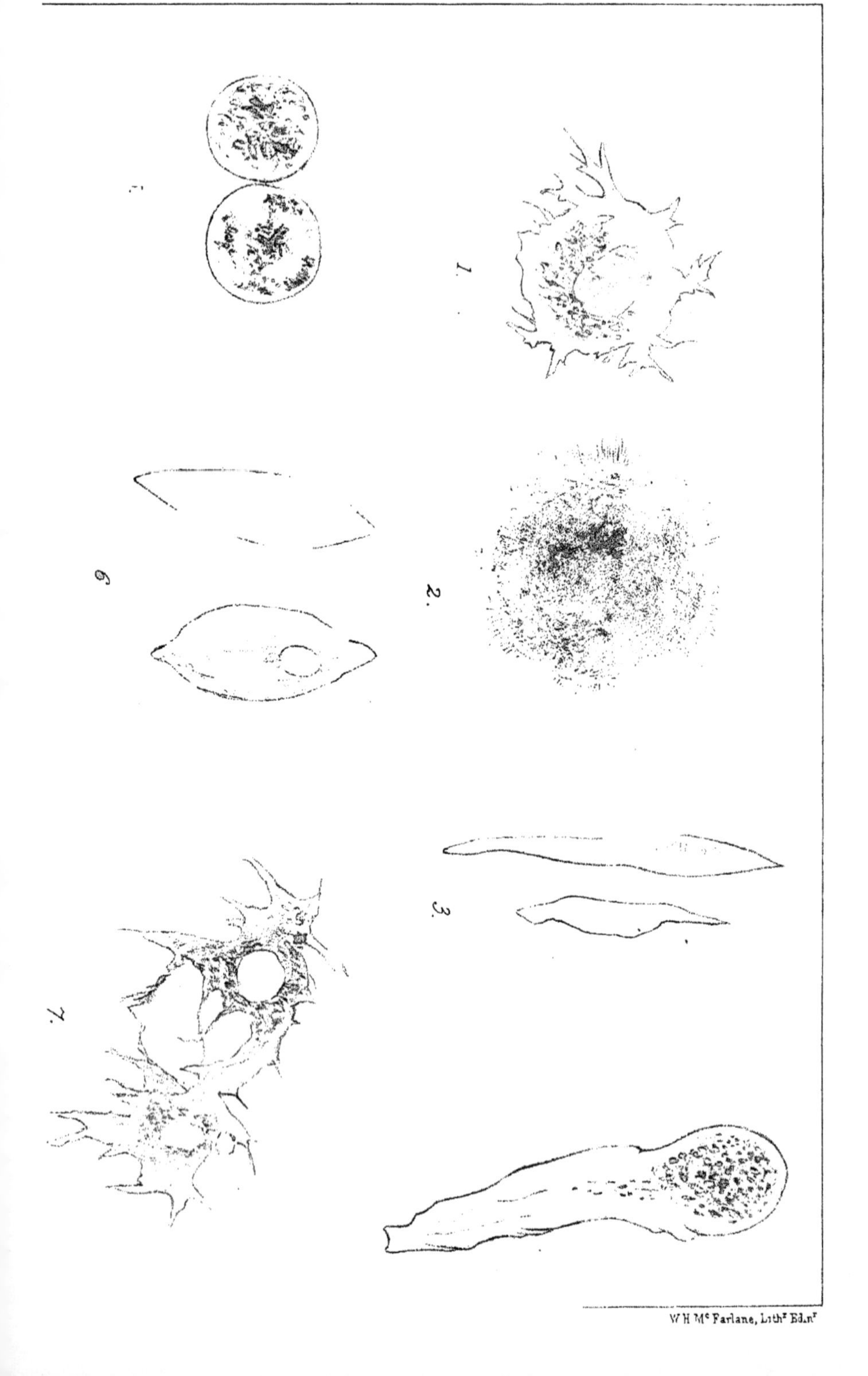

W H Mc Farlane, Lithr Edinr

DESCRIPTION OF PLATE XVII,

Illustrating Mr. E. Ray Lankester's notes on some Migrations of Cells.

Fig.

1.—Cell with pseudopodial processes from perivisceral cavity of *Lumbriculus* (probably migrated from the intestine).

2.—Arrested sperm-cell from the perivisceral cavity of *Tubifex* (in October).

3.—Migrated muscular-fibre elements found in the perivisceral cavity of *Tubifex* and *Lumbriculus*.

4.—Goblet-cell from the intestine-wall of *Limnodrilus*.

5.—Migrated globular cells of intestine-wall from perivisceral cavity of *Tubifex*.

6.—Oat-shaped perivisceral cells of *Enchytræus*, migrated from intestine.

7. Branched yellow cells of the perivisceral endothelium of a species of *Limnodrilus*.

JOURNAL OF MICROSCOPICAL SCIENCE.

DESCRIPTION OF PLATE XVIII,

Illustrating Mr. C. Stewart's paper on a New Sponge, *Tethyopsis columnifer*.

Fig.

1.—The sponge of the natural size.

2.—Ditto, with the basal portion and lower part of the stem in section.

3.—Under surface.

4.—Spiculum of the lower portion.

5.—Spiculum of the axis of the stem.

6.—Stellate spicula of the dermal membrane.

7.—Stellate spicula of the sarcode.

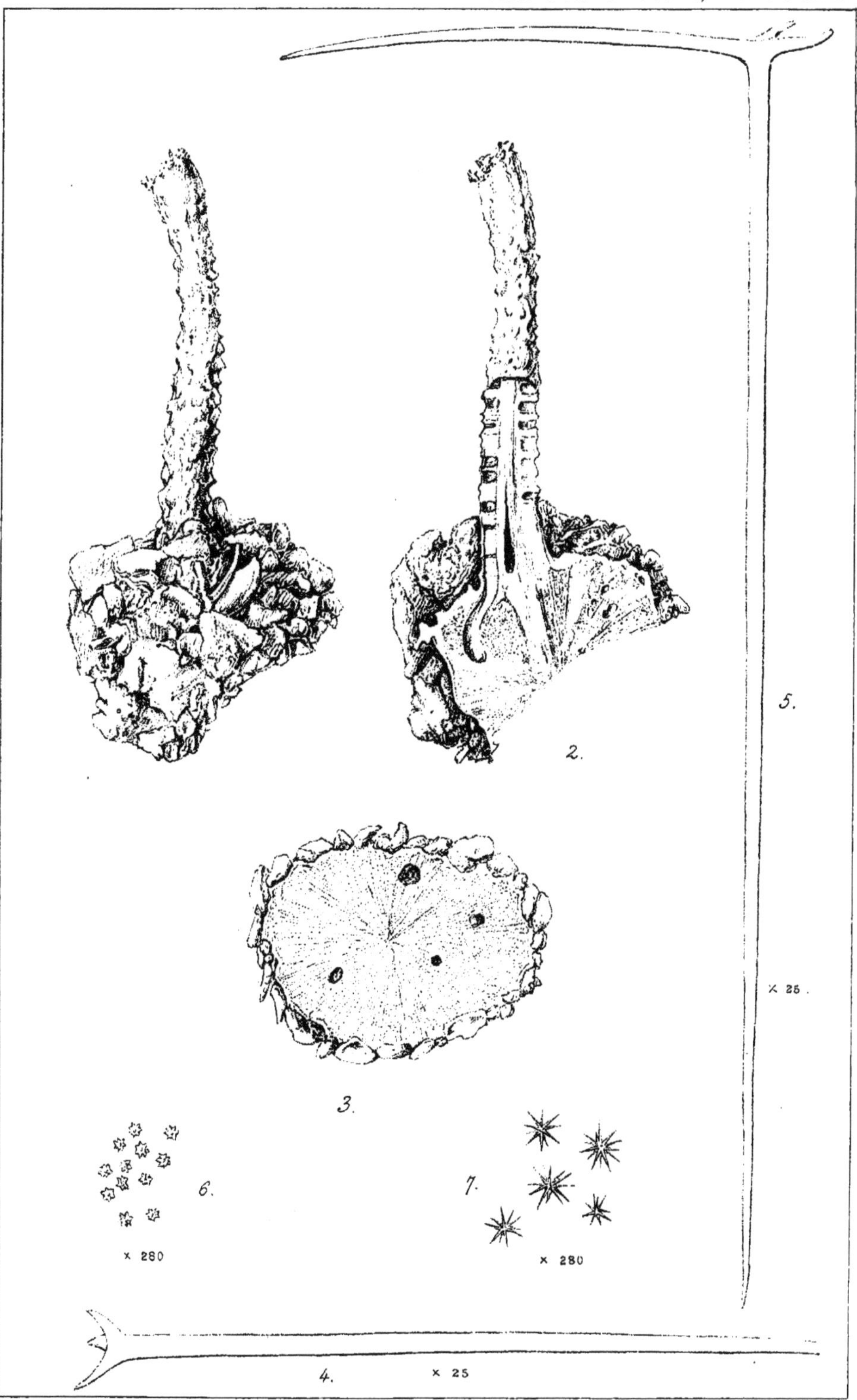

W H Mc Farlane, Lithr Edinr

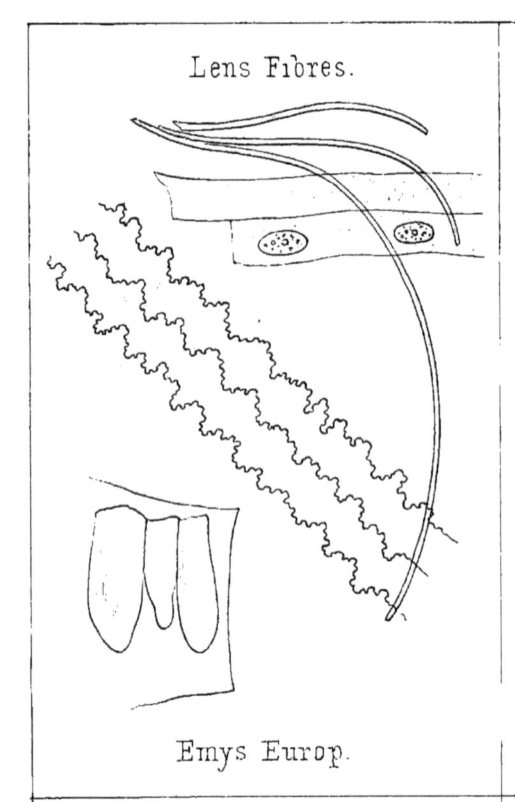

Lens-fibres & Capsular Epithelu

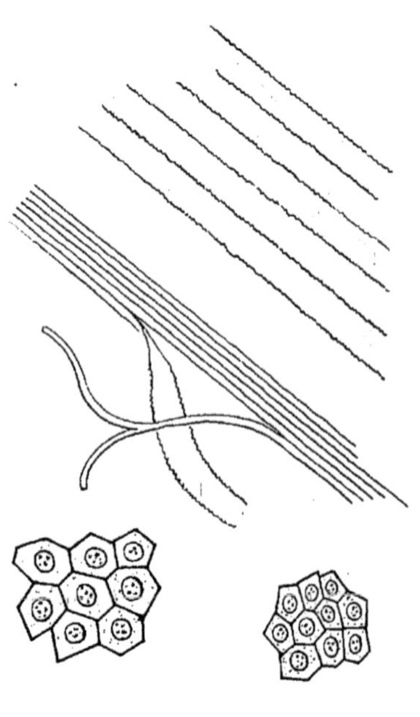

Rana temporaria.

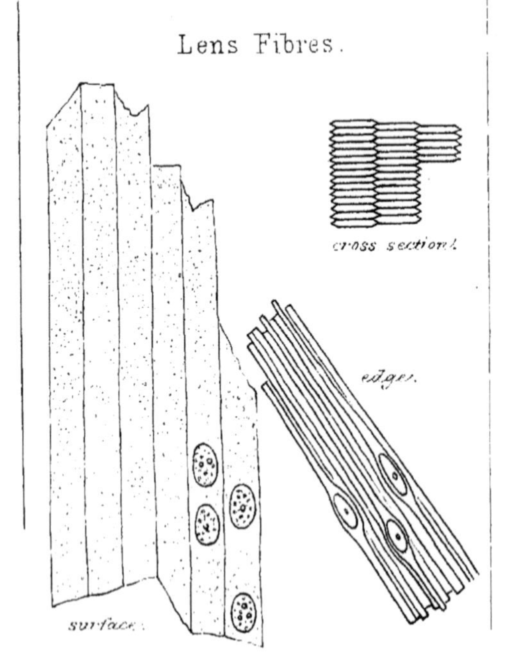

Chelone Mydas.

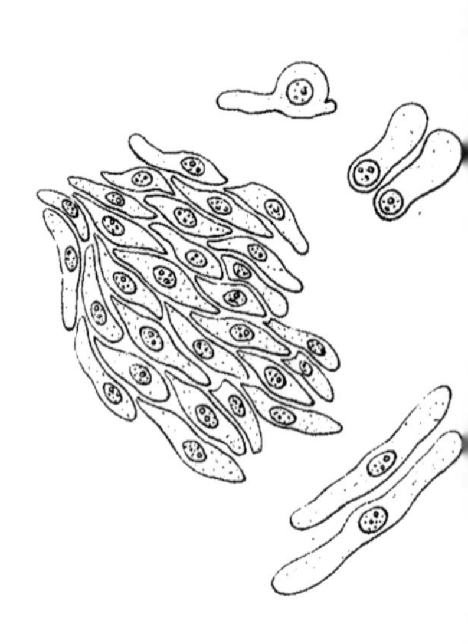

Development of lens fibres. Hum

J Hulke, F.R.S. del.

W H Mc Farlane l

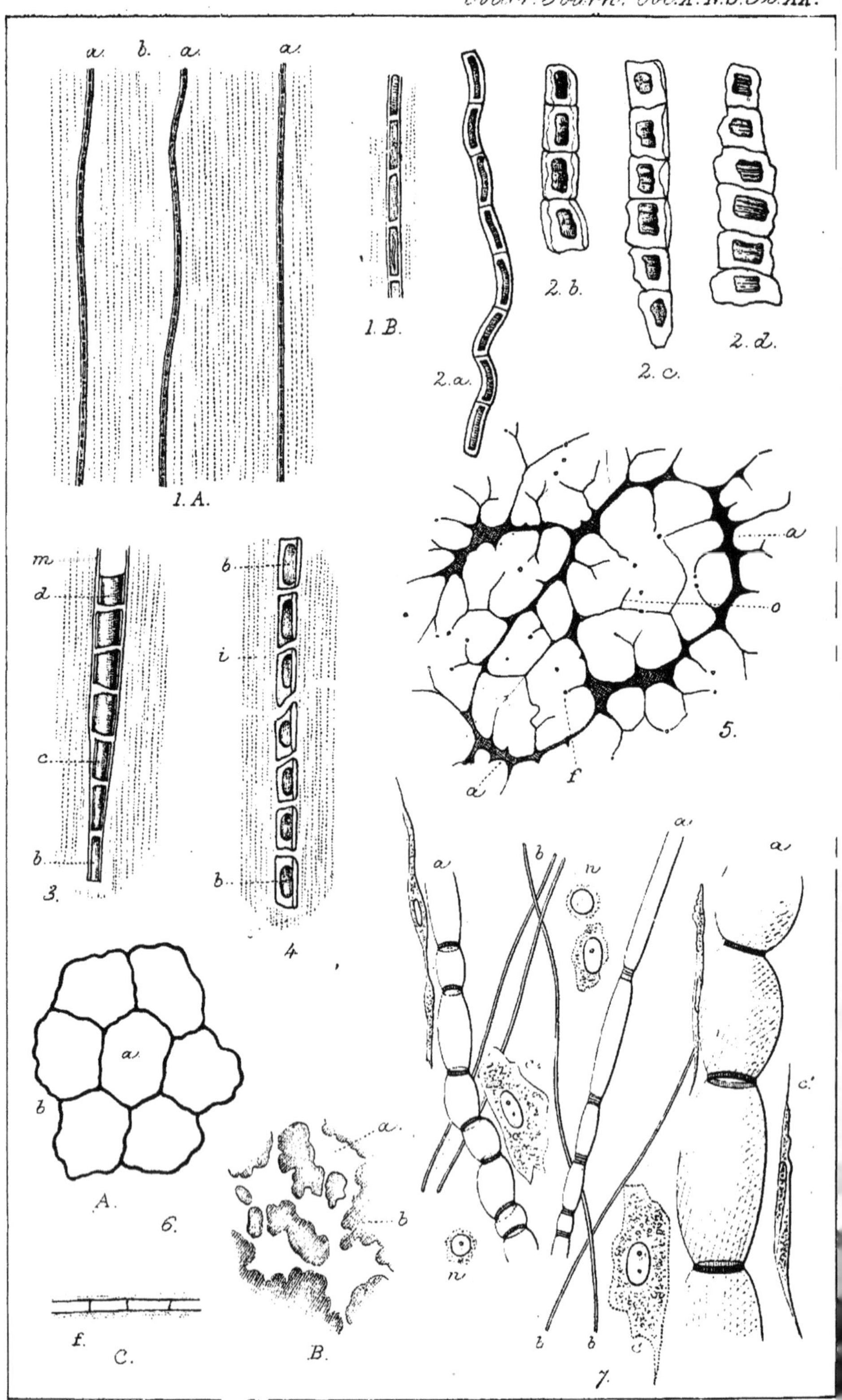

L. Ranvier, delt. W. H. McFarlane, Lithr. Edinr.

EXPLANATION OF PLATE XIX,

Illustrating Mr. Hulke's observations on the Histology of the Eye.

Four diagrams of lens fibres.

EXPLANATION OF PLATE XX,

Illustrating Dr. Ranvier's paper on the Cellular Elements of Connective Tissue.

Fig.

1.—Tendon from the tail of an albino rat.

A. Tendon kept stretched, × 100 diameters; *a*, cellular tubes; *b*, bundles of connective tissue rendered transparent by acetic acid.

B. One of the cellular tubes under a power of 400 diameters.

2.—Cellular tubes from the tendons of the tail of the adult mole, × 400 diameters; *a*, unopened tube; *b*, tube partly opened; *c*, tube nearly completely opened; *d*, tube compressed by the contraction of the connective substance; the nuclei are plicated.

3.—Tendon from the tail of an adult albino rat, × 400 diameters; *b*, rolled-up nucleus; *c* and *d*, opened nuclei; *m*, membrane of the common sheath.

4.—Tendon from the tail of an adult albino rat after 48 hours of inflammation, × 400 diameters; *b*, nucleus of the cell rendered apparent; *i*, substance of the connective-tissue bundles.

5.—Transverse section of a tendon from the tail of an albino rat, aged one month; *a*, septa giving rise to finer septa appearing like fibrills *f*.

6.—Impregnation of the tendons by a solution of nitrate of silver of 2 to the 100.

A. Epithelium of the surface. B. Layer of sub-epithelial connective tissue. C. Tube with the separation of the cells marked by the deposit of silver.

7.—Sub-cutaneous cellular tissue from the inguinal region of the dog; *a*, connective bundles treated by formic acid, and presenting annular fibres; *b*, elastic fibres; *c*, flat cells of the connective tissue seen in front; *c*, the same seen in profile; *n*, cells similar to embryonic cells and to the white corpuscles of blood and lymph.

EXPLANATION OF PLATE XXI,

Illustrating Mr. W. Saville Kent's paper on two New Genera of Alcyonoid Corals.

Fig.

1.—*Gymnosarca bathybius,* nat. size, attached to and partly investing a fragment of *Lophohelia prolifera.*

2.—A detached piece of the same, enlarged.

3.—A single calyx, × 10 diameters, and illustrating the arrangement of the spicula.

4.—Spicula of the general cœnenchyma, × 40 diameters; *a,* a slender arcuate spiculum from the tentacular region.

5.—*Cereopsis Bocagei,* nat. size.

6.—Two calices of the same, the one fully and the other partially expanded, and showing the disposition of the spicula, × 12 diameters.

7.—A single calyx in another condition of contraction.

8.—A calyx contracted to its greatest possible extent.

9.—Coloured spicula from the basal tentacular region, showing their triangular disposition.

10.—Isolated ones from the same region, × 50 linear.

11.—Minute coloured spicular from the upper portion of the tentacles, similarly enlarged.

12.—Spicula of the general cœnenchyma.

13.—Half of one of the pinnate tentaculæ, viewed superiorly and enlarged 2 diameters.